APPLIED ENERGY

AN INTRODUCTION

APPLIED ENERGY

AN INTRODUCTION

Mohammad Omar Abdullah

CRC Press
Taylor & Francis Group
Boca Raton London New York

CRC Press is an imprint of the
Taylor & Francis Group, an **informa** business

CRC Press
Taylor & Francis Group
6000 Broken Sound Parkway NW, Suite 300
Boca Raton, FL 33487-2742

First issued in paperback 2019

© 2013 by Taylor & Francis Group, LLC
CRC Press is an imprint of Taylor & Francis Group, an Informa business

Library of Congress Cataloging-in-Publication Data

Abdullah, Mohammad Omar.
Applied energy : an introduction / Mohammad Omar Abdullah.
p. cm.
Includes bibliographical references and index.
ISBN 978-1-4398-7157-7 (hardback)
1. Power resources. 2. Energy transfer. 3. Industries--Power supply. I. Title.

TJ163.2.A23 2012
621.31--dc23   2012015650

Visit the Taylor & Francis Web site at
http://www.taylorandfrancis.com

and the CRC Press Web site at
http://www.crcpress.com

Dedication

"Verily! In the creation of the heavens and the earth, and in the alternation of night and day, there are indeed signs for men of understanding..."

Rough translation from Al-Qur'an (3:190)

This book is dedidated to:

My beloved parents — who educated me and brought me up;
my loving family & relatives;
all of my teachers — wherever they are;
all of my students; and
colleagues & friends around the globe.

Contents

Preface   xvii

Acknowledgments   xix

About the Author   xxi

1 Introduction to Applied Energy   1
   1.1 General Introduction   1
   1.2 Basic Concepts: Energy and Power   1
        1.2.1 What Is Energy?   1
               1.2.1.1 What Are the Basic Forms of Energy?   1
        1.2.2 What Is Power?   5
        1.2.3 Energy Resources   6
               1.2.3.1 What Are the Nonrenewable Energy Resources?   6
               1.2.3.2 What Are the Alternative or Renewable Energy Sources?   7
        1.2.4 World Energy Consumption and Demand   7
        1.2.5 The Conservation of Energy   8
        1.2.6 The Availability of Energy   8
   1.3 Energy Equation   8
        1.3.1 The Energy Conservation Equation (First Law of Thermodynamics)   8
        1.3.2 The Energy Availability Equation (Second Law of Thermodynamics)   9
        1.3.3 General Mechanical Energy Equation   9
        1.3.4 Gibbs Free-Energy Relation   12
        1.3.5 Energy Performance Curve Equation in Terms of Useful Output for Hybrid Energy System   12
   1.4 Energy Conversion Systems   13
        1.4.1 Types of Energy Conversion Systems   13
               1.4.1.1 Conventional Energy Conversion Systems   14
               1.4.1.2 Fossil Fuel Conversion System   15
               1.4.1.3 Alternative or Renewable Energy Conversion Systems   17
               1.4.1.4 Peat and Other Energy Conversion Systems   23
   1.5 Energy Storage, Methods and Safety   23
   1.6 Energy Efficiencies and Losses   25
   1.7 *Application Projects, Assignments, Questions, and Problems*   29

2 Energy Industry and Energy Applications   37
   2.1 Energy Power Industry   37
        2.1.1 What Is Energy Power Industry?   37
        2.1.2 Energy Production from Energy Power Industry   38
               2.1.2.1 Improving Energy Production from Oil and Gas Industries   39
               2.1.2.2 Improving Coal Production Efficiency by Coal Industries   39
               2.1.2.3 Improving Energy Production Efficiency in Nuclear Industries   40
               2.1.2.4 Improving Energy Production Efficiency in Renewable Energy Industries . . .   40
               2.1.2.5 Improving Electrical Production Efficiency in Electric Power Industry 41
   2.2 Energy Applications in Energy-Intensive Industry   41
        2.2.1 What Is Energy-Intensive Industry?   41
        2.2.2 Energy Application in Energy-Intensive Industries   42
               2.2.2.1 Aluminum, Iron, and Steel Industries   42

		2.2.2.2	Chemical and Petrochemical Industry	46
		2.2.2.3	Petroleum Refinery	48
		2.2.2.4	Cement and Concrete Industries	49
		2.2.2.5	Textiles, Pulp, and Paper Industries	50
		2.2.2.6	Metal Processing Industry	52
		2.2.2.7	Mining Industry	53
	2.2.3	Energy Efficiency in Energy-Intensive Industry		53
	2.2.4	*Case Study 1*: Comparing Aluminum Production Methods and Energy Consumption		55
2.3	Energy Applications in SME Energy Industries			56
	2.3.1	What Are SME Energy Industries?		56
	2.3.2	What Is the Difference between SMEs and Other Industries?		56
	2.3.3	What Are the Types of SME Energy-Related Industries?		57
	2.3.4	Energy Efficiency and Applications in SME Energy Industry		57
	2.3.5	*Case Study 2*: Energy Consumption Saving Opportunity in Melting Process for Aluminum Casting		58
		2.3.5.1	Calculation of Energy Consumption of New Casting Method	59
2.4	Energy Applications in Building Industries			60
	2.4.1	What Is the Building Industry and Its Applications?		60
	2.4.2	Energy Efficiency and Applications in Building Industries		60
		2.4.2.1	Energy Consumed in Building Construction	60
		2.4.2.2	On-Site Energy Resources	61
		2.4.2.3	Near Future Energy Application Trend in Built Environments	62
2.5	Energy Applications in Transportation Industries			63
	2.5.1	What Are the Transportation Industries?		63
	2.5.2	Energy Consumption in Transport Industry		64
		2.5.2.1	Impacts of Transport on Energy Consumption	64
	2.5.3	Examples of Automotive Transportation Industry and Common Usages		65
	2.5.4	Examples of Air Transportation Industry and Common Applications		65
		2.5.4.1	Air Transport and Travel Industry	66
	2.5.5	Examples of Sea Transportation Industry and Common Applications		66
	2.5.6	Brief Overview of Climate Change and Greenhouse Emissions Mitigation from Transport Industry		67
	2.5.7	Energy Efficiency and Applications in Transport Industry: Current and Future		68
		2.5.7.1	Road Transportation Scenario	68
		2.5.7.2	Sea (Marine) Transportation Industry Scenario	68
2.6	Residential Energy Applications			68
	2.6.1	Residential Rnergy Applications		68
2.7	*Application Projects, Assignments, Questions, and Problems*			69

3	**Energy Resources, Supply, and Demand**			**79**
3.1	Introduction to Energy Resources, Supply, and Demand			79
3.2	Energy Resources			80
	3.2.1	What Is Energy Resources?		80
		3.2.1.1	Energy Resources Comparison	80
		3.2.1.2	Reserves to Production [R/P] Ratio	80
	3.2.2	How Much Energy Reserves from the Vast Amount of Energy Resources Do We Have?		81
3.3	Energy Supply and Energy Demand			83
	3.3.1	What Is *Energy Supply* and *Energy Demand*?		83
	3.3.2	Energy Supply and Demand from Our World Energy Resources		83
	3.3.3	Energy Supply and Demand in Some Selected Countries		88
		3.3.3.1	Asia Pacific Region	88
		3.3.3.2	Europe Region	92
		3.3.3.3	Middle East Region	95
		3.3.3.4	North America Region	97
	3.3.4	How to Cope with Our Current and Future World Energy Supply and Demand Problems		98

      3.3.4.1   *1. Finding and developing conventional and new sources of fossil fuels.* 98

      3.3.4.2   *2. Using energy wisely, efficiently, and in an environmentally friendly way.* 99

      3.3.4.3   *3. Developing Alternative Energy and other energy resources.* 99

3.4   Energy Flow Visualization and Sankey Diagram 99

    3.4.1   What Is a Sankey Diagram and Energy Flow Visualization? 100

    3.4.2   Applications Benefits of Sankey Diagram and Flow Visualization Analysis 100

    3.4.3   Basics of Sankey Diagram 100

    3.4.4   Examples of Sankey Diagram Applications 100

      3.4.4.1   Energy Application 1. Building industry — Go green to save the total energy loss. 100

      3.4.4.2   Energy Application 2. General: Energy usage for United States in 2006 and 2009. 101

3.5   *Application Projects, Assignments, Questions, and Problems* 103

4 Energy Management and Analysis   111

4.1   Energy Audits 111

    4.1.1   What Are Energy Audits? 111

    4.1.2   Basic of Energy Audits 111

    4.1.3   Benefits and Applications of Energy Audits 112

4.2   Energy Use and Fuel Consumption Study 112

    4.2.1   Primary Energy — Secondary Energy Use Relationship and Overall Efficiency 113

      4.2.1.1   Example 4.2 113

    4.2.2   Fuel Consumption Study 115

      4.2.2.1   *Fuel Consumption Rate* 115

      4.2.2.2   *Specific Fuel Consumption* 115

      4.2.2.3   Example 4.3 115

      4.2.2.4   Econometrics modeling study on fuel consumption 116

4.3   Energy Life-Cycle Analysis 116

    4.3.1   What Is Energy Life-Cycle Analysis? 116

    4.3.2   Basis and Type of Energy Life-Cycle Analysis 117

      4.3.2.1   Case Study 1: Life-Cycle Study of Soy Bean Diesel Oil 117

      4.3.2.2   Case Study 2: Life-Cycle Study of a Coal Power Plant 118

      4.3.2.3   Case Study 3: Life-Cycle Analysis of an Adsorption Air-Conditioner Powered by

            Three Different Energy Resources 120

4.4   Energy, Environment, and Health 125

4.5   Energy Pollutants, Safety and Controls 129

    4.5.1   What Are the Energy Pollutants? 129

      4.5.1.1   Pollutants from Energy Power Industry (Oil and Gas, Nuclear) 129

      4.5.1.2   Pollutants from Power Plants (Energy-Intensive Industry) 129

      4.5.1.3   Pollutants from SMEs 130

      4.5.1.4   Pollutants from Automotive/Transportation Industries 130

      4.5.1.5   Pollutants from Industrial and Residential Wastes 131

4.6   Impact of Energy Use on Our Society and the Environment 131

    4.6.1   Ways to Mitigate Impact of Energy Use 132

4.7   Energy Policy, Planning, and Statistics (Energy Target, Production and Consumption) 133

    4.7.1   Energy Policy 133

      4.7.1.1   Financial Incentive or Fiscal Policy 133

      4.7.1.2   Energy Regulations 133

      4.7.1.3   Energy Standards 134

      4.7.1.4   Example of Energy Policy Implemented 134

    4.7.2   Energy Planning and Statistics 134

      4.7.2.1   Energy Planning and Statistics for the United States 135

      4.7.2.2   Energy Planning and Statistics for Japan 135

      4.7.2.3   Energy Planning and Statistics for China 136

      4.7.2.4   Energy Planning and Statistics for Europe 137

4.8   Applied Energy Modeling and Simulation 138

    4.8.1   Modeling Energy Supply and Demand 138

4.8.2   Energy Modeling and Simulation of Power Plants and Their Applications 138
4.8.3   Energy Modeling and Simulation of Building Environment 138
4.8.4   Energy Modeling and Simulation of Transport and Vehicle Applications 138

5   Energy Saving, Recovery, and Storage   147
5.1   Introduction to Energy Saving Technologies and Energy Storage 147
5.2   Energy Recovery 147
     5.2.1   Energy Recovery in Oil and Gas (O&G) Industry 147
         5.2.1.1   Enhanced Oil Recovery (EOR) 147
         5.2.1.2   Horizontal Wells and Related Drilling Technology for Maximum Oil Production/Recovery 149
         5.2.1.3   Shale Oil or Shale Gas Production/Recovery 150
     5.2.2   Energy Recovery (ER) in Energy-Intensive Industries 150
         5.2.2.1   Energy Recovery in the Aluminum Industry 150
         5.2.2.2   Energy Recovery in the Chemical and Petrochemical Industries 151
         5.2.2.3   Energy Recovery in Pulp and Paper (Forest Products) Industry 152
         5.2.2.4   Energy Recovery in Glass Industry 152
         5.2.2.5   Energy Recovery in Metal Processing Industry 153
         5.2.2.6   Energy Recovery in Steel Industries 154
         5.2.2.7   Energy Recovery in Cement Industry 154
         5.2.2.8   Summary of Energy Recovery Requirements for Energy-Intensive Industry . . . 155
     5.2.3   Energy Recovery in SME Energy-Related Industry 155
5.3   Process Integration: Pinch Technology and Energy Optimization 155
     5.3.1   What Is Pinch Technology and Process Integration? 156
     5.3.2   Use of Process Integration 156
     5.3.3   General Application Advantages of Process Integration 156
     5.3.4   Basis of Pinch Technology and Analysis 156
5.4   Energy Saving in Combined Cycles Power Plants 157
     5.4.1   Binary Vapor Cycle 158
     5.4.2   Combined Gas and Steam Turbine Cycles 158
5.5   Energy Saving in Combined Heat and Power (CHP) Plants 160
     5.5.1   What Is Combined Heat and Power (Cogeneration)? 160
     5.5.2   What Are the Similarities and Differences between Combined Cycle and CHP? 160
     5.5.3   Basis and Types of CHP 161
     5.5.4   Application Advantages of CHP 161
     5.5.5   *Case Study 1*: Concentrated Solar Power–Desalination CHP 163
     5.5.6   Future of CHP (Cogeneration) and Applications 163
5.6   Energy Saving in Energy-Intensive and SMEs Energy-Driven Industries 163
     5.6.1   Energy Saving in Steel and Related Industries 164
         5.6.1.1   Example 5.2. 164
     5.6.2   Energy Saving in Cement and Related Industries 165
     5.6.3   Other Energy-Intensive and Related Industries 165
     5.6.4   Energy Savings in SME Energy-Driven Industries 165
5.7   Energy Saving in Building, Heating, Refrigerating and Air-Conditioning Systems 166
5.8   Energy Storage 167
     5.8.1   Fossil Fuel (Petroleum and Gas) Storage 167
     5.8.2   Ammonia Storage 167
     5.8.3   Nuclear Storage and Safety 168
         5.8.3.1   Short-Term or Temporary Storage 168
         5.8.3.2   Long-Term Storage 168
     5.8.4   Battery Storage 169
         5.8.4.1   Battery for Everyday Applications 169
     5.8.5   Flywheel Storage 170
     5.8.6   Solar Energy Storage 171
     5.8.7   Wind Energy Storage 172

5.8.8   Pump Hydroelectric Storage 172

5.8.9   Fuel Cell Storage 173

5.8.10   Hydrogen Storage 174

      5.8.10.1 Application Advantages 174

      5.8.10.2 Application Disadvantages 174

5.8.11   Thermal Energy Storage 174

5.8.12   Inductor, Capacitor, and Supercapacitor Storage 175

      5.8.12.1 Inductor 175

      5.8.12.2 Capacitor 175

      5.8.12.3 Supercapacitor or Ultracapacitor 175

5.8.13   Superconducting Magnetic Energy Storage (SMES) and Superconducting Coil 177

      5.8.13.1 Application Advantages of SMES Systems 177

      5.8.13.2 Disadvantages 177

5.8.14   Compressed Air Storage 178

5.9   Energy Storage Comparison and Energy Density 178

5.10   *Application Projects, Assignments, Questions, and Problems* 179

6 Energy from Fossil Fuels versus Alternative Energy   187

6.1   Introduction to Fossil Fuels (Petroleum, Natural Gas, and Coal) 187

6.1.1   Petroleum 187

      6.1.1.1 Volumetric Equation of Oil in Place (OIP) 188

      6.1.1.2 Example 6.1 188

      6.1.1.3 Solution 189

      6.1.1.4 Example 6.2 189

      6.1.1.5 Solution 189

      6.1.1.6 Chemical Composition of Petroleum 189

      6.1.1.7 General Classifications of Petroleum 190

      6.1.1.8 Example 6.3 191

      6.1.1.9 Solution 191

      6.1.1.10 Petroleum and Refining Process 191

      6.1.1.11 How Is Petroleum Processed? 192

6.1.2   Petroleum Fuel or Products' Properties Required for Energy Applications 192

      6.1.2.1 Energy from Combustion of Petroleum 193

6.1.3   Natural Gas (Gaseous Petroleum) 194

      6.1.3.1 Gas Originally in Place (GOIP) 194

      6.1.3.2 Gas Composition 194

      6.1.3.3 The Ideal Gas Equation and the Gas Law 195

      6.1.3.4 Example 6.4. 195

      6.1.3.5 Solution 196

      6.1.3.6 Natural Gas and Refining Process 196

      6.1.3.7 How Is Natural Gas Processed? 196

      6.1.3.8 Energy from Combustion of Natural Gas 196

      6.1.3.9 Uses and Application Advantages of Natural Gas . . 197

6.1.4   Coal 198

      6.1.4.1 Classification of Coal 198

      6.1.4.2 How Is Coal Processed? 199

      6.1.4.3 Energy and Combustion of Coal 199

      6.1.4.4 Uses of Coal 199

      6.1.4.5 Emission from Combustion of Coal 200

6.1.5   Other Fossil Fuels 201

      6.1.5.1 Oil Shales 201

      6.1.5.2 Natural Bitumen and Extra-Heavy Oil 202

6.2   Environmental Concerns of Fossil Fuels 202

6.2.1   Technologies to Tackle Environmental Challenges from Fossil Fuels 202

      6.2.1.1 Coal Pollution 203

6.3   Introduction to Alternative/Renewable Energy Resources 205
      6.3.1   What Is Alternative or "Renewable" Energy? 205
6.4   Fossil Fuels versus Alternative/Renewable Fuels for Energy Applications 206
6.5   Main Advantages of Using Alternative/Renewable Energy Sources 207
6.6   Present Limitations of Alternative/Renewable Energy 207
6.7   How Can Alternative Energy Replace Fossil Fuels for Our Near Future Energy Applications? . . . 207
6.8   *Application Projects, Questions, and Problems* 208

7   Solar and Electrochemical Energy Conversion   215
7.1   Introduction to Solar Energy 215
7.2   Solar Photovoltaic (PV) 215
      7.2.1   What Is Solar Photovoltaic? 215
            7.2.1.1   What Is a Solar PV Cell? 215
            7.2.1.2   What Is a Solar PV Module or "Solar Array"? 216
      7.2.2   The Working Principle of a Basic Solar Photovoltaic (PV) Generator 216
            7.2.2.1   Brief Description of Components 217
      7.2.3   General Classifications and Applications of PV Systems 217
      7.2.4   Application Advantages of a Solar Photovoltaic (PV) Power System 217
      7.2.5   Recent Advancements and Future Applications of PV Power Systems 217
7.3   Solar Thermal Energy 218
      7.3.1   Classification of Solar Thermal Energy Based on Motion Type of the Solar Collector 218
      7.3.2   Classifications of Solar Thermal Energy Collector/Solar Thermal Power Plant Systems . . 218
            7.3.2.1   Evacuated-Tube Solar Collector 218
            7.3.2.2   Solar Thermal Collecting Systems 219
            7.3.2.3   Solar Power Plants 219
      7.3.3   Application Advantages of Solar Thermal Energy Conversion Systems 222
      7.3.4   Application Disadvantages and Technical Concerns of Solar Thermal Energy 222
      7.3.5   How to Estimate Solar Energy Output 223
            7.3.5.1   Example 7.2 223
            7.3.5.2   Solution: 223
      7.3.6   *Case Study 1.* Solar Energy Application: A Simple Solar Clothes Dryer 224
      7.3.7   Recent Advancement and Future Applications of Solar Thermal Power 225
            7.3.7.1   Application Examples of Solar Thermal Power Plants and PV-Powered Generators
                   around Our World 227
7.4   Introduction to Electrochemical Energy Conversion 227
7.5   Battery 227
      7.5.1   What Is a Battery? 227
            7.5.1.1   What Are the Basic Components of a Battery as an Energy Generator? 227
      7.5.2   Classifications of Batteries 228
            7.5.2.1   Dry cell battery 228
            7.5.2.2   Wet Cell Battery 230
      7.5.3   Cell Performance 230
      7.5.4   Application Advantages and Disadvantages of Batteries 232
      7.5.5   How to Estimate Batteries' Energy Output 232
7.6   Fuel Cell 232
      7.6.1   What Is a Fuel Cell? 232
      7.6.2   Fuel Cell Classifications 232
            7.6.2.1   Fuel Cell Classification I: Based on Electrolyte 232
            7.6.2.2   Other Fuel Cell Types 237
            7.6.2.3   Fuel Cell Classification II: Based on Power Size 238
      7.6.3   Example Fuel Cell Systems 239
      7.6.4   Fuel Cell Voltage and Fuel Cell Efficiency 240
      7.6.5   Fuel Cell Energy Output Estimation 240
      7.6.6   Application Advantages of Fuel Cell as Energy Generator 241
      7.6.7   Application Disadvantages or Limitations of Fuel Cells 242

7.6.8 *Case Study 2.* Parametric Study of an Alkaline Fuel Cell (AFC) 242
7.7 Summary 246
7.8 *Application Projects, Questions, and Problems* 247

8 Hydro, Wind, and Geothermal Energy **253**
8.1 Hydro Energy 253
    8.1.1 Introduction to Hydro Energy 253
    8.1.2 The Basic Components of a Hydro-Electric Power Plant 254
    8.1.3 Site Selection for a Hydro-Electric Power Plant 255
    8.1.4 Hydro-Electric Turbine Classifications 255
        8.1.4.1 What Is *Specific Speed* of Turbine? 257
        8.1.4.2 What Are the Common Prime-Movers or Turbines? 258
        8.1.4.3 Dams Versus Weirs 258
    8.1.5 Turbine Efficiency: Full Load Efficiency versus Part Load Efficiency 258
    8.1.6 Micro-Hydro Applications 260
        8.1.6.1 Plant Factor 260
        8.1.6.2 Example 8.2 260
        8.1.6.3 Load Factor 260
        8.1.6.4 Example 8.3 260
        8.1.6.5 Unit Energy Cost 260
        8.1.6.6 Example 8.4 261
    8.1.7 Application Advantages of Hydro Energy and Hydro Power Plant 261
    8.1.8 Examples of Hydro Energy Power Plants around the World 261
8.2 Ocean Energy (Tidal, Wave, Thermal) 262
    8.2.1 Tidal Energy 262
        8.2.1.1 What Is Tidal Energy? 262
        8.2.1.2 What Are the Application Advantages of Tidal Energy? 262
        8.2.1.3 How Tidal Energy Is Used for Electric Energy Production? 263
        8.2.1.4 Classification of Tidal Power Plants and Water Turbines 263
        8.2.1.5 Examples of Tidal Applications around the World 263
        8.2.1.6 Some Recent Advancements of Tidal Technologies 264
    8.2.2 Wave Energy 265
        8.2.2.1 How Is Wave Energy Used for Electric Energy Production? 265
        8.2.2.2 What Are the Advantages and Disadvantages of Wave Energy Applications? . . . 266
        8.2.2.3 Example of Wave Energy Applications around the World 266
    8.2.3 Ocean Thermal Energy 266
        8.2.3.1 What Is Ocean Thermal Energy? 267
        8.2.3.2 Ocean Thermal Energy Conversion Plant (OTEC) 267
8.3 Wind Energy 268
    8.3.1 Introduction to Wind Energy and Its Applications 268
        8.3.1.1 Some Applications of Windmills 268
    8.3.2 How to Estimate Wind Energy Output 268
    8.3.3 Wind Energy Extraction and Aerodynamic Effects on Windmill Power Output 270
        8.3.3.1 *Tip Speed Ratio* 270
        8.3.3.2 *Solidity* 270
    8.3.4 Offshore Windmills 272
    8.3.5 Some Examples of WindMill Plants 272
8.4 Geothermal Energy 273
    8.4.1 What Is Geothermal Energy? 275
    8.4.2 Classification of Geothermal Energy Resources and Applications 275
        8.4.2.1 Vapor-Dominated Steam ("Dry Steam") 275
        8.4.2.2 Water-Dominated Steam ("Wet Steam") 276
        8.4.2.3 Hot Dry Rock (HDR) Resources. 276
    8.4.3 How to Estimate Geothermal Energy Output 276
        8.4.3.1 What Are the Application Advantages of Geothermal Energy? 278

		8.4.4	Type and Classifications of Geothermal Energy Applications	279
		8.4.5	Geothermal Power Plant	279
			8.4.5.1 Dry Steam Power Plant (with Vapor Dominated)	280
			8.4.5.2 Flash Steam Power Plant (with Liquid Dominated)	280
			8.4.5.3 Binary Cycle Geothermal Power Plant	280
		8.4.6	Direct Geothermal Heat Applications	281
		8.4.7	World Geothermal Energy Production	282
		8.4.8	Some Examples of Geothermal Power Plants Around the World	282
		8.4.9	Future of Geothermal Technology and Applications	285
	8.5	*Application Projects, Assignments, Questions, and Problems*		286

9 Nuclear Energy and Energy from Biomass — **295**

	9.1	Nuclear Energy and Nuclear Power Plant		295
		9.1.1	What Are Nuclear Power and Nuclear Power Plants?	295
		9.1.2	What Are the Application Advantages of Nuclear Energy and Nuclear Power Plants?	296
			9.1.2.1 Example 9.1	297
		9.1.3	What Is the Working Principle of a Nuclear Power Plant?	298
			9.1.3.1 The Science of Nuclear Fuel: A Quick Overview	298
			9.1.3.2 The Technology of Nuclear Fuel: An Overview of the Nuclear Fuel and Technology Brief	298
		9.1.4	The Basic Nuclear Power Plant	298
		9.1.5	Classification of Nuclear Power Plants	299
			9.1.5.1 Pressurized Water Reactor (PWR)	299
			9.1.5.2 Boiling Water Reactor (BWR)	301
			9.1.5.3 Gas-Cooled Reactor (GCR)	302
			9.1.5.4 Advanced Gas-Cooled Reactor (AGR)	303
			9.1.5.5 Liquid Metal Cooled Reactor (LMCR)	303
			9.1.5.6 Liquid Metal Fast Breeder Reactor (LMFBR)	304
			9.1.5.7 Pressurized Heavy Water Reactor (PHWR, Better Known as CANDU)	305
			9.1.5.8 Other Reactor Designs	306
		9.1.6	How to Estimate Nuclear Energy Output	306
		9.1.7	Examples of Nuclear Power Plants around the Globe	307
		9.1.8	Recent Advancements and Future of Nuclear Energy	308
			9.1.8.1 Future of Nuclear Storage and Safety	308
			9.1.8.2 Future of Nuclear Energy for Sustainability	309
			9.1.8.3 Some Possible or Likely Future Scenarios of Nuclear Energy Applications	309
	9.2	Energy from Biomass (Bioenergy)		309
		9.2.1	What Are Biomass and Bioenergy?	309
		9.2.2	Biomass Energy Classifications	310
			9.2.2.1 Digestion	311
			9.2.2.2 Fermentation	311
			9.2.2.3 Gasification	311
			9.2.2.4 Pyrolysis	312
			9.2.2.5 Combustion	312
		9.2.3	Application Advantages/Disadvantages of Bioenergy	316
			9.2.3.1 Application Advantages	316
			9.2.3.2 Disadvantages	316
		9.2.4	*Case Study 1*: Biodiesel for Oil-Based Mud Drilling	316
		9.2.5	*Case Study 2*: Economics of Producing Microalgal Biodiesel (Chisti, 2008 [21])	317
		9.2.6	Examples of Biomass Power Plant Applications around the Globe	318
		9.2.7	Recent Advancement and Future of Bioenergy	318
	9.3	*Assignments, Questions, and Problems*		319

10 Hybrid Energy **327**
  10.1 Hybrid Energy and Hybrid Energy Methods 327
      10.1.1 What Is Hybrid Energy? 327
  10.2 Hybrid Geothermal/Fossil System 327
      10.2.1 Hybrid Geothermal Preheat/Fossil Energy System 328
      10.2.2 Hybrid Geothermal Fossil/Superheat Energy System 328
      10.2.3 One-Stage Compound Hybrid System 329
  10.3 Integrated Gasification Combined-Cycle (IGCC) 331
      10.3.1 The Advantages of IGCC Compared with Conventional Coal-Based Power Generation Systems 331
      10.3.2 The Disadvantages of IGCC 332
  10.4 Hybrid Fuel Cell/Turbine and Fuel Cell/Microturbine System 332
      10.4.1 Hybrid Fuel Cell/Turbine (FCT) 333
      10.4.2 Hybrid Fuel Cell and Microturbine (FCMT) 333
          10.4.2.1 What Are Microturbines and Micropower? 333
          10.4.2.2 Introduction to FCMT 333
      10.4.3 Hybrid SOFC-PEM/Turbine System 334
      10.4.4 Hybrid Fuel Cell/Battery System 335
  10.5 Hybrid Electric Vehicles (HEVs) 335
      10.5.1 Hybrid Electric Vehicles Classification 335
      10.5.2 Energy Storage Device Requirements for Hybrid Electric Vehicles 337
  10.6 Hybrid Energy Systems for Rural Application 337
      10.6.1 Overview of Hybrid Energy Systems for Rural Applications 338
      10.6.2 Energy Sizing and Control of Hybrid Energy Systems for Electricity Production 338
          10.6.2.1 PV–Diesel Hybrid Energy System Configurations 339
      10.6.3 Hybrid Energy Systems and System Optimization 340
      10.6.4 Application Advantages of Renewable Hybrids for Rural Village Electrification 341
          10.6.4.1 Future Hybrid Energy Systems 341
  10.7 Other Hybrid and Integrated Energy Application Systems 342
  10.8 Fuel Efficiency of Hybrid Energy Systems: Case Study 342
  10.9 *Application Projects, Questions, and Problems* 343

11 Other Energy Conversion Methods **349**
  11.1 Absorption Energy Conversion 349
      11.1.1 Absorption Energy Conversion 349
      11.1.2 What Is an Absorption Thermal Energy Convertor? 349
          11.1.2.1 Working Principle of an Absorption Energy System 349
          11.1.2.2 Materials Selection and Working Pair 350
          11.1.2.3 Application Advantages of Absorption Energy Systems 350
          11.1.2.4 Industrial Design and Commercial Absorption Systems 351
          11.1.2.5 Some Advanced Absorption Cycles 351
  11.2 Adsorption Energy Conversion 354
      11.2.1 Basics of Adsorption 354
          11.2.1.1 The Adsorption System 354
          11.2.1.2 Adsorbent–Adsorbate Pairs Selection 355
          11.2.1.3 Dubinin–Astakhov (D–A) Equation and Heat of Adsorption 357
          11.2.1.4 Thermodynamic Analysis of a Typical Adsorption Cycle 358
      11.2.2 Energy Performance of Adsorption Energy System 359
      11.2.3 Performance of Adsorption Systems for Different Applications 359
      11.2.4 Application Advantages of Adsorption Energy System 359
  11.3 Thermoelectric and Thermionic Energy Conversion 360
      11.3.1 Thermoelectric Energy Conversion 360
          11.3.1.1 What Is a Thermoelectric Energy Convertor? 360
          11.3.1.2 The Working Principle of a Thermoelectric Generator 360
          11.3.1.3 What is the Estimated Energy Output of a Thermoelectric Module? 360

          11.3.1.4 Materials Selection for Thermoelectric Generator 361

          11.3.1.5 Application Advantages of a Thermoelectric Generator 363

          11.3.1.6 What Is the Potential Application Area of a Thermoelectric Energy Convertor? . 363

          11.3.1.7 Application Examples 363

      11.3.2 Thermionic Energy Conversion 363

          11.3.2.1 What Is a Thermionic Energy Convertor? 363

          11.3.2.2 Working Principle of a Thermionic Generator 364

          11.3.2.3 Materials Selection for Electrodes of a Thermionic Generator 364

          11.3.2.4 Applications of a Thermionic Generator 364

          11.3.2.5 Thermoelectric Generator and Thermionic Generator: Application Comparisons . 364

      11.3.3 Combined Thermionic–Thermoelectric Generator 365

  11.4 Magnetohydrodynamic (MHD) Power Conversion 366

      11.4.1 Working Principle of an MHD 366

      11.4.2 Classification of MHD Power Systems 367

      11.4.3 Application Advantages of MHD 367

      11.4.4 Some Current Usage and Future Applications of MHD 370

          11.4.4.1 *CHP (Co-Generation) with MHD Applications* 370

          11.4.4.2 Nuclear Powered MHD Energy Systems 370

          11.4.4.3 Coal Fired MHD Technology 371

          11.4.4.4 MHD Enhanced Solar Energy Applications 371

          11.4.4.5 Thermo-Acoustic–MHD Electrical Generator 371

          11.4.4.6 MHD Plasma Devices for Vehicle Applications 371

  11.5 *Questions and Problems* 373

12 Some Applied Energy Related Issues   **379**

  12.1 "Energy Storage" Issues 379

      12.1.1 Fossil Fuel Storage Issues 379

          12.1.1.1 Ammonia Storage and Ammonia Storage Tanks Issues 379

          12.1.1.2 Nuclear Energy Storage Issues 380

      12.1.2 Renewable Energy Storage Issues 380

          12.1.2.1 Battery Storage Issues 380

          12.1.2.2 Compressed-Air Storage Issues 383

          12.1.2.3 Pumped Hydroelectricity Storage Issues 383

          12.1.2.4 Flywheel and Kinetic Energy Storage Issues 383

          12.1.2.5 Geothermal Energy Storage Issues 384

          12.1.2.6 Solar Energy Storage Issues 384

          12.1.2.7 Fuel Cells and Hydrogen Storage Issues 385

  12.2 "Energy Sustainability" Issues 387

  12.3 Waste and "Waste-to-Energy (WTE)" Issues 388

  12.4 "Zero-energy" Building (ZEB) Issues 389

  12.5 "Biofuels from Crops" Issues 391

      12.5.1 *Algae* 393

      12.5.2 *Switch Grass* 394

      12.5.3 *Prairie Grass* 394

      12.5.4 Biofuel from Waste Fats (Biodiesel) 394

  12.6 Adsorption Technology from Vehicle Waste Heat 395

  12.7 *Assignments, Questions, and Problems* 395

Epilogue   **403**

Appendix   **405**

Glossary   **407**

Index   **409**

Preface

Applied energy refers to any energy conversion when applied to energy systems. The subject matter and the associated scope are therefore very vast, and include energy conversion and conservation, energy resources and sustainability, energy processes, and environment applications analysis. Thus this book derived its benefits largely based on numerous useful resources, publications, and industrial experiences, such as that of books, lecture notes, industrial literature, as well as the latest journals related to applied energy available to the author.

Is this book for you? This book is specially written in the simplest possible manner yet it is a comprehensive textbook for an applied energy courses. This book assumes that you *do not have* prior knowledge of applied energy at all and provides an introduction to applied energy. Especially if you are a beginner who wants to study applied energy related subjects, you will find this book very helpful in giving a concise overview of the key aspects of energy and related studies. It is also for those at the intermediate level in certain areas of applied energy, who want to widen their scope in other areas of applied energy. Or, it is even suitable for those who merely want to test their fundamental understanding in relation to the vast areas of applied energy — especially energy resources, management, and applications.

How is this book designed, and what are its special features? This book is designed in such a way that it is geared towards practical- and application-based learning, presented in various forms such as essential notes followed by practical projects, assignments, and objective and practical questions with respect to vast applied energy areas. In each chapter, a small section is devoted to introducing some elements of applied energy design and innovation as a practice; we named it *Application Project*, linking knowledge with applied energy design and practice. By applying these comprehensive ways of delivery, the readers can increase their understanding of the various applied energy subjects and the so-called life long learning in a quick and enjoyable way. This book does not cover every minor detail of energy or applied energy, but it gives you a grasp of all the important aspects of applied energy study.

How is this book organized? This book can be used as a textbook, tutorial, or review, where you start at the beginning and work through to the end. This book is organized into sections covering specific topics so that you can concentrate on the materials you want. The book is subdivided into 12 sections, each dealing with a separate but related topic. The philosophy of this book is to cover a broad range of the important questions of applied energy.

It is important for students to have good motivation for studying any subject and to be able to place into context the concepts presented. Thus, in this book, for the convenience of the readers, the chapters and contents are carefully organized as follows: Chapter 1 is an introduction that describes various basic concepts and definitions used throughout the book. Chapter 2, which focuses on energy industry and energy applications, is essential to all students. In Chapter 3, energy sources and supply and demand are introduced. Chapter 4 presents energy management, energy policy, energy plan, and analysis. This chapter includes some of the most useful concepts in energy management. Chapter 5 focuses on various energy saving technologies and energy storage methods. Chapter 6 is devoted to describing the comparison of fossil fuels versus alternative energy; Chapter 7 through Chapter 9 discuss the various types and applications of alternative energies; while Chapter 10 discusses the hybrid energy technology and hybrid energy schemes. Chapter 11 presents some other energy conversion methods. Chapter 12 concludes with a discussion on various applied energy issues.

At the end of each chapter, sample assignments and application projects are given aimed to enhance knowledge and practice. Multiple choice questions and theoretical questions are also provided for review purposes.

Suggestions for using this textbook Depending on the course credits offered, a 2-term/semester course can be conducted with Chapters 1-6 during the first term or semester, followed by the remaining chapters (Chapters 7–12) in the second term/semester. The book can also be covered quickly by *compressing* accordingly to only one term/semester — this mode of teaching is practically suitable for programs that adopt, say, 3 or 4 credits in energy discipline but that require a comprehensive coverage.

What are the limitations of this book?   This textbook is aiming for a wide-ranging introduction on applied energy, and this may indeed be of interest to those students who wish to add an applied energy and environment focus to their first degree. The proposed book will be useful reading material for post-graduate students as well especially during team and individual project work. Its main weakness, however, is lack of an in-depth treatment of topics; while the generality and wide scope of this book are its strength, they are also its weakness.

Finally, the author would like to thank readers in advance for comments, suggestions, advice, and guidance for further improvement of this book. The author can be reached by e-mail, at amomar@feng.unimas.my or amomar13@gmail.com.

Mohammad Omar Abdullah

Acknowledgments

The author very much appreciates all CRC staff; in particular; Leong Li-Ming, Amber Donley, Jennifer Ahringer, Michele Dimont, Simon Bates, and Florence Kizza for all their understanding and assistance throughout the preparation of this textbook. The author would like to acknowledge Mohamad Nazim Jambli (UNIMAS) and Shashi Kumar (CRC) for their valuable tips and advice on the use of LyX and LaTeX software for writing this book. Finally, the author would like to thank his students, especially Harunal Rejan and Tian Chuan Min, for helping with some of the figures in this book.

About the Author

Dr. Mohammad Omar Abdullah currently is an Associate Professor at the Department of Chemical Engineering & Energy Sustainability, Universiti Malaysia Sarawak (UNIMAS), Malaysia. He has delivered more than 15 years of lectures, short courses, and seminars around the world and has published numerous papers on energy and related fields in many professional journals and conferences. He was appointed as a Senior Member of the Academy of Malaysian SMEs (S.M.A.M.S) in March 2009 during the IKS2009 Presentation and Exhibition at the Putra World Trade Center (PWTC), Kuala Lumpur, Malaysia. He is a member of the American Chemical Society (ACS), USA; member of the American Society of Heating, Refrigerating and Air-conditioning Engineers, Inc. (ASHRAE), USA; Member of the Institue of Mechanical Engineers (IMechE) and CEng, United Kingdom. He has bachelor's and master's degrees, both in Petroleum Engineering, from UTM, Malaysia, and a Ph.D. degree in Mechanical and Manufacturing Engineering from the Univerisity of Hertfordshire (UH), United Kingdom. He is also a board member of the the Board of Study Panel and the Industrial Panel for the faculty of Petrolatum and Renewable Energy Engineering (FPREE), UTM.

1

Introduction to Applied Energy

1.1 General Introduction

Energy systems are integral parts of our own bodies and our society. Each day, every one of us changes energy from one form to another and uses it to do work for us, within us and around us! In order to live more comfortably, we build and improve numerous energy systems for our own good, from air-conditioning and power plants, to vehicle waste heat usage and zero-energy homes.

Applied Energy refers to any energy conversion when applied to energy systems (see Figure 1.1). The subject matter and the associated scope are therefore very vast — it refers primarily to our daily domestic and industrial energy usage and applications in various areas of energy conversion and conservation, energy resources, energy processes, energy storage, safety, and transport, environmental energy pollutants, and sustainable energy systems, as well as techno-economics, energy innovation, and applied energy issues around the globe.

In this book, *Applied Energy: An Introduction*, we shall endeavor to come up with an introduction to various applied energy studies, relating to six main areas of energy applications, namely (1) Energy (power) industry, (2) Energy intensive industry and manufacturing, (3) Transportation applications, (4) Building industry applications, (5) Small and medium enterprise (SME) industry as well as (6) residential and other general energy applications. Focusing on energy conversion, and conservation in the energy industry, we will also attempt to discuss energy applications in small-medium enterprises, solar energy, hydro and wind energy, biomass energy, hybrid energy, and energy sustainability issues.

In addition to illustrations, we shall cover numerous worked examples, application projects, case studies, assignments and homework problems to empower our comprehension and confidence. By going through the materials presented in this introductory but somewhat comprehensive book on applied energy, you can determine which key ideas and useful knowledge are worth the effort during the last week of the term or semester.

1.2 Basic Concepts: Energy and Power

1.2.1 What Is Energy?

The energy of a body is its capacity to do work, and makes change possible. Each day, we all change energy from one form to another and use it to do work for us and in order to live more comfortably. Energy is the total amount of work done, such that mathematically, energy or work done E is essentially the integral of power P, i.e.

$$Energy, E = workdone = \int P dt \qquad (1.1)$$

There are many different energy resources available for us to do work.

1.2.1.1 What Are the Basic Forms of Energy?

Generally, energy forms are either potential or kinetic. Potential energy comes in forms that are stored including chemical, gravitational, mechanical, and nuclear energy. Kinetic energy forms are used for doing a variety of work, for instance, electrical, chemical, electrochemical energy, thermal (heat), electromagnetic energy (light), motion, and

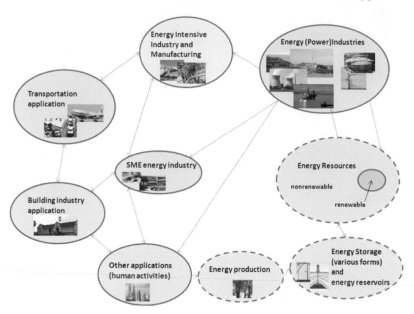

Figure 1.1
Applied energy study diagram.

vibration (sound) energy. Here are some of the brief descriptions of various forms of energy,* together with their application advantages.

Electrical Energy or Electricity Electrical energy is the energy developed due to flow of electrons among the atoms of matter of a conduit when an electric field is applied. Unlike other energy sources, electricity is a secondary source of energy. We must use another form of energy (examples coal) to produce electricity. A power station is a place where other forms of energy, e.g. coal, natural gas, hydro energy, and nuclear energy, are turned into electrical energy for transmission to places that use electrical energy. Electricity is sometimes called an energy carrier because it is a well established, efficient, and safe way to move energy from one place to another. Also, it can be conveniently used to perform many tasks. As the world population grows, we use more energy for our daily activities, and more technology and innovations for numerous energy applications. As a result the demand for global electricity grows continuously. Electrical Power, $P_{electrical}$ and Electrical Energy $E_{electrical}$ can be calculated from

$$P_{electrical} = V \times I \tag{1.2}$$

$$E_{electrical} = P_{electrical} \times T \tag{1.3}$$

where
  $P_{electrical}$ = electrical power (Watts);
  V = voltage (Volt);
  I = current (ampere);
  $E_{electrical}$ = electrical energy (kWh); and
  T = Time (seconds).

Kinetic Energy The kinetic energy of an object is the energy that it possesses due to its motion [3]. We can consider an object such as a bullet flying through the air. The bullet has "kinetic energy" due to the fact that it is in motion relative to another bullet that is stationary [4].

---

*For more detailed treatment of the subjects in relation to various energy forms and energy conversions, readers can refer to standard physics books (e.g., Serway & John [31]) and energy handbooks (e.g., Goswami & Kreith [2]).

Kinetic energy, $E_{kinetic}$ is calculated as follows:

$$E_{kinetic} = \frac{1}{2}mV^2 \tag{1.4}$$

We use this energy for two primary application advantages:
1. It is easily available in moving objects as all that is needed is motion.
2. Kinetic energy is clean and it does not pollute our environment.

Potential Energy Potential energy, or stored energy, is the ability of a system to do work due to its position or internal elastic structure. For example, gravitational potential energy is a stored energy determined by an object's position in a gravitational field. Our Earth's gravity is necessary for the potential gravitational energy. Spring potential energy is the energy stored in a spring.

Gravitational potential energy,

$$E_{potential} = mgh \tag{1.5}$$

where
$E_{potential}$ = potential energy in joules;
m = mass (kg);
g = gravity (m/s^2);
h = height (m).

Spring potential energy,

$$U = \frac{1}{2}k\,x^2 \tag{1.6}$$

where
U = spring potential energy;
k = spring force constant;
x = distance from equilibrium.
We use potential energy for three primary application advantages:
1. The potential energy is virtually free (spring energy, for instance)
2. Much more constant and reliable than wind, solar, or wave power
3. No waste or pollution produced.

Thermal or heat energy Thermal or heat energy is the internal energy present in a system by virtue of its temperature difference with surroundings. Temperature of a system is a measure of how much thermal energy it has. The higher the temperature, the faster the molecules are moving around and "vibrating." Thermal energy can be produced by burning fossil fuels (coal, oil, natural gas) or biomass (e.g, wood). It can also be derived from steam in a geothermal or through nuclear reactions in a nuclear plant.

Thermal energy, Q

$$Q = m\,Cp\,(\triangle T) \tag{1.7}$$

We use thermal energy for the following primary application advantages:
1. It is used for industrial power generation and available globally in most areas.
2. It can provide continuous, reliable energy that is not dependent on the weather (with the exception of solar thermal).

Chemical Energy Chemical energy is a form of potential energy related to the structural arrangement of atoms or molecules, which exists because of the forces of attraction (chemical bond) exerted between the different parts of each molecule. Chemical energy of a chemical substance can be transformed to other forms of energy by a process called chemical reaction. For instance, the glucose in our own body has chemical energy because the glucose releases energy when chemically reacted with oxygen. We all use this energy to generate force and release heat. The batteries that power all of our mobile phones, the fossil fuels we consume every day in our vehicles and power plants, are all associated with chemical energy applications.

We use chemical energy for the following primary advantages:

1. Chemical energy is used by our own bodies every day.
2. Chemical energy is one of the most efficient energy sources to store and utilize.
3. Recent developments in chemical energy technology have led to long-lasting rechargeable batteries and fuel cells.

Electrochemical Energy or electrochemical potential　　Electrochemical energy is the potential energy stored in a battery or electric cell, where both chemical energy and electricity are also involved; hence "electrochemical" or electrochemical potential. IUPAC [5] defined electrochemical potential as the partial molar Gibbs energy of a substance at the specified electric potential, where the substance is in a specified phase.

Electrochemical potential can be expressed as,

$$\bar{\mu}_i = \mu_i + z_i F \Phi \tag{1.8}$$

where

$\bar{\mu}_i$ is the electrochemical potential of species i, J/mol;
μ_i is the chemical potential of the species i, J/mol;
z_i is the valency charge of the ion i, dimensionless;
F is Faraday's Constant, C/mol; and
Φ is the local electrostatic potential, V.

Electrochemical potential is important in industrial applications, especially for effective energy storage systems such as batteries, supercapacitors, and fuel cells. They are used for numerous starting, lighting, and ignition applications for computers, cordless tools, emergency power and lighting, vehicles and aircraft, remote monitoring stations, toys, missiles, pacemakers, satellites, hearing aids, portable communication devices, electric vehicles, industrial controls, spacecraft, traction electronics, etc.

We use electrochemical energy for the following primary application advantages:

1. Electrochemical energy provides a very effective energy storage system.
2. Electrical engineering is suitable for starting, lighting and ignition in numerous appliances.
3. It has additional applications advantages, such as decentralized peak power-shaving and load leveling.

Electromagnetic energy and Light energy　　Electromagnetic Energy or "radiant" energy can be defined as the energy in the form of transverse magnetic and electric waves. Light is the radiant energy of electromagnetic waves or radiation in the visible portion of the electromagnetic spectrum. Light may also be thought of as photons or energy particles. The word "photon" derives from the word "photo" which means "light." Therefore, electromagnetic energy usually refers to systems that transfer electrical power "wirelessly." Electromagnetic energy was a great discovery of the nineteenth century, the energy application areas of which are, for instance, radio waves, x-rays, and gamma-rays. And here are some of its application in our daily activities: The Sun transfers radiant energy to our Earth in the form of infrared energy, visible light, and ultraviolet rays; light bulbs transfer radiant energy to our eyes in the form of visible light. Microwave ovens use radiant energy to cook food. And radio waves transfer information to our radios and televisions also via radiant energy.

We use electromagnetic energy for the following primary applications advantages:

1. Electromagnetic energy is clean and applicable in many implements.
2. It is easy to generate electricity and can be made to work on an extremely small scale such as in microchips.
3. Unlike nuclear energy, it has no radioactive components that can explode violently.

Sound energy or acoustic energy　　Sound energy or acoustic energy is a mechanical wave energy produced by vibrating objects, associated with the vibration movement of air molecules, and within the hearing frequencies. The telephone and mobile devices convert sound energy into electrical energy, and back into sound energy again for our daily uses. Ultrasonic testing uses high frequency sound energy to conduct ultrasonic inspection, do examinations (for example, to detect cracks and leaks in industrial tanks), and make measurements (example: for thickness measurements). Sonic and ultrasonic weapons are used by the military. We use sound energy for the following application advantages:

1. Every day, we use sound energy in order to hear things.
2. Sound energy is used for ultrasonic testing inspection.

Nuclear energy Nuclear energy is an energy that is generated through the use of uranium (a natural metal that is mined all over the world) via nuclear reactions. Nuclear energy is created through chemical reactions that involve the splitting or merging of the atoms of nuclei together. The process of splitting an atom's nucleus is termed fission, and the process of merging the nuclei of atoms is termed fusion, which releases energy. Converting nuclear masses into energy forms is known through the popular chemical equation discovered by Einstein:

$$E = mc^2 \qquad (1.9)$$

where
  E is known as the amount of energy released,
  m is known as the mass of the nuclei; and
  c is the value of the speed of light.

We use nuclear energy for the following primary advantages:
  1. Nuclear energy has very high energy density content and uses relatively much less fuel in the electricity-generating power plants.*
  2. It is a clean energy that also produces less waste, and does not produce carbon dioxide or smoke. Therefore it does not contribute to the greenhouse effect.
  Next, we will discuss power, and various topics associated with energy and power.

1.2.2   What Is Power?

Power, is defined as the rate of energy change per length of time. It tells us the quantity of energy that changed during a certain period of time. Mathematically, power or rate of work can be expressed as

$$P = \frac{dU}{dt} = \frac{dw}{dt} \qquad (1.10)$$

where
  P is power;
  U is energy;
  t is time; and
  w is work.

1.2.2.0.1   Units of Energy and Power
The measure of energy in SI unit is the Joule (J). A Joule is the amount of energy required to exert a force of 1 Newton through a distance of 1 meter, so that 1 J = 1 N-m.

For thermal energy, energy units like Btu's and calories are generally used, e.g., energy from food. A Btu is approximately the amount of energy needed to increase the thermal energy of 1 pound of water enough to raise its temperature by 1 °F. A calorie is the amount of energy needed to raise the temperature of 1 g of water by 1 °C. In other words, it is the amount of energy required to increase the temperature of 1 kg of water by 1 °C. 1 Btu = 1055 Joules. Now, the SI unit of power is Watt (W) or 1 Joule per second (1 J/s = 1 Watt). Other units are Btu per hour, horsepower (hp), etc. 1 hp = 745.7 W.

In our residential electrical appliances and electric industrial applications, for convenience purposes, the units of electric energy are often given in Watt-hours (Wh) or kilowatt-hours (kWh). Application for power plants and other large-quantity electricity powered plants are generally expressed in megawatt-hours (MWh). World energy consumption usually expressed in tetrawatt-hours (TWh). Sometimes, it is also expressed in total tonne of oil equivalent† (toe) where the International Energy Agency (IEA) defined one (1) toe to be equal to 11.63 MWh or around 41.868 GJ. Therefore 1 Mtoe= 11,63GWh.

Example 1.1

A local agro-based SME company currently employs a typical type of motor for fruit-cutting processes. The 2.5 hp motor operates with an efficiency of **70%**.

(a) What is the power drawn from the supply in kW?

---

*As a rule of thumb, 3 kg of uranium U-235 = 8,640 tonnes of coal!

†Tonne of oil equivalent (toe) is defined as the amount of energy released by burning one metric ton (1000 kg) of crude oil.

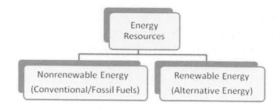

Figure 1.2
Energy resources classification.

(b) If the supply voltage is 240V, what is the input current?

(c) What is the possible energy saving per year, if Management decided to change to a new motor of the same capacity but with higher efficiency of 75% (assuming operating hours = 3,920 hours/year, and 1 hp = 746 W)?

Solution:

(a) Since efficiency η of an energy system is defined as the power output P_{out} per energy input P_{in},

$$\eta = \frac{P_{out}}{P_{in}} \times 100\% \Rightarrow 0.7 = \frac{2.5hp \times \frac{746W}{hp}}{P_{in}}$$

$$\therefore P_{in} = 2664.3W = 2.67kW \,\#$$

(b) The input current I is,

$$I = \frac{P}{V} = \frac{2664.3}{240} = 11\,A\,\#$$

(c) New motor power input requirement:

$$\eta = \frac{P_{out}}{P_{in}} \times 100\% \Rightarrow 0.75 = \frac{2.5hp \times \frac{746W}{hp}}{P_{in,new}} \Rightarrow P_{in,new} = 2486.7W = 2.49kW$$

Therefore, total saving is,

$$(2.67 - 2.49)kW \times 3,920\,hours/year = 706kWh/year\,\#$$

1.2.3 Energy Resources

Our energy resources on Earth can be broadly classified into two categories (Figure 1.2): nonrenewable (also known as "conventional" and "fossil fuel") and renewable (or "alternative") energy. We normally call fossil fuels *nonrenewable* because their supplies are limited and we cannot produce as much fossil fuel in a short time as we like. Unlike fossil fuels, renewable energy can be replenished close to or at the same rate that they are used — thus at times we also call it *sustainable* energy.

1.2.3.1 What Are the Nonrenewable Energy Resources?

Around the world, most of our electricity energy is generated based on nonrenewable energy resources. Fossil fuel (coal, petroleum, natural gas) and uranium are the primary nonrenewable energy sources. They are used to make electricity and provide thermal heat for heating/cooling and transport, as well as for manufacturing all kinds of industrial products. Other fossil fuels incluse bituminous sands (tar sands), oil shale, and lignite, which need expensive processing before we can use them.

Figure 1.3
World energy consumption overview (From Enerdata, 2010, Yearbook Statistical Energy Review 2010 http://yearbook.enerdata.net/. With permission.)

1.2.3.2 What Are the Alternative or Renewable Energy Sources?

Alternative or renewable energy sources include biomass, geothermal energy, hydro power, solar energy, ocean energy, and wind energy. They are called renewable* because they are *replenished* or *regenerated* in a short time. We use renewable energy sources mainly to make electricity and provide thermal heat for applications.

1.2.4 World Energy Consumption and Demand

Figure 1.3 presents an overall view of our current energy consumption pattern around the world. The figure indicates that the largest demand is in Asia, Asia Pacific, Europe, and North America [6].

China became the world's largest energy consumer, about 18% of the world total, as seen in Table 1.1, since its consumption surged by 8% during 2009, from only 4% in 2008. This is followed closely by the USA (2,201 Mtoe); and then India (655 Mtoe), Russia (621 Mtoe), Japan (459 Mtoe), and Germany (315 Mtoe).

Table 1.1 Total energy consumption by country for Year 2009

Country by ranking	Unit: Mtoe
China	2,234
USA	2,201
India	655
Russia	621
Japan	459
Germany	315
France	254
Canada	244
Brazil	238
South Korea	233

(From Enerdata, 2010, Yearbook Statistical Energy Review 2010 http://yearbook.enerdata.net/. With permission. [6])

---

*It is to be noted that the term "alternative" energy is preferred by some as energy can not actually be "renewed" but converted from one form to another.

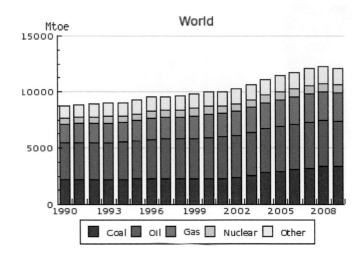

Figure 1.4
Total world energy consumption from 1990–2008 (From Enerdata, 2010, Yearbook Statistical Energy Review 2010 http://yearbook.enerdata.net/. With permission.)

Figure 1.4 shows our total world energy consumption from 1990–2008. Oil remained the largest energy source (33%) despite its share decreasing over time. Coal has shown a growing role in the world's energy consumption, where in 2009 it accounted for 27% of the total consumption. It is expected that worldwide energy demand will increase by about 30% between 2010 and 2030, with hydro and other renewables increasing at the highest rates. We shall further discuss these in subsections 1.4.1.2 and 1.4.1.3, under fossil fuels conversion and renewable energy conversion, respectively.

1.2.5   The Conservation of Energy

The conservation of energy or First Law of Thermodynamics tells us that energy changes form and it moves from place to another place but the total amount of energy in a system remains unchanged. In other words, we can say that "energy input" equals "energy stored" plus "energy output." It tells us nothing about loss energy, idle energy or waste energy.

1.2.6   The Availability of Energy

The availability of energy or the Second Law of Thermodynamics deals with the availability of energy, which means "energy is conserved but *not necessary* that its availability or usefulness." It recognizes that some of the energy of a fuel is not available for conversion to work in an energy conversion system, due to losses. In other words, it means that in the real world no process is essentially 100% efficient. We shall further discuss some equations in relation to energy conservation and energy availability (First and Second Law of Thermodynamics) in the next section.

1.3   Energy Equation

1.3.1   The Energy Conservation Equation (First Law of Thermodynamics)

We can write the energy conservation equation in a simple manner,

$$\text{(Energy stored)} = \text{(Energy input)} - \text{(Energy output)} \tag{1.11}$$

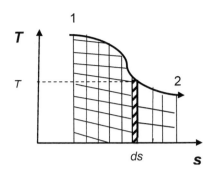

Figure 1.5
T-s diagram.

$$dU = \delta Q - \delta W \tag{1.12}$$

Where,
dU is the infinitesimal increase in internal energy of the system,
δQ is the infinitesimal heat flow into the system, and δW is the infinitesimal work done by the system.

1.3.2   The Energy Availability Equation (Second Law of Thermodynamics)

The energy availability equation can be expressed in a simple equation as follows:

$$ds \geq 0 \tag{1.13}$$

where s is the entropy for an isolated system.

A simple definition of entropy s is that it is the amount of energy that is not available for work for a system. In other words it is the energy form of a system that relates to its internal state of disorder* — which contributes to loss or disturbance. The higher entropy denotes bigger disordered states, thus more energy loss from the system.

Just for a quick review revision on entropy s, we shall already note and are quite familiar with the property s that appears in many charts and tables of properties; and we also have read lines of constant entropy on many graphs. For example, in a *T-s* diagram, e.g., Figure 1.5, the shaded area or $\int_1^2 T\,ds$ represents the heat supply usually denoted with Q.

1.3.3   General Mechanical Energy Equation

The General Mechanical Energy Equation is a statement of the conservation of energy principle. It involves energy, heat transfer, and work. Here are some of the energy equations in various forms, with relation to common energy applications, and with their respective importance.

Energy content per unit mass of a closed system in rate form   Applying the energy conservation law, the energy content per unit mass of a closed system can be changed by energy input applied on the system, i.e., heat transfer (Q) and work transfer (W). Therefore, we can write conservation of energy for a closed system, expressed in rate form, as

$$\frac{dE_{system}}{dt} = \dot{Q}_{net,in} + \dot{W}_{net,in} \tag{1.14}$$

---

*This includes excitation of quantum states at the microscopic level.

Where the net rate of heat transfer to the system is

$$\dot{Q}_{net,in} = \sum \dot{Q}_{in} + \sum \dot{Q}_{out} \tag{1.15}$$

And the net power input to the system:

$$\dot{W}_{net,in} = \sum \dot{W}_{in} + \sum \dot{W}_{out} \tag{1.16}$$

Since we are considering only steady flow processes in most of our applied energy systems, there is no change of E system with respect to time t, thus the $\frac{dE_{system}}{dt}$ term reduced to zero. This results in the following form,

$$\dot{Q}_{net,in} = -\dot{W}_{net,in} \tag{1.17}$$

Or, the usual form of the steady flow energy equations, with inlet *1* and outlet *2*, of the form,

$$\dot{m}\left(\frac{V_1^2}{2} + h_1 + gZ_1\right) + q = \dot{m}\left(\frac{V_2^2}{2} + h_2 + gZ_2\right) + w \tag{1.18}$$

The kinetic energy term $\frac{V^2}{2}$ and potential energy term gz is usually negligible compared with other energy terms, i.e., the change of enthalpy h, the heat transferred q, or the work done. If no work is done ($w=0$), the energy equation becomes simply,

$$q = \dot{m}\left(h_2 - h_1\right) \tag{1.19}$$

Two of the most common mechanical energy system components used in energy-related industries is pump and turbines. We can compare the differences between pump and turbines in an easy way as follows:

Notes: Pumps (and other devices, such as fans, compressors, and propellers)   Pumps are machines that supply energy to fluid. A pump moves fluids (liquids or gases) from lower pressure to higher pressure, and overcomes this difference in pressure by adding energy to the system. This results in an increase of pressure in pumps. The energy conversion involved is: Mechanical Energy → Hydraulic energy.

Turbines   Turbines are machines that capture (extract) energy from a moving fluid — liquid or gas. A decrease in pressure takes place in turbines. Turbines convert the energy of fluid into mechanical energy. Hydraulic energy → Mechanical Energy.

The following are some of the common energy conservation equations widely used in pump, fan, turbine, and other energy-related systems.

Energy per unit mass equation for a pump, fan, or similar devices   Now, considering incompressible flow,[*] the mechanical energy equation for a pump or a fan can be written, with inlet *1* and outlet *2*, in terms of energy per unit mass, as follows:

$$\frac{V_1^2}{2} + \frac{P_1}{\rho} + gZ_1 + w_{shaft} = \frac{V_2^2}{2} + \frac{P_2}{\rho} + gZ_2 + w_{loss} \tag{1.20}$$

Where
V = flow velocity;
ρ = density;
P = pressure;
g = acceleration of gravity;
Z = elevation height;
w_{shaft} = net shaft energy per unit mass for a pump, fan, or similar devices;
w_{loss} = loss due to friction.

---

[*]Flow with fluid motion of negligible changes in density.

Energy per unit mass equation for a turbine or similar devices Unlike a pump, a turbine produces works, so that now the shaft works *Wshaft* moves to the right hand of the above equation,

$$\frac{V_1^2}{2} + \frac{P_1}{\rho} + gZ_1 = \frac{V_2^2}{2} + \frac{P_2}{\rho} + gZ_2 + w_{shaft} + w_{loss} \tag{1.21}$$

Where *Wshaft* = net shaft energy output per unit mass for a turbine or similar devices

Energy per unit volume equation for a pump, fan, or similar devices The mechanical energy equation for a pump or a fan (1) can also be written in terms of energy per unit volume (Unit: N/m2) by multiplying Eq. (1.20) with fluid density ρ:

$$\frac{\rho V_1^2}{2} + P_1 + \gamma Z_1 + \rho w_{shaft} = \frac{\rho V_2^2}{2} + P_2 + \gamma Z_2 + \rho w_{loss} \tag{1.22}$$

where $\gamma = \rho\,g$ = specific weight.

Energy per unit weight (or head) (unit: N.m/N = m) equation for a pump, fan, or similar devices We can also write the mechanical energy equation for a pump or a fan in terms of energy per unit weight (or head) by dividing Eq. (1.20) with gravity g,

$$\frac{V_1^2}{2g} + \frac{P_1}{\rho g} + Z_1 + \frac{w_{shaft}}{g} = \frac{V_2^2}{2g} + \frac{P_2}{\rho g} + Z_2 + \frac{w_{loss}}{g} \tag{1.23}$$

Or in the following form,

$$\frac{V_1^2}{2g} + \frac{P_1}{\gamma} + Z_1 + h_{shaft} = \frac{V_2^2}{2g} + \frac{P_2}{\gamma} + Z_2 + h_{loss} \tag{1.24}$$

where
$\gamma = \rho g$ = specific weight,
$h_{shaft} = \frac{W_{shaft}}{g}$ = net shaft energy head per unit mass,
$h_{loss} = \frac{W_{loss}}{g}$ = headloss.

Example 1.2 Consider a condenser of an industrial refrigerator unit located at a petrochemical plant. A flow of steam enters the condenser at a velocity of 110 m/s with specific enthalpy of 1438 kJ/kg . The condensate leaves the condenser with specific enthalpy of 292kJ/kg. (A) If the heat rejected at the condenser is 1150 kJ/kg, what is the velocity of the condensate? (B) What is the rate of heat removal if the mean mass flow rate is 1.2 kg/s?

Solution:

$$\dot{m}\left(\frac{V_1^2}{2} + h_1 + gZ_1\right) + q = \dot{m}\left(\frac{V_2^2}{2} + h_2 + gZ_2\right) + w$$

$$\Rightarrow q = (h_2 - h_1) + \frac{(V_2^2 - V_1^2)}{2} + g(Z_2 - Z_1)$$

(A) Therefore,

$$\Rightarrow -1150 kJ/kg = (292 - 1438)kJ/kg + \frac{(V_2^2 - 110)\frac{m}{s^2}^2 \cdot \frac{kg}{kg}}{2} \cdot \frac{1}{10^3 \frac{Nm}{kJ}} + 0$$

$$\therefore V_2 = 64\,m/s \,\#$$

(B) The rate of heat removal,

$$Q = \dot{m}(h_2 - h_1) = 1.2\ \text{kg/s}(160 - 2300)\text{kJ/kg} = -2,568\text{kW}\#$$

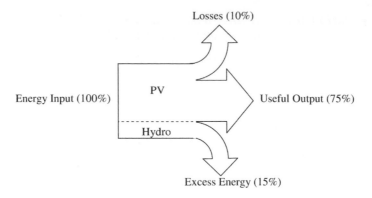

Figure 1.6
A typical hybrid energy system [7].

1.3.4   Gibbs Free-Energy Relation

The Gibbs free energy of a system at any moment in time is defined as the enthalpy of the system minus the product of the temperature times the entropy of the system. Gibbs free energy,

$$G = H - TS \tag{1.25}$$

Gibbs free-energy is a thermodynamic potential that measures the "useful" or process-initiating work obtainable from an isothermal, isobaric thermodynamic system. The change in the Gibbs free energy of the system that occurs during a reaction is therefore equal to the change in the enthalpy of the system minus the change in the product of the temperature times the entropy of the system. The change of Gibbs free energy can be expressed as,

$$\Delta G = \Delta H - \Delta(TS) \tag{1.26}$$

And, at constant temperature, the change of Gibbs free energy is,

$$\Delta G = \Delta H - T\Delta S \tag{1.27}$$

1.3.5   Energy Performance Curve Equation in Terms of Useful Output for Hybrid Energy System*

Let us consider a typical combined PV/Hydro hybrid energy system as shown in Figure 1.6, where the total energy input (100 %) is a combination of power range derived from PV (85–95 %) and hydro (5–15 %) hybrid system.
We can write the energy balance as follows:

$$E_I = E_o + E_{Loss} + E_{Excess} \tag{1.28}$$

$$(Energy Input) = (Useful Output + System Losses + Excess Energy) \tag{1.29}$$

System operational efficiency, η

$$\eta = \frac{useful\ output}{total\ input} = \frac{E_o}{E_I} = \frac{75}{100} = 75\% \tag{1.30}$$

Notice that the numbers at the right side of the above equation are typical example values only.
Introducing an inactive (idle) energy input, E_0^I into Eq. (1.28), at $E_0 = 0$, and expressing in terms of E_{Loss}^0 and E_{Excess}^0, as follows:

$$E_0^I = E_{Loss}^0 + E_{Excess}^0 \tag{1.31}$$

---

*Adopted from Abdullah et al.[7]

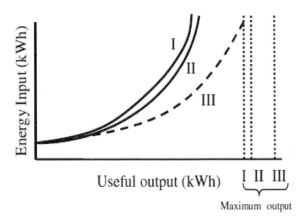

Figure 1.7
The energy performance curve, i.e., energy input vs. useful output for three typical hybrid energy systems [7].

The net energy input efficiency, ε_{net}, therefore is,

$$\varepsilon_{net} = \frac{useful\ output}{total\ net\ input} = \frac{E_0}{E_I - E_I^0} \equiv \varepsilon^0 \left(1 - \frac{E_0}{E_{0,max}}\right) \tag{1.32}$$

Now, the energy performance equation can be established as in Eq. (1.33),

$$E_I = E_I^0 + \frac{E_0}{\varepsilon^0 \left(1 - \frac{E_0}{E_{0,max}}\right)} \tag{1.33}$$

Subsequently, we can draw typical energy performance curves as in Figure 1.7, in terms of energy input versus useful energy output, based on the equations above.

Note: The dashed lines indicating maximum energy output curves show that the useful output of the typical hybrid energy system above could be well improved by utilizing higher combined total energy inputs. More discussion on hybrid energy will be covered in Chapter 10.

Next, we shall first introduce various energy conversion systems.

1.4 Energy Conversion Systems

An energy conversion system is any energy system that involves a process of changing one form of energy into another, and produces desired work. As discussed in the previous sections, energy in a system may be transformed or converted so that it resides in a new state. This conversion results in many varieties of our physical work. For instance, a pump converts mechanical energy into hydraulic energy; a turbine converts the energy of fluid into mechanical energy and drive shaft, an internal combustion engine converts the potential (chemical) energy in gasoline and oxygen into heat, which is then transformed into the kinetic energy of propulson that moves a vehicle. A solar photovoltaic cell converts solar radiation into electrical energy to light a bulb or power a solar calculator.

1.4.1 Types of Energy Conversion Systems

Broadly, we can classify the energy conversion systems into two categories i.e. Conventional Energy Conversion and Alternative (Non-conventional) Energy Conversion (Figure 1.8).

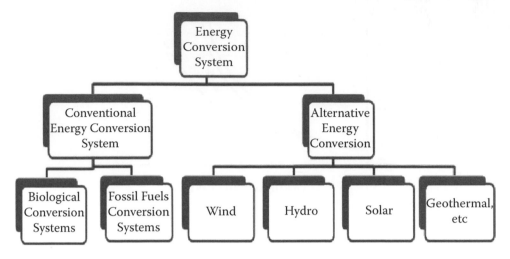

Figure 1.8
Energy conversion classification.

1.4.1.1   Conventional Energy Conversion Systems

1.4.1.1.1   *Biological Conversion Systems*   What are biological conversion systems? Biological conversion systems are the conversion systems that an available in human and other living organisms (notably animals and plants) in our Earth's ecosystem. The energy flow through living organisms first occur with sunlight and photosynthesis, where the plants, as primary producers, are at the first level in the energy food chain. Subsequently, the energy flows through the rest of the food chain, i.e., primary consumers, secondary consumers, tertiary consumers, and so on (Figure 1.9); and finally the decomposers or "detritivores" (dead cells) act as the bottom part of the food chain, so that energy goes to the soil and back to the plants and the cycle starts again (see Figure 1.10).

During the above cycle processes, energy is consumed by each level and some energy is lost in the form of heat energy. Similarly, nutrients, water, and other materials are continually recirculated within and among ecosystems.

Primary producers, such as *autotrophs*, "self-feeding" organisms, are capable of storing biochemical energy by synthesizing organic molecules from inorganic precursors. Primary produces manufacture complex organic molecules

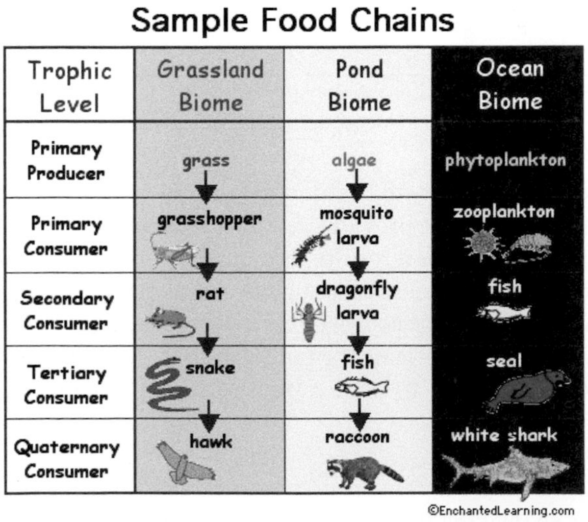

Sample Food Chains

Trophic Level	Grassland Biome	Pond Biome	Ocean Biome
Primary Producer	grass	algae	phytoplankton
Primary Consumer	grasshopper	mosquito larva	zooplankton
Secondary Consumer	rat	dragonfly larva	fish
Tertiary Consumer	snake	fish	seal
Quaternary Consumer	hawk	raccoon	white shark

©EnchantedLearning.com

Figure 1.9
Sample food chains. (Courtesy of enchantedlearning.com [9].)

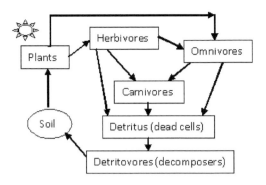

Figure 1.10
Simplified food-energy cycle.

from simple inorganic compounds, namely water (H_20), carbon dioxide (CO_2), and nutrients through the photosynthesis process together with light energy:

$$6CO_2 + 6H_2O \stackrel{photosythesis}{\longrightarrow} C_6H_{12}O_6 + 6O_2 equation \tag{1.34}$$

The photosynthesis process reduces carbon compounds and produces oxygen.

What is a trophic level? This is a feeding level, often represented in a food chain just described. Primary producers comprise the bottom trophic level, followed by primary consumers (herbivores), then secondary consumers (carnivores feeding on herbivores), and so on.

What happens to the primary producers? Generally, primary producers are either consumed or decomposed. From the equation for the aerobic respiration process,

$$C_6H_{12}O_6 + 6O_2 \longrightarrow 6CO_2 + 6H_2O + heat\ energy \tag{1.35}$$

In the process, metabolic work is done and the associated energy in the chemical bonds is converted to heat energy. However, if a primary producer was not consumed, it would be covered by layers of silt, sand, and other rock formations. It is believed that during the prehistoric period, e.g., the Carboniferous period, about 360 to 286 million years ago, enormous amounts of primary producers in excess of consumption accumulated in swamps. It was buried and compressed to form coal, oil deposits, and gas, under suitable high pressure and temperature (HPHT) conditions. Today, we drill down through layers of silt, sand, and rock formations to reach the reservoirs containing oil and gas.

What is an ecology pyramid or ecology cone? Figure 1.11 represents quantitatively the efficiency of energy transfer in a food chain (Science Aid [9]). From an applied energy point of view, we can perhaps estimate that, due to energy loss, only about 8 to 10% of the energy is actually transferred from one stage to the next level of the food chain. In other words, the energy conversion efficiency is around 8 to 10%. We construct Figure 1.12 in the form of an ecological cone, to conveniently represent energy transfer in the food chain. However, we shall notice that the energy availability of the Earth essentially is 100%! We shall discuss biomass technology and its applications in Chapter 9.

1.4.1.2 Fossil Fuel Conversion System

The fossil fuel (oil, gas, and coal) conversion system involves few conversion stages. It starts with stored chemical energy in the fossil fuel and ends with electricity production; usually it is a multistage, chemical-thermal-mechanical-electrical energy conversion system. There are two types of electricity energy utilization; one is fixed (e.g., for our residential applications) and the other is mobile (transport applications such as our vehicles). Figure 1.13 (top figure)

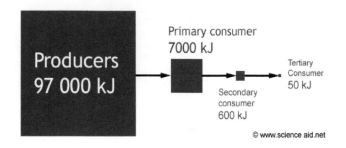

Figure 1.11
Energy transfer block diagram. A typical energy consumption block diagram for the ecological food chain. (From Science Aid, ©www.scienceaid.net [9].)

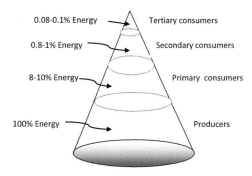

Figure 1.12
Ecological cone and energy transfer in food chain.

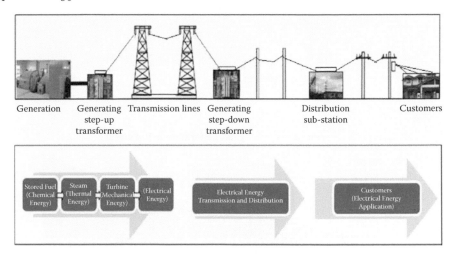

Figure 1.13
Conventional electric energy power generation from fossil fuel, its transmission, and distribution system. Steam turbine application driven by fossil fuels is about 80% of our total world electricity.

show conventional electric energy power generation from fossil fuel, together with its transmission and distribution system (usually known as energy from grid). Most power stations burn coal, oil, or natural gas to run the generators, and normally the energy conversion processes is as follows: Burn fossil fuel → heat water to steam → steam turns turbine → turbine turns generator → electric power production through induction. The energy conversion system therefore is a chemical-thermal-mechanical-electrical conversion system (Figure 1.13, bottom figure).

Now, for fixed electricity utilities, electricity is normally transmitted using high-voltage power lines. Nearly all of the power we use comes from large power stations, although diesel generators are also used usually as backup and rural power applications* (see, e.g., [7]) Biomass-driven generators are also used in some other remote places. Relatively few are using fuel cell generators, as fuel cell technology has relatively high cost implications.

Now, as far as energy resources are concerned, in 2008, fossil fuels provide around 67% of our world's electrical power (Table 1.2). Around 95% of the world's total energy demands including heating, transport, electricity generation and other uses depending on fossil fuels. Of all the fossil fuel resource, coal provides the highest at around 41% of our energy, and oil provides 5%. Natural gas provides around 21% of the world's consumption of energy, and as well as being burnt in gas power stations, is used by many people to heat their homes in cold-climate countries. Other fossil fuels include bituminous sands (tar sands) and oil shale, which need expensive processing before we can use them. Annual electric net generation in the world from 1980 through 2008 is shown in Figure 1.14, where the gradient of net electricity generation from fossil fuels (coal/natural gas/ petroleum) is clearly shown to be less than the gradient of net electricity by nuclear energy; renewable energy has the highest gradient of net electricity generation.

Table 1.2 Source of electricity world total — 2008

Resources	Coal	Oil	Natural gas	Nuclear	Hydro	Other	Total
TWh/year	8,263	1,111	4,301	2,731	3,288	568	20,261
Proportion	41%	5%	21%	13%	16%	3%	100%

(Data source: IEA [10])

We shall cover fossil energy conversions and electricity generation in more detail in Chapter 6.

1.4.1.3 Alternative or Renewable Energy Conversion Systems

Due to the fossil energy shortage around the globe and the increasing associated environmental pollution problems, apart from energy efficiency of energy conversion systems, development of reliable renewable energy systems is becoming more and more encouraging.

---

*It is estimated that a quarter of the world's population, around 2 billions in numbers, still lack access to electric energy from grid.

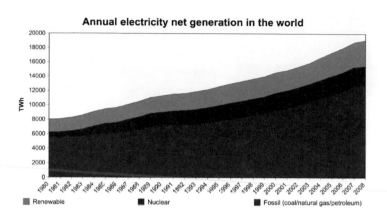

Figure 1.14
Annual electric net generation in the world (IEA) [10].

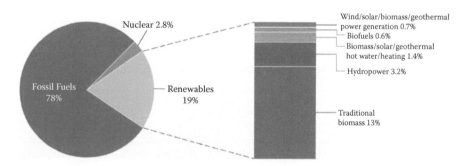

Figure 1.15
Renewable energy share of final energy consumption in 2008 [11].

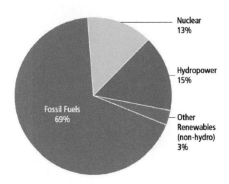

Figure 1.16
Share of global electricity from renewable energy in 2008 [11].

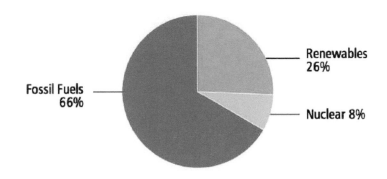

Figure 1.17
World Generating Capacity by Source in 2009 [11].

According to the Renewables 2010 Global Status Report [11] published recently, we can see that the renewable energy share had been cited to be about 19% in 2008 (Figure 1.15). In 2008 as well too, hydro energy share had reached 15% while other renewables amount to only about 3% (Figure 1.16). In 2009, according to World Generating Capacity by Source, renewable energy (hydro + non-hydro renewable) had increased by 7% to reach about 26% (see Figure 1.17) in total total share of global electricity!

According to the World Energy Council Report, also published in 2010, the share of renewable energies in electricity production, excluding hydro energy, increased from 3% in 2008 to 5.1% in 2010 [12].

As to further "most-likely" energy trend estimation (see Figure 1.18), according to BP2030 [13] and the World Energy Council Report [12], the contribution of renewables (excluding hydro energy) to energy growth would further increase from 5% (1990–2010) to 18% (2010–2030). The contribution of fossil fuels to primary energy growth is projected to fall from 83% (1990–2010) to 64% (2010–2030).

Below is a brief description of the common renewable conversion systems.

Hydro* energy *conversion system The suitable turbine types used for a hydro power plant depends on many influencing parameters, with two primary parameters, i.e., the location (usually called water head) and flow characteristics of the available water source. Turbine types generally can be classified into two main types, i.e., Impulse Turbines and Reaction Turbines (Figure 1.19).

The power generated from a hydro turbine can be estimated from,

$$\text{Power } P = \eta g h Q \tag{1.36}$$

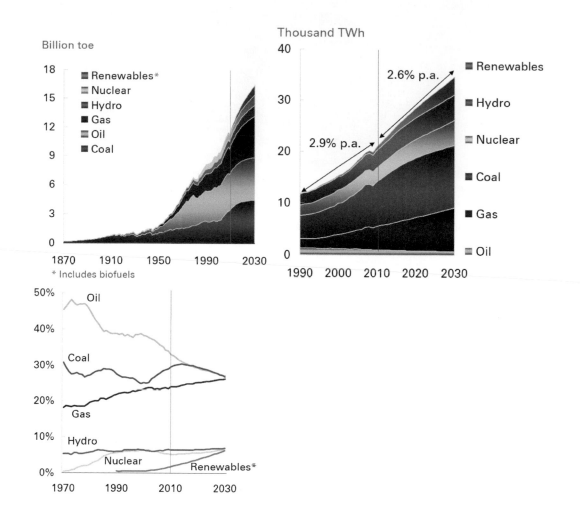

Figure 1.18
World commercial energy use (top left), shares of world primary energy (top right), and world power generation (bottom left). (Courtesy of BP Energy Outlook 2030 [13].)

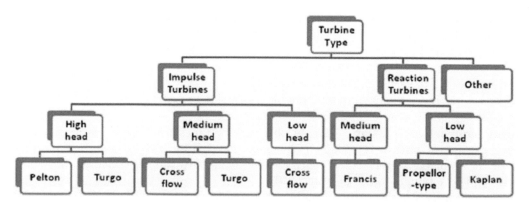

Figure 1.19
Turbine type classifications.

Figure 1.20
The Three Gorges Dam provides the world's largest hydro-electric power station, with a total capacity of 22,500 MW. The total length of the dam axis is 2309.47m with a crest elevation of 185m and maximum dam height of 181m. The normal reservoir storage water level is 175 m, the total reservoir storage capacity is 39.3 billion m^3. (Courtesy of Three Gorges Project [14].)

Hydro power driven generators are used for electricity generation applications, supplying power to grid, for both residential and industrial applications. Apart from being environmentally-friendly, it basically requires zero fuel input. Today, it is the most mature conversion technology that could produce big scale electricity energy generation applications. A typical example of successful hydro conversion systems is the well known "Three Gorges Dam" in China (Figure 1.20).

Wind energy conversion system In a wind energy power plant, wind turbines convert the kinetic energy in the wind into mechanical and electrical power. The power output P from a wind turbine is given by the well-known wind power equation,

$$P = \tfrac{1}{2}\, Cp\rho AV^3 \tag{1.37}$$

where
ρ is the density of air (1.225 kg/m^3)
Cp is the power coefficient,
A is the rotor sweep area, and
V is the wind speed.

Wind farms are of two types — on land and offshore. Wind farms can have a lot of individual turbines, up to a series of 100 turbines or more for generating a huge amount of electricity production (for example, the offshore wind farm at Kent, Figure 1.21) The electricity generated from wind turbines can be used for many applications, power to grid, for residential and commercial applications, etc.

Application advantages: Apart from its being environmentally-friendly, it is basically zero fuel input, and suitable for remote and rural electrification.

Solar energy conversion system Solar energy is considered by Scientists to be the most important available energy, where all fossil and renewable energies are derived from. It is universally and freely available in abundance everywhere; thus the total actual energy capacity is extremely large. Solar energy can be broadly divided into two types, i.e., photo energy and heat energy. Although solar energy available in abundance, until today, unlike with hydro energy, the techno-economical aspects of the solar photovoltaic collection techniques have not fully reached a stage that is economically viable for wide electricity production as compared to conventional fossil energy driven or hydro power. Two primary techno-economical concerns with relation to solar energy are how to collect it efficiently and how to store it, due to low efficiency of solar PV and the requirement of big battery banks for energy storage. Solar thermal energy, on the other hand, is widely used around the globe for drying, and for space- and water-heating.

Figure 1.21
The world's largest offshore wind farm. The 300 MW Thanet Wind Farm is installed offshore, off the coast of Thanet in Kent, United Kingdom. It consists of 100 turbines in total. The wind farms project covers an area of 35 km². [Courtesy of BBC News [15] and The Press Association (PA) Photos Limited.)

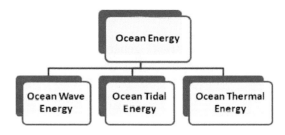

Figure 1.22
A simple classification of ocean energy.

Application advantages: Apart from its being environmentally-friendly, it is basically zero fuel input, suitable for thermal energy production and small scale rural electrification.

Geothermal energy conversion system   Geothermal energy is our Earth's high grade "interior" temperature i.e., it uses the heat of magma below the Earth's crust, due to radioactive decay. The energy capacity of geothermal energy is estimated to be around 50 mW per m^2, and with a temperature gradient of around 30 K per km depth.

Application advantages: The application of geothermal energy conversion systems is limited to areas near tectonic plate boundaries. Three general usages of geothermal energy are: (a) Electricity generation; (b) Direct use; and (c) use as geothermal heat pumps.

Tidal, wave, and ocean energy conversion systems   Our ocean can produces 3 types of energy, i.e., ocean wave energy, ocean tidal energy, and ocean thermal energy (see Figure 1.22). Oceans cover $\geq 70\%$ of our Earth's surface, and act as the World's largest areas of (tidal/wave) energy providers and thermal collectors.

Advantages: As with most other renewable energies, apart from being environmentally-friendly, they basically require zero fuel input, and are sustainable and free.

The primary disadvantage is in relation to the high saline content of sea water, which can contribute corrosion related problems. Tidal power is subject to unsteady energy generation (due to variation in tidal head), sedimentation, and erosion problems. Ocean wave energy is subject to the variations of wave frequency and amplitude and thus requires good inconsistency control in energy generation.

Biomass energy conversion system   Biomass refers to biological materials derived from living organisms such as corn and wood. It is used to produce either electricity or heat. Biomass energy application includes: Ethanol

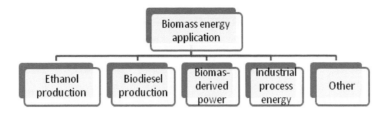

Figure 1.23
Biomass energy applications classification.

production, bio-diesel production, biomass-derived power and often use in numerous industrial process energy plants worldwide (Figure 1.23).

Advantages of biomass conversion are: (a) Unlike fossil fuels, biomass generally is regarded as an inexhaustible fuel source; (b) It is available worldwide even in rural areas.

However, biomass conversion currently is still considered an expensive conversion method, in terms of producing the biomass, converting it to fuel (ethanol, bio-diesel, etc.), and for reliable electricity energy generation. We shall cover renewable energy conversions in more detail in Chapter 7 through Chapter 11. We shall also endeavor to find out the latest issues in relation to biomass energy applications, covered in Chapter 12.

1.4.1.4   Peat and Other Energy Conversion Systems

Peat and peat energy conversion systems   Peatlands are areas of landscape, with or without vegetation, that have a naturally accumulated peat layer at the surface. Globally, peatlands are major stores of carbon. The organic component of peat deposits has, however, a fairly constant anhydrous, ash-free calorific value of 20-22 MJ/kg. Peat is used as fuel for electricity/heat generation, and directly as a source of heat for industrial, residential, and other purposes. Peat is used in combined heat and power plants (CHP) (45% of the total use), in condensing power generation (CP) (38%), district heating (DH) (10%), and residential heating (RH) (8%) (World Energy Council [13]). The world's annual peat harvest is equivalent, according to Joosten and Clarke [16], to about 15 million metric ton of carbon. From a reserve point at view, it is clear that the world possesses huge reserves of peat overall, with undisturbed peatlands reported to be around 3,500,000 km^2. According to Strack [17], the global peat carbon pool is in the area of 500 billion metric tons.

Other energy conversion systems   Absorption and adsorption energy conversion, thermoelectric and thermionic energy conversion, magnetohydrodynamic (MHD) power conversion, as well as other emerging energy conversion methods, will be covered in Chapter 6.

1.5   Energy Storage, Methods and Safety

What is energy storage?   Energy storage is refers to a space or system devices that we can use to store some form of energy to perform some useful operations at a later time. An oil reservoir, for instance, stores petroleum while a gas reservoir store natural gas. Even our own human body also passesses excellent means of storage, such that the food that we consume can be utilized for useful energy production and work! Storing energy allows humans to balance the supply and demand of energy, and most importantly, to achieve energy sustainability.

How our fossil fuels — (oil and gas are) stored   Oil and gas were stored originally in reservoirs formed as oil or gas formations. We extract them by using suitable drilling and production technologies for our applications. Once extracted, the fuels are transported in trucks and pipelines for delivery. Petroleum oil is usually stored in large-capacity storage sites equipped with specially designed tanks, either above ground or below ground (like underground caverns). Similar to oil, natural gas can be stored in either above-ground or underground reservoirs/storage tanks or

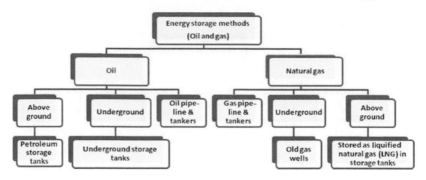

Figure 1.24
A simple classification of oil and gas storage.

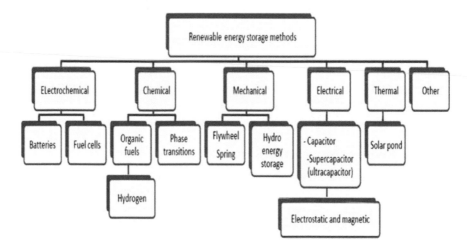

Figure 1.25
Classification of renewable energy storage methods.

in old gas wells. Natural gas can be stored as liquefied natural gas (LNG) in LNG storage tanks (Figure 1.24). LNG can be converted into other products, for example, ammonia, by a process called steam reforming, and subsequently stored in huge industrial refrigerated storage tanks.

Renewable energy storage methods   Renewable energy storage systems in both residential and commercial use today can be broadly categorized into a few methods, such as mechanical, electrical, chemical, biological, and thermal (see Figure 1.25).

What are the problems and general concerns of renewable energy storage?   Unlike fossil fuel storage, the technology of which is quite matured and established, the problem of storing large amounts of accessible energy in an efficient and cost effective way has still remained one of the most difficult tasks for our renewable energy systems [18]. Energy storage in the form of hydro energy, wind energy and pumped hydroelectric are perhaps three of the most successful high capacity utility applications (can generate up to 2 GW power or more). The limitation of pumped hydroelectric is the big reservoir requirement and the need of a suitable site and pumping provision for such big installation. Wind energy requires high enough wind potential areas such as wind farms situated offshore, and the batteries available today are not designed for large systems — mainly due to cost limitations. Beside most batteries available are short in lifetime, this can results in disposal problems. For average capacity energy applications, mechanically-operated flywheels, superconducting magnetic devices, and super-capacitors are usually used and applied for storing energy. However, they are generally very expensive for the same energy density content that batteries can afford to store.

Figure 1.26
A typical double-walled anhydrous ammonia tank at Bintulu. The tank is 22.4 min diameter and 24.6 min height — about the size of a quarter of a football field. (From *Corrosion Journal* [21].)

Why safety of energy storage? To ensure our human life, health, and environments are well protected, safety measures is always one of the main topics for good industrial practices, an energy storage safety is one of the main concerns. Oil and gas industries, for instance, have very stringent energy storage tanks standards, e.g., Society of Petroleum Engineers (SPE), American Society of Mechanical Engineers (ASME), International Organization for Standardization (ISO), etc., and need to comply at all times. The application of high pressure, compressed air, and toxic energy fuels all have safety concerns in relation to the possibility of catastrophic tank ruptures, corrosion, or even explosions. Nuclear waste and radioactive storage require very high standard of storage, handling, and disposal site procedures (for example, see Ronald Allen Knief [19]). Commercial storage designs adopted, for example, the ISO 11439 Standard, for storage of natural gas, as fuel for automotive vehicles [20]. ISO 23273-2 describes protection against hydrogen hazards for vehicles fueled with compressed hydrogen, in particular, for fuel cell vehicles. A further example is ammonia tank storage, which normally require the storage vessel (see, e.g., Figure 1.26) to be periodically tested for application integrity, and saved from possibilities of tank failure, for instance, stress corrosion cracking [21]. Renewable energy conversion systems, likewise, require high energy storage capability options, for instance superconductive magnetic energy storage (SMES) techniques (see, e.g., Johnson et al. [22], Tam et al.[23], and Ali and Dougal [24]). We shall cover in more detail the various energy storage systems, safety, method, energy storage comparisons, and their associated applications, in Chapter 5.

1.6   Energy Efficiencies and Losses

What is energy conversion efficiency? Energy conversion efficiency is the ratio between the useful output of an energy conversion system and the energy input to the system. Depending on the output desired, the useful output can be electric power, mechanical work, heat, and so on.

$$\text{Energy conversion efficiency} = (\text{useful energy output})/(\text{total energy input}) \qquad (1.38)$$

If a series of energy conversions is required to produce the final form of energy output, we will find losses in each step of the conversion process. The overall efficiency of such systems will be essentially reduced compared to direct or shorter conversion. The overall efficiency $\eta_{overall}$ is the product of the efficiencies of each step in the conversion process.

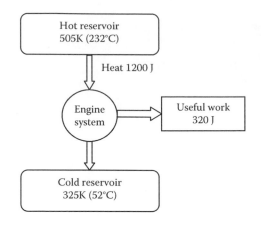

Figure 1.27
Schematic energy flow of a typical automobile engine (Case I).

$$\eta_{overall} = \sum (\eta_i) = (\eta_1) \times (\eta_2) \times (\eta_3) \times \ldots \tag{1.39}$$

Example 1.3 A typical electric-powered car having a 2A, 24V battery was charged at constant current from 8am till 10pm. Determine: (A) How much electric energy is used? (B) What is the cost of electricity to charge the battery? (C) Calculte the total energy conversion loss. (D) What is the actual efficiency of the battery? [Data given: electric cost =\$0.28/kWh,*and assuming 90% battery (chemical energy) to electricity conversion, and 60% electrical to mechanical energy conversion].

Solution:

(A) $E_{electric}$= (V x I) x T = (24 volts x 0.2 A) x 14 hour = 67.2 Wh = 0.672 kWh.
(B) The cost incurred is 0.672kWh x $\frac{\$0.28}{kWh}$= \$0.19.
(C) Total energy conversion loss = 0.9x0.6 = 0.54 = 54%.
(D) Actual battery efficiency = 100%- 54% = 46%.

Energy conversion efficiency can be given in dimensionless numbers between 0 and 1.0, or in percentage form, 0 to 100%. Due to energy losses, efficiencies for most energy conversion application are less than 100%. However, heat pump, refrigerator and other devices that "transfer" heat rather than convert it can have efficiency more than 100% and usually their efficiencies are expressed in coefficient of performance (COP).

Further as to our introduction of the First Law and Second Law of Thermodynamics, i.e., in Subsections 1.2.5, 1.2.6, 1.3.1, and 1.3.2, we shall now study the efficiency relationships with regard to both of the mentioned laws, in terms of the First Law of Efficiency (Conversion Efficiency), the Carnot Efficiency (Reversible Efficiency), and the Second-Law Efficiency.

Suppose we are considering an automobile engine as in Figure 1.27. The hot reservoir as shown is the cylinder chamber where ignition takes place. The heat energy tranfer Q_H results from the fuel combustion process. And the cold reservoir as shown is the cooling water that circulates through the engine and loses heat Q_C through the radiator, supplemented by a ventilation fan. The useful work done, W_{useful}, is the work done for turning the crankshaft. We shall now proceed to give 3 examples as an easy way to show the different efficiency relationships.

Example 1.4 *Engine I* Referring to the schematic energy flow diagram of a typical automobile engine as shown in Figure 1.27. (A) What is the heat lost to the cold reservoir? (B) Calculate the entropy changes in the hot reservoir, the cold reservoir, and the engine, respectively. (C) What is the thermal efficiency of the engine? (D) Compute the Reversible (Carnot) efficiency and the Second-Law Efficiency.

---

*Note that this is the average local consumers' residential electric rate as of June 2011, and this figure will be adopted throughout the text. Also, we must note that electric price varies between countries, localities, and applications.

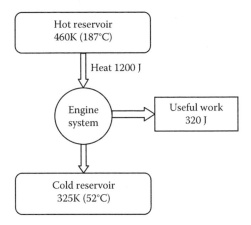

Figure 1.28
Schematic energy flow of a typical automobile engine (Case II).

Solution:
(A) Heat loss supply to the cold reservoir,

$$Q_{cold} = Q_{Hot} - W_{useful} = 1200 \text{ J} - 320 \text{ J} = 880 \text{ J}\#$$

(B) Entropy changes,

$$\triangle S_{Hot} = \frac{Q_{Hot}}{T_{Hot}} = \frac{-1200 \text{ J}}{505 K} = -2.38 \text{ J/K}\#$$

$$\triangle S_{Cold} = \frac{Q_{cold}}{T_{cold}} = \frac{880 J}{325 K} = 2.71 \text{ J/K}\#$$

$$\triangle S_{Net} = \triangle S_{Hot} + \triangle S_{Cold} = 0.33 \text{ J/K}\#$$

(C) The thermal efficiency of the engine is,

$$\eta_{thermal} = \frac{Q_{Hot} - Q_{Cold}}{Q_{Hot}} = 1 - \frac{Q_{Cold}}{Q_{Hot}} = 1 - \frac{880}{1200} / = 0.27\#$$

(D) The reversible (Carnot) efficiency of the engine is,

$$\eta_{Carnot} = \frac{T_{Hot} - T_{Cold}}{T_{Hot}} = 1 - \frac{T_{Cold}}{T_{Hot}} = 1 - \frac{325}{505} / = 0.36\#$$

And, the Second-Law Efficiency is,

$$\therefore \eta_{II} = \frac{\eta_{thermal}}{\eta_{Carnot}} = \frac{0.27}{0.36} / = 0.75 \text{ or } 75\%\#$$

Example 1.5 *Engine II* Now, let us consider another engine with a lower hot reservoir of 460K as shown in Figure 1.28. For this case, (A) What is the heat lost to the cold reservoir? (B) Calculate the entropy changes in the hot reservoir, the cold reservoir, and the engine. (C) What is the thermal efficiency of the engine? (D) Compute the Revessible (Carnot) efficiency and thus the Second-Law Efficiency.

Solution:
(A) Heat loss supply to the cold reservoir,

$$Q_{cold} = Q_H - W_{useful} = 1200 \text{ J} - 320 \text{ J} = 880 \text{ J}\#$$

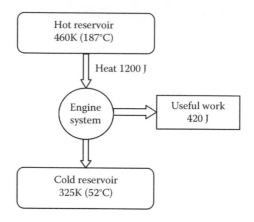

Figure 1.29
Schematic energy flow of a typical automobile engine (Case III).

(B) Entropy changes,

$$\triangle S_{Hot} = \frac{Q_{hot}}{T_{hot}} = \frac{-1200J}{460K} = -2.60 \text{ J/K} \#$$

$$\triangle S_{Cold} = \frac{Q_{cold}}{T_{cold}} = \frac{880J}{325K} = 2.71 \text{ J/K} \#$$

$$\triangle S_{Net} = \triangle S_{Hot} + \triangle S_{Cold} = 0.11 \text{ J/K} \#$$

(C) The thermal efficiency of the engine is,

$$\eta_{thermal} = \frac{Q_H - Q_C}{Q_H} = 1 - \frac{Q_C}{Q_H} = 1 - \frac{880}{1200} / = 0.27 \#$$

(D) The Revessible (Carnot) efficiency of the engine is,

$$\eta_{Carnot} = \frac{T_H - T_C}{T_H} = 1 - \frac{T_C}{T_H} = 1 - \frac{325}{460} / = 0.29 \#$$

And, the Second-Law Efficiency is,

$$\therefore \eta_{II} = \frac{\eta_{thermal}}{\eta_{Carnot}} = \frac{0.27}{0.29} / = 0.93 \text{ or } 93\% \#$$

Example 1.6

Engine III Refer to another conceptual engine as shown in Figure 1.29. (A) Determine the heat lost supply to the cold reservoir. (B) Calculate the entropy changes in the hot reservoir, the cold reservoir, and the engine, respectively. (C) What can we infer from (B) above?
   Solution:
(A) Heat lost supply to the cold reservoir,

$$Q_{cold} = Q_H - W_{useful} = 1200 \text{ J} - 420 \text{ J} = 780 \text{ J} \#$$

(B) Entropy changes,

$$\triangle S_{Hot} = \frac{Q_{hot}}{T_{hot}} = \frac{-1200J}{460K} = -2.61 \text{ J/K} \#$$

$$\triangle S_{Cold} = \frac{Q_{cold}}{T_{cold}} = \frac{780 J}{325 K} = 2.40 \text{ J/K\#}$$

$$\triangle S_{Total} = \triangle S_{Hot} + \triangle S_{Cold} = -0.21 \text{ J/K\#}$$

(C) We notice that the net entropy is a negative value, $ds \leq 0$, which implys that this engine violates the Second Law of Thermodynamics, i.e., application impossible.

Table 1.3 summarizes calculation results of the three examples above for Engines I, II, and III. Comparing Engine I and II, we can observe that the First Law Efficiency (thermal efficiency $\eta_{Thermal}$) for both engines is similar, i.e., 27%, although both engines are supplied with different heat source levels of 505K and 460K for Engine I and II, respectively. However, we found out that Second-Law Efficiency η_{II} for Engine II is superior than Engine I. Engine II is converting 93% of the available work to useful work, while it is only converting 75% for Engine II. As to the third engine, Engine III, we notice that the net entropy is a negative value, $ds \leq 0$, which implys that this engine violates the Second Law of Thermodynamics, which renders its application impossible.

Table 1.3 Comparison results of different energy schemes of the engine

Engine	T_{Hot}	Q_{cold}	W_{useful}	$\triangle S_{Hot}$	$\triangle S_{Cold}$	$\triangle S_{Net}$	$\eta_{Thermal}$	η_{Carnot}	η_{II}
I	505K	880J	320J	-2.38	2.71	0.33	27%	36%	75%
II	460K	880J	320J	-2.61	2.71	0.11	27%	29%	93%
III	460K	780J	420	-2.61	2.40	-0.21	-	-	-

$(T_{Cold}= 325K;\ Q_{Hot}=1200J)$

For further treatment of Second-Law Efficiency, readers can find more detaiedl discussions in thermodynamics books; for example, Cengel and Boles [26].

We will cover the energy industry and energy applications in the next chapter.

1.7 *Application Projects, Assignments, Questions, and Problems*

Application Project

Linking knowledge with applied energy design and practice

1. You are required to design, build, and test a simple toy car. The following are given: a small motor, shaft, four wheels, car body, and a battery. Describe various energy applications in your toy car, and the associated energy losses in a typical toy car that you have built. What is the power capacity of your car? How efficient is your car, and what is the likely losses range in your toy car? (Optional: Also, draw a Sankey Diagram to show the energy flows in your system) Note: Report format: abstract, introduction, experiment method and materials, results, conclusion, and recommendation, references, appendices. (Group project: 2–3 persons per group.)

2. You are required to design, build, and conduct a simple "hydro" generator test at the energy laboratory. The following are given: a multimeter with electrical wiring, an alternator (or small generator), timing pulleys, turbine shaft, and blades. It is understood that under different operating conditions, we would obtain different energy outputs, i.e., power measured in Watts (W). Your group is required to conduct simple experiments at various operating conditions, for example, at different water resource flow rates and with different sizes of pulleys used. Measure the power output at six different flow rates, and at two different elevations, usually called

Figure 1.30
A sample micro turbine. This typical turbine was designed and built using marine grade stainless steel 316L to resist corrosion specially for high-salinity water sensitive environment usage. (Courtesy of author.)

"head." What is the energy efficiency of the total system? Discuss your result obtained. A sample custom-made micro-turbine is given in Figure 1.3. (Group Project: 3–4 persons per group.) (Adopted from Group Project #2, 2008, [25].)

Assignments

1. Analyze and discuss the latest energy resources available in your country (in percentage distribution) and the associated energy utilizations. Also, find out the latest related government policies; choose any one of the policies and discuss in less than three pages. (Adopted from Assignment #3, 2007 [25])

2. What are the possible energy losses in the building you are reciting now? List them, explain and discuss how the losses can be possibly minimized (adopted from Assignment#2, 2007 [25].)

Questions and Problems

Multiple choice questions

1. Which one of the following is *not* an energy storage method?

    (A) Fuel cell
    (B) Power transformer
    (C) Ultracapacitor
    (D) Spring
    (E) Flywheel
    (F) Battery
    (G) Compressed air

2. Which one of the following statements does not correctly describe fossil fuel (oil and gas) storage?

    (A) Oil and gas are stored originally in the oil or gas reservoirs between rock and sand formations.

(B) The fossil fuels are transported in trucks and pipelines for temporary storage and delivery.

(C) Petroleum oil is usually stored in large-capacity storage sites above ground because it is not suitable to be stored below ground.

(D) Natural gas can be stored in storage tanks or in old gas wells.

(E) Natural gas can be stored as liquefied natural gas in huge storage tanks.

3. What is the efficiency of an energy conversion machine if it requires 1,500 joules of energy input and produces 900 joules of unwanted waste heat energy?

(A) 40%

(B) 50%

(C) 60%

(D) 75%

(E) 85%

4. What is the efficiency of a coal-fired boiler if feed water supplied per hour is 250 kg while coal fired per hour is 27kg? Enthalpy rise for water-to-steam conversion is 145 kJ per kg; coal calorific value = 2050 kJ/kg.

(A) 80.5%

(B) 70.5%

(C) 65.5%

(D) 60%

(E) 59%

5. Which of the following statements is/are true in relation to energy and power generation?

I. Electricity is useful for providing heat, light, and power for human activities.

II The Electricity process can be described in three stages, i.e., electricity generation, power transmission, and distribution.

III Alternating current electric power lines can transport electricity at low costs across great distances by taking advantage of the ability to transform the voltage using power transformers.

IV Around the globe, most power stations are burning biomass and other renewable resources to run the generators for producing electricity.

(A) I only;

(B) I and II only;

(C) I, II, and III only;

(D) I and III only;

(E) II and IV only.

6. Which one of the following is a renewable energy conversion system?

I Hydro energy conversion system

II Wind energy conversion system

III Solar energy conversion system

IV Natural gas conversion system

(A) I only;

(B) I and II only;

(C) I, II, and III;

(D) I and III only;

(E) II and IV only.

7. Which of the following are upcoming technologies for sustainable electricity production?

I Nuclear power generation

II Wind power generation

III Solar power generation

IV Biomass power generation

V Ocean and tidal energy generation

(A) I and II only;

(B) I and V only;

(C) I, II, and III only,

(D) I and III only.

(E) All except I.

8. The following are the primary techno-economical concerns/questions with relation to solar PV energy:

I How to collect efficiently?

II How to store?

III How to use?

IV How to sell?

V What is the total amount of energy available?

(A) I and II only;

(B) II and V only;

(C) I, II, and III only;

(D) I and III only

(E) All except IV.

9. Peat is used for the following:

I As fuel for electricity generation

II As fuel for heat generation

III Used in combined heat and power (CHP) plants

IV Use in condensing power generation (CP)

V Use in district/residential heating

(A) I and II only;

(B) II and V only;

(C) I, II, and III;

(D) I and III only;

(E) All the above.

Theoretical questions

1. Define energy and power.

2. How are the energy storage methods classified?

3. What is the similarity and difference between a battery and a fuel cell? Answer: A fuel cell can store energy chemically similar to a battery; but unlike battery, as long as there is a flow of chemical fuels into the cell, a fuel cell can produce electricity continuously.

4. Describe three of the basic forms of energy, and list their application advantages.

5. Describe biological conversion systems. Draw an ecology pyramid or ecology cone to show the approximate quantity of energy consumption at tropich levels of a food chain. What is the energy conversion efficiency of a typical food chain of a biological conversion system?

6. Briefly discuss the following alternative energy conversion systems:

   (a) Wind energy conversion system

   (b) Hydro energy conversion system

   (c) Solar energy conversion system

7. What is the difference between a pump and a turbine? Based on energy conservation principle, derive mechanical energy equations for both devices.

8. What is energy conversion efficiency? Defined the overall energy efficiency of a conversion system if it has undergone a series of energy conversions to produce the final form of energy output.

9. Describe current fossil energy scenarios around the globe. What do you think about the application of the fossil energy in the near future, say, 10–30 years to come? And, explain why you think that.

10. What are currently the main problems of renewable energy application as compared to fossil fuel currently?

11. Discuss the energy storage problems or concerns in relation to the following energy systems:

    (a) Pump hydraulic

    (b) Batteries

    (c) Wind energy

    (c) Super-capacitor

12. In your opinion, what will be the best forms of energy resources in the next 10–30 years to come? What are the likely energy application scenarios around the globe? Give examples and discuss.

Homework problems

1. A typical electrical generator has power output of 90MW with an efficiency of 75%.

   (a) What is the rate of energy supplied to the generator per hour?

   (b) What is the rate of energy input requirement if the system efficiency increases to 79%?
   (**Answer:** *120 MWh or* $432 \times 10^9 J$; *113.9 MWh or* $410 \ x \ 10^9 \ J$).

2. In a hydro energy system, a dam is constructed to provide and sustain suitable water head, and a hydroturbine is employed to generate the power required from the potential and avalable kinetic energy.

   (a) Applying the energy conservation equation, solve for the net shaft energy head per unit mass generated in terms of v_1, v_2, p_1, p_2, z_1 and z_2, where *1* is the inlet conditon and *2* the outlet conditon; and

   (b) Solve for the power generated as a function of Q, z_1, and z_2. Ignore any system friction losses.

(**Answer:** $\dot{W}_{shaft} = \rho g Q(Z_2 - Z_1)$)

3. Water flows into a boiler at 85°C, at an enthalpy of 355.9 kJ/kg. The water then leaves the boiler as steam at 100°C, If heat transfer added by the boiler is 137.9 kW, what is the actual flow rate of the water entering the boiler? Assume constant-pressure process.

    (**Answer:** $\dot{m} = 0.059\,kg/s$)

4. A 90MW power generator with capital cost of $700,000 is closing for one (1) month every year for maintenance.

    (**a**) What is the net earning per year if the cost of electricity supplied to a commercial area is $ 0.3 per kWh?

    (**b**) If the electricity is supplied to a residential area at a lower tariff electric charge of only $ 0.15 per kWh, and with a reported monthly operation and management of 2% of the capital cost due to the need for extra maintenance and other operational costs, what is the fate of this new energy scenario? Is the electricity provider making a profit?

    (**c**) What is the possible earning if the electricity provider employs a 30MW photovoltaic, with capital cost of $300,000, as an energy supplement to the residential area in (**b**) above? Assume that the latest system can work the whole year without the need of maintenance.

    (**Answer:** $213,840,000, $106,752,000,$1.45800 \times 10^5$)

Answers to multiple choice questions:

1. B.

2. C.

3. (A) efficiency = (1500-900)/1500 = 40%

4. (C) Boiler efficiency = heat supplied/heat input by coal = (250 x145)/(27x2050)= 65.5%

5. (C). I, II, and III only. IV is incorrect. Around the globe, at the present, most power plants/stations are still burning fossil fuels (about 67%) to run the generators for producing electricity.

6. (c) All except (IV); natural gas is regarded as a fossil fuel.

7. (E) All except I. Note: Unlike fossil fuels and nuclear power, biomass generally is regarded as an inexhaustible fuel source for power generation.

8. Answer: (A) I and II only. Note: The primary problem with PV is low efficiency and solar PV storage.

9. Answer: E.

Bibliography

[1] Serway, R. A., Jewett, J. W. 2004. *Physics for Scientists and Engineers* (6th ed.) Brooks/Cole, Belmont, CA.

[2] Goswami, D. Y. and Kreith, F., editors. 2008. *Energy Conversion*. CRC Press, Taylor & Francis Group, Boca Raton, FL.

[3] Jain, M. C. 2009. *Textbook of Engineering Physics* (Part I). PHI Learning Pvt. Ltd., New Delhi.

[4] Sears, F. W., Brehme, R. W. 1968. *Introduction to the Theory of Relativity*. Addison-Wesley, Redding, MA.

[5] IUPAC. 2010. *Compendium of Chemical Terminology*, 2nd ed. (the "Gold Book"). Compiled by A. D. Mc-Naught and A. Wilkinson. Blackwell Scientific Publications, Oxford (1997). XML on-line corrected version: http://goldbook.iupac.org (2006) created by M. Nic, J. Jirat, B. Kosata; updates compiled by A. Jenkins. ISBN 0-9678550-9-8. doi:10.1351/goldbook. Last update: 12-22, 2010; version: 2.2.

[6] Enerdata. 2010. *Yearbook Statistical Energy Review 2010* http://yearbook.enerdata.net/

[7] Abdullah, M. O., Yung, V. C., Anyi, M., Othman, A.K., Ab. Hamid, K.B., Tarawe, J. 2010. Review and Comparison Study of Hybrid Diesel/Solar/Hydro/Fuel Cell Energy Schemes for a Rural ICT Telecenter, *Energy* 35 : 639–646.

[8] EnhancedLearning.com, Food Chains and Food Webs, 2010.http://www.enchantedlearning.com/subjects/foodchain/, Retrieved June 4, 2011.

[9] Science Aid. 2008. Food and Energy, http://scienceaid.co.uk/biology/ecology/food.htm, Retrieved June 4, 2011.

[10] World Energy Outlook. 2010. International Energy Agency (IEA). Also online at http://www.iea.org/weo/2010.asp

[11] Renewables 2010 Global Status Report. Renewable Energy Policy Network for the 21st Century.

[12] World Energy Council, 2010 Survey of Energy Resources, London, United Kingdom.

[13] BP Energy Outlook 2030, London, January 2011.

[14] "Three Gorges Project." Chinese National Committee on Large Dams. http://www.chincold.org.cn/dams/rootfiles/2010/07/20/1279253974143251-1279253974145520.pdf. Retrieved May 15, 2011.

[15] BBC Web news, http://www.bbc.co.uk/news/uk-england-kent-11395964, Retrieved May 16, 2011.

[16] Joosten, H., Clarke, D., 2002. *Wise Use of Mires and Peatlands*, International Mire Conservation Group and International Peat Society, Jyväskylä, Finland.

[17] Strack, M. (Ed.), 2008. *Peatlands and Climate Change*, International Peat Society, Jyväskylä, Finland.

[18] Huggins, R. A. *Energy Storage*, Springer.

[19] Knief, R. A. 1985. *Nuclear Criticality Safety : Theory and Practice*. USA: American Nuclear Society.

[20] ISO 11439 (2006). Gas cylinders - High pressure cylinders for the on-board storage of natural gas as a fuel for automotive vehicles. ISO 11439:2000 Stage: 90.92 (2006-07-18)

[21] Abdullah, M. O., Zen, J., Yusof, M. 2011, On the Stress Corrosion Cracking, Crack Growth Prediction and Risk-based Inspection of Industrial Refrigerated Ammonia Tanks, *Corrosion*, 67 (4) National Association of Corrosion Engineers (NACE) International, Houston, Texas.

[22] Johnson, B.K., Law, J.D., Saw, G.P., 2001. Using a superconducting magnetic energy storage coil to improve efficiency of a gas turbine powered high speed rail locomotive. *IEEE Transactions on Applied Superconductivity*, 2001, 11(1) pp. 1900–1903.

[23] Tam, K.-S.; Kumar, P., and Foreman, M., Using SMES (superconductive magnetic energy storage) to support large-scale PV (photovoltaics) power generation, *Journal of Solar Energy Science and Engineering*, 1990, Volume: 45:1.

[24] Ali, M.H., Dougal, R.A. Comparison of SMES and SFCL for transient stability enhancement of wind generator system, Energy Conversion Congress and Exposition (ECCE), 2010 IEEE, pp. 3382–3387.

[25] Abdullah, M.O. Lecture notes, Energy Resources and Application, 1996–2010.

[26] A.Ç. Yunus and Boles, M. A. 1989. *Thermodynamics: An engineering approach*, New York: McGraw-Hill, International edition.

2

Energy Industry and Energy Applications

In this chapter, we will consider the energy industry and the associated energy applications used in the industry. Although we are familiar with the term "industry" as a manufacturing or technically productive enterprise often named after its primary product, for example, the chemical or steel industry, it is not easy to give an exact definition of "energy industry."

Here, we define energy industry as *all* the industries that are involved in energy applications, either directly or indirectly, comprised of the following activities (Figure 2.1):

- *Production* of primary energy (energy power industry), including fuel mining, extraction, and fuel refining;

- *Generation and supply of electrical energy* (electrical power industry), including electricity generation, electric power distribution, and sales;

- *Conversion of energy and application* (energy-intensive industry, small and medium enterprise (SME) energy-related industry, building industry, transport industry), including all energy applications for products output and human comfort.

Around the world today, we use a huge amount of fuels; for oil alone: we consume more than 80 million barrels of oil per day (bpd).* Therefore, the energy industry has big responsibility in energy generation, energy supply, energy usage, and application, as well as producing numerous energy products for our everyday use. It is indeed a crucial infrastructure for our human civilization, means of livelihood and comfort.

The energy industry comprises the following six main categories:

1. Energy power industry

2. Electrical power industry

3. Energy-intensive industry

4. Small and medium enterprise (SME) energy-related industry

5. Building industry

6. Transport industry

Figure 2.1 is a chart showing the energy industry classification given above. We will now begin to discuss each and every class of the industry. We will first cover the *energy power industry* as this is the class of industry that encounter and captures energy resources in a productive way for various energy applications, and it is a *mediator* between energy resources and energy users.

2.1 Energy Power Industry

2.1.1 What Is Energy Power Industry?

The energy power industry can be defined as the industries involved in the production of primary energy including fuel mining, extraction, and fuel refining. And, the energy power industry is one of the most crucial parts of the energy infrastructure, which acts to retrieve useful fuels available in our Earth.

---

*World oil supply is around 88.5 million bpd for January 2011.

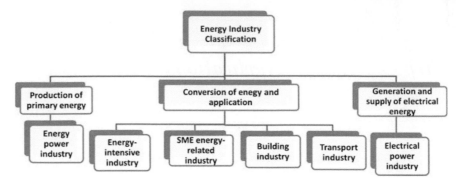

Figure 2.1
Energy industry classification.

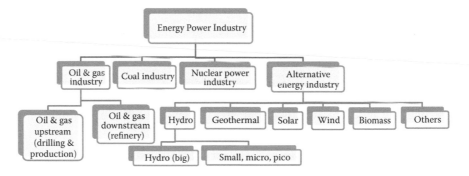

Figure 2.2
Energy power industry.

Energy power industry comprises the following four main categories of industries (see Figure 2.2):

1. *Oil and gas industry.* Oil and gas industry deals with upstream, which includes geological survey, and drilling and production for crude oil and gas activities. Oil and gas downstream activities basically deals with the post-production, i.e., the refinery[†] process of the oil or petroleum.

2. *Coal industry*

3. *Nuclear power industry*

4. *Alternative/renewable energy industry.* The alternative energy industry, better known as renewable energy industry, comprises industries that deal with power production from hydro, geothermal, solar, wind, biomass, and so forth. We can further classify hydro power into two types based on their capacity. The big capacity hydro industry involves big scale energy production at the MWs range,[*] while the the rest of the hydro systems (small, micro, and pico-hydro) are in the kWs power scale when working individually.

2.1.2   Energy Production from Energy Power Industry

To further the discussion given in Chapter 1, Section 1.24 ("World Energy Consumption and Demand)" we notice that there is great concern about current energy and fuel production rate, as compared to the energy demand rate globally. Thus our main worry and concern is threefold:

[†]Refinery industry is consider under the energy-intensive industry category.

[*]A typical example is the well known "The Three Gorges Dam" hydro system in China with power production capacity of 22,500 MW (Figure 1.20). The big hydro power industry is contributing the biggest share in respect of the alternative/renewable energy for primary energy production.

1. Effective fuel production or energy production;

2. Energy and fuels' continuous supply and sustainability; and

3. Clean energy supply and applications.

In this subsection, we will describe how the energy power industries endeavor to address, to some extent, the first concern. We will leave the second and third concerns on energy sustainability, energy supply, and applications to the rest of the sections and other chapters of this book.

2.1.2.1 Improving Energy Production from Oil and Gas Industries

At present, the following are the methods practiced by the oil and gas industries for improving energy production efficiency:

1. *Finding new wells and improving current wells' operations.* In addition to the search for new potential reservoirs via exploration, and preparing new oil and gas wells for production, drilling teams from the oil and gas industries also revitalize existing wells by:

   - Well repairing and stimulating, e.g., replacing of production tubings to restore and enhance oil and gas production;
   - Well stimulation, e.g., (a) deepening of existing wells to enhance oil and gas production, (b) acids and fracturing fluids are pumped into the well under high pressure to fracture, clean and stimulate the rock formation and thus increase production.

2. *Horizontal well, directional, and extended-reach drilling.* Horizontal drilling is developed to allow wider access into areas of reservoirs including those areas that were previously unreachable. Directional and extended-reach drilling allow operators to access reservoirs far from the drilling location. Figure 2.3 is a photo showing a typical oil well drilling platform where drilling and production operations are taking place.

3. *Improvement in completions technology.* Completions technology also has evolved, securing wells in less time and with greater protection by:

   - Cased-hole completion, which involves making small holes in the well casing and formation so that oil or gas can flow effectively into production tubing.
   - Open-hole completion, where sand screens and gravel packs are inserted into the reservoir to stabilize the hole and maintain the flow of oil from the reservoir into the well bore.

4. *Smart well technology and control.* This includes directional drilling with remote control ability by the above-ground operator in sending commands to direct drilling operations, which receives real-time log data about the oil-bearing rocks.

5. *Enhanced oil recovery (EOR).* EOR is a practice to improve oil and gas production. Further discussions on EOR applications will be given in Chapter 5.2, "Energy Recovery."

6. *Efficient total power management.* Efficient power management aims to optimize all operations for oil and production, as well as ensuring safety and environmental aspects of the wells and surroundings.

2.1.2.2 Improving Coal Production Efficiency by Coal Industries

Coal is mined by two methods: surface or "opencast" mining, and underground or "deep" mining.

At present the following are the methods practiced by coal industries for improving energy production efficiency:

1. *Finding new reserves through improved exploration activities.* The exploration activities comprised of geological, geochemical, and geophysical surveys, followed by exploration drilling. Improving exploration techniques ensures new reserves, thus and an extended coal supply.

2. *Improve in coal drilling technology.* Over the years, many mining techniques has evolved to allow, for instance, previously inaccessible reserves to be reached.

Figure 2.3 (SEE COLOR INSERT)
An offshore drilling platform. An offshore structure where typical drilling and associated production activities are carried out in the oil and gas industries. (From NOAA, public domain.)

3. *Improved coal mining operations.* One of the typical examples is the recent improvement of spoil removal during mining operations. By the use of a newly designed lightweight dragline bucket, in the South Walker Creek Mine, Australia, the energy required to fill the new lightweight bucket has been reported to the reduced by 20% compared to the traditional bucket used (World Coal Institute [3]).

4. *Enhancing safety and control.* Improvements in safety levels in both underground and opencast mining.

5. *Efficient coal power management.* Efficient coal power management aims to optimize all operations for coal production, as well as ensuring safety and environmental aspects of coal mines and their surroundings.

2.1.2.3 Improving Energy Production Efficiency in Nuclear Industries

At present, the following are the methods practiced by nuclear industries for improving energy production efficiency:

1. Improve operating performance, which includes possible reduction of capital cost, operation and maintenance costs, and construction time.

2. Considering co-generation (combined heat and power) applications.

3. Improvement in nuclear waste management.

4. Improvement in safety, storage, and disposal.

2.1.2.4 Improving Energy Production Efficiency in Renewable Energy Industries

As presented in Chapter 1, renewable energy is one of the feasible methods for energy sustainability and the way forwards for future energy supply.

At present, the following are the methods for improving energy production from renewable energy industries:

1. Improved system efficiency.

2. Improved energy production performance while reducing costs.

3. Natural resources enhancement.

4. *Considering hybrid energy systems.* Hybrid energy systems enhance fuel saving, energy recovery, and increase the overall system efficiency. We will cover hybrid energy systems in Chapter 10.

(a)   (b)

Figure 2.4 (SEE COLOR INSERT)
Electrical transmission and distribution. (a) A typical electrical power substation, a subsidiary station for electrical power transmission and distribution at which required voltage is stepped up or stepped down via a transformer. (b) A typical electrical transmission tower. (Courtesy of author.)

2.1.2.5   Improving Electrical Production Efficiency in Electric Power Industry

The following are the four main steps for further improvements:

1. Improvement of electrical energy generation.

2. Improvement of electrical energy transmission (see Figure 2.4).

3. Improvement of large scale energy storage, including the use of smart grids to deliver efficient and economical electricity services.

4. Improvement of electrical energy distribution.

Next, we shall start discussing energy applications in energy-intensive industries.

2.2   Energy Applications in Energy-Intensive Industry

2.2.1   What Is Energy-Intensive Industry?

Energy-intensive industries are the industries that use large amounts of energy to transform energy processes or materials for our everyday use. Collectively, and depending on the countries, they supply 60–90% of the energy processes and materials vital to our world economy and applications.

The US Department of Energy under its Industrial Technologies Program (ITP) had identified eight priority energy-intensive industries, i.e., aluminum, chemicals, forest products, glass, metal casting, mining, petroleum refining, and steel industries (see US DOE, 2011 [4]).

The following is a simple classification of energy-intensive industries (Figure 2.5) together with their brief descriptions:

1. Metal processing industry. Metal processing industries consist of various high-energy intensive application processes involving casting, rolling, forging, machining, drilling, milling, and so on.

2. Mining industry

3. Petroleum refining

4. Chemical industry

5. Aluminum, iron and steel industry

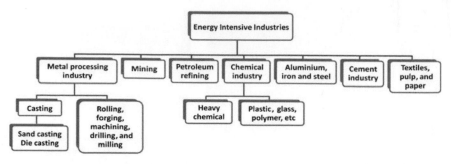

Figure 2.5
Energy-intensive industry classifications.

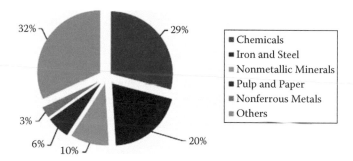

Figure 2.6
World industrial sector energy consumption by major energy-intensive industry shares in 2006. (From US EIA, 2009 [6].)

6. Cement industry

7. Textiles, pulp, and paper industry

2.2.2 Energy Application in Energy-Intensive Industries

Figure 2.6 shows the 2006 world industrial sector energy consumption by major energy-intensive industry shares. In 2006, five industries accounted for 68% of all energy used in the industrial sector: chemicals (29%), iron and steel (20%), nonmetallic minerals (10%), pulp and paper (6%), and nonferrous metals (3%). According to EIA, the quantity and fuel mix of future industrial energy consumption will be determined largely by energy use in these five industries (US EIA, 2009 [6]).

2.2.2.1 Aluminum, Iron, and Steel Industries

The iron, aluminum, and steel industries are one of the largest energy-consuming and energy-intensive industrial sectors in the world, notably in many developed countries; for example the United States, China, and Japan.

Aluminum and steel are two essential metals to modern manufacturing. Their special corrosion resistance ability makes them the preferred metals for many manufactured products, such as automotive, cans and packaging, spacecraft components, and building industries.

Aluminum metal is made from two sources:

1. *Primary aluminum* produced from ore. Primary aluminum is extracted from ore with typical steps that includes bauxite mining, bauxite refining, and electrochemical reduction of alumina in a smelter [5].

2. *Secondary aluminum* produced from scrap. Secondary aluminum is recovered from recycled scrap material, where the scrap should have aluminum content > 20 wt.%. Scrap with aluminum content < 20 wt.% is usually discarded. About 0.39 kWh of energy is needed in processing scrap for production of 1 kg of aluminum (Hiraki and Akiyama 2009) [1].

Figure 2.7
Typical aluminum smelter plant. (From Sohar Aluminum Smelter.)

On average it takes ~ 16 kWh and > 18 kWh of electricity to produce 1 kg aluminum from alumina (aluminum oxide), for normal and refined quality (99.999%) of aluminum, respectively. Further details and breakdown with respect to aluminum production energy consumption is given in Sect. 2.3.5, Case Study 1. Thus, aluminum smelting is energy intensive, that is why the world's smelters are located in areas that have access to abundant economical power resources, such as hydro-electric, natural gas, and coal zones. A photo of a typical aluminum smelting plant is given in Figure 2.7.

What is aluminum smelting and how is aluminum steel processed?   Bauxite is the main ore used for aluminum production. It contains 30–60 wt.% alumina, and is extracted by the open-pit mining method, where the layer of soil on top of the alumina deposit, usually called *overburden* ,is removed. Conventional drilling or blasting operations are not involved due to the soft nature of bauxite deposits. The richer ores are used as mined. The lower grade ores may be of benefit by crushing, washing, and subsequent drying to remove clay and silica waste. The mined bauxite is then transported to refineries. The production of the aluminum metal comprises two basic steps:

1. *Refining.* Production of alumina from bauxite by the Bayer process in which bauxite is digested at high temperature and pressure in a strong solution of caustic soda. The resulting hydrate is crystallized and calcined to the oxide in a kiln or fluid bed calciner.

2. *Reduction via electrolytic refining.* Aluminum is produced by the electrolysis of molten aluminum oxide. Reduction of alumina to virgin aluminum metal employs the Hall–Heroult electrolytic process using carbon electrodes and cryolite flux, for producing aluminum of high purity (>99.99%).

Figure 2.8 shows block diagrams describing various processes of aluminum smelting, which are all energy-intensive.*
  The aluminum oxide was first extracted from bauxite, i.e., the aluminum ore. This process is called the *Bayer process*, in which the bauxite is digested under pressure with caustic soda (sodium hydroxide). The aluminum oxide dissolves to form sodium aluminate, which is filtered off. Aluminum hydroxide is filtered off and then heated. This process is called *calcination*, which produces alumina (aluminum oxide).
  In aluminum smelting, the electrolysis process occurs in the furnace:
  A mixture of alumina and cryolite is fed into the furnace. In the furnace, carbon rods act as the anodes, are lowered into the molten mixture. Electricity passes between them and the carbon furnace lining acts as the cathode.The electricity splits up the aluminum oxide into aluminum metal, which collects as a molten layer on the floor of the furnace. Oxygen combines with the carbon anodes to from carbon monoxide, which burns off as CO_2.
  Aluminum, made by an electrolytic process, is now widely used in transportation, packaging, building construction, and electrical transmission.
  As a result of increasing cell size and better process control, energy consumption has improved with time as shown in Figure 2.9. [9].

---

*For consumption rate see also Sect.2.3.5, Case Study 2.

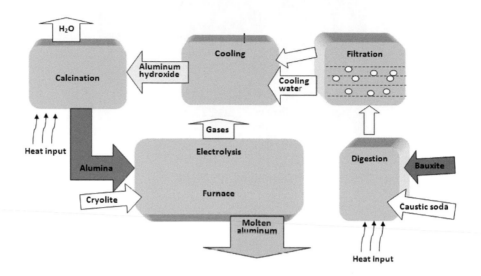

Figure 2.8
Aluminum smelting process block diagram.

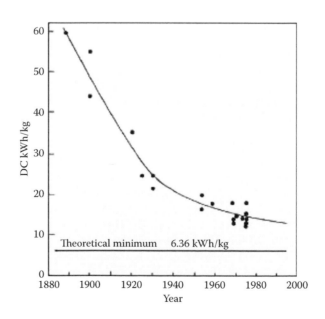

Figure 2.9
Energy consumption for aluminum production. (From the *Electrochemistry Encyclopedia*. With permission.)

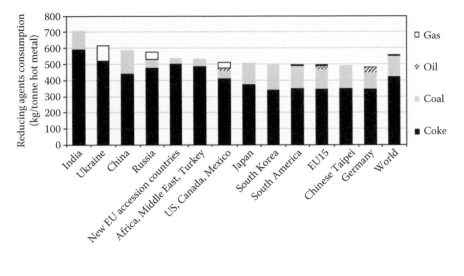

Figure 2.10
Blast furnace reductant use, 2005. (From IEA, 2007 [7].)

2.2.2.1.1 Production and Uses of Aluminum Worldwide aluminum capacity in December 1997 was 24,302,000 metric tons/yr, of which 1,861,900 tons/yr was shut down mainly due to high energy cost. The United States capacity was 4,215,000 metric tons/yr, of which 654,000 tons/yr was shut down. The United States once had a more dominant position, but new reduction plants have been moving to countries with less expensive hydro and coal power, such as Norway, Brazil, Venezuela, and Australia [9].

Next we will cover the production of steel and associated energy applications.

2.2.2.1.2 Production of Steel and Energy Usage Like aluminum, steel production involves many high-energy demand processes such as heating, cooling, melting, and solidification. This includes the wide use of electrical-driven arc furnaces. Common fuels used are coking coal, fuel oils, and gas. Coal and electricity are the main energy and feedstock sources.

The production of steel can be described briefly as follows:

1. *Produced from iron ore*: Coke and agglomerated iron ore is reacted in a blast furnace to yield liquid iron. Thereafter, the liquid iron is converted into steel in a basic oxygen furnace.

2. *Produced from steel scrap*: This is an alternative production process that uses steel scrap and electricity to generate liquid steel in an electric arc furnace.

3. *Direct Reduced Iron (DRI)*: The third option of production is via the Direct Reduced Iron (DRI) method, also called the *Sponge Iron* method, produced from iron ore, and the solid iron product is used as a feedstock for electric arc furnaces.

The blast furnace* is the most energy consuming step in the steel making process, accounting for > 50% of total energy use in blast furnace steel making, i.e., production from iron ore (Gielen and Taylor, 2009) [22]. Figure 2.10 shows the use of coke, coal, and other reducing agents in blast furnaces. It is an indication of total energy use for steel making. The best performing country, Germany, uses about 485 kg per metric ton of hot metal.

The application advantages of steel include the following:

1. High-strength, lightweight steel has contributed to bridges, pipelines, railways, structural utilities, and so on.

---

*A blast furnace is a type of metallurgical furnace used for smelting. Fuel and ore are continuously supplied through the top of the furnace, while air or O_2 is blown into the bottom of the chamber, so that the chemical reactions take place throughout the furnace as the material moves downward. The end products are usually molten metal and slag phases tapped from the bottom, and flue gases exiting from the top of the furnace. This type of furnace is typically used for smelting iron ore to produce hot metal (pig iron), an intermediate material used in the production of commercial iron and steel (IEA, 2007).

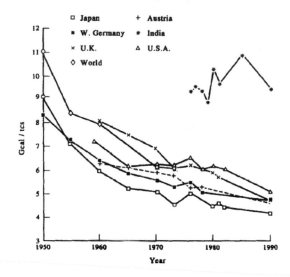

Figure 2.11
SEC of steel in selected countries and the world. (Courtesy of International Iron and Steel [10].)

2. Steel is the most recycled product among all metals.

3. In general, steel has shown the highest decline of total air emissions among major manufacturing sectors.

4. Steel has one of the highest specific energy consumption (SEC) reductions among metals in the world (see Figure 2.11).

The specific energy consumption of steel in developed countries ranges from 4 to 6 Gcal/tcs. As presented in Figure 2.11, over the 1950–1990 period, the SEC for the world and most countries decreased due to energy application improvements.

And, most recently according to the American Iron and Steel Institute,* "steel has voluntarily reduced energy consumption by 27% since 1990."

In the next subsection, we will consider another high energy consumption industry — the chemical and petrochemical industry.

2.2.2.2 Chemical and Petrochemical Industry

According to recent IEA statistics, the chemical and petrochemical industry accounts for more than 30% of the total industry energy use worldwide.

What are the chemical and petrochemical industries? The chemical and petrochemical industries are industries or companies that produce industrial chemicals and petrochemicals.

Chemicals are the fundamental building blocks of many products that meet our basic requirements for food, daily usages, health, and environments, the products of which are related to the manufacture or production technologies with respect to our foods and drinks, medical and medicine, biotechnology, computing, telecommunications, and other advanced chemical-related technologies.

The chemical industries produce a vast range of industrial chemicals. But a relatively small number of chemicals account for the bulk of chemical production usually called *heavy chemical* due to the large amount of production. The processes are uniform worldwide (IEA). Most heavy chemicals are made from salts, minerals or gases in the air. Examples are sulfuric acid, ammonia, sodium hydroxide, and sodium carbonate.

---

*American Iron and Steel Institute, The North American Steel Industries reduces energy intensity, Available at www.steel.org/en/Sustainability/Energy Reduction.aspx, accessed on April 16, 2012.

We call chemicals that are produced in relatively small amounts *fine chemicals*, examples of which are agrochemicals, dyes, and pharmaceuticals..

The chemical and petrochemical industries contribute to about 30% of world industrial energy consumption. The energy intensive processes include:

1. *Electrolysis.* For example, a large chemical plant producing caustic soda through electrolysis of brine consumes huge amounts of electricity, supplied by a power plant, with more than 30 MW generating capacity.

2. Refining and related processes, e.g., steam cracking.

3. Catalytic reforming.

Ten of the world's most energy-intensive petrochemical and plastic commodities are:

1. Ammonia

2. Ethylene

3. Propylene

4. Benzene

5. Toluene

6. Xylene

7. Thylene

8. Polyethylene

9. Propylene

10. Styrene

The classification of petrochemical products, i.e., hydrocarbon series, is detailed in Chapter 6. Generally, the HC series can be classify into paraffins, cycloparaffins, and aromatics. For instance, the olefins group includes ethylene and propylene (generally derived from hydrocarbon feed stocks using steam cracking). The aromatics e.g., benzene, toluene, and xylene, are produced via steam cracking of catalytic reforming.

The raw materials for most plastics are hydrocarbons obtained from oil refining. To produce various products and different forms, plastics are shaped by moulding, extrusion and laminating processes.

A photo of a typical petrochemical plant is given in Figure 2.12.

Figure 2.12
Typical chemical plant. (Courtesy of author and photographer Mak Koon Kong.)

Figure 2.13 (SEE COLOR INSERT)
The Melaka refinery. The 926-acre refinery complex has a maximum production rate range of 265,000 to 300,000 barrels per day (bpd), supplying about 30% of the Malaysian petroleum product needs. (Courtesy of PETRONAS.)

What are the world's 4 primary industrial chemicals (heavy chemicals)? The four main industrial chemicals are ammonia, sulfuric acid, caustic soda, and sodium carbonate.

1. *Ammonia.* The major use of ammonia is to make fertilizer (75%), wood pulp (15%), nylon (5%), and nitric acid (5%).

2. *Sulfuric acid.* The uses of sulfuric acid are many; for instance, agricultural chemicals and fertilizer (25%), chemicals and plastics (25%), paints and pigments (20%), detergents and soaps (20%), fiber (5%), other (5%).

3. *Caustic soda (sodium hydroxide).* Caustic soda is important in the manufacture of chemicals (30%), pulp and paper (30%), soap and detergents (10%), and alumina (5%), etc.

4. *Sodium carbonate.* One of the notable application of sodium carbonate is in the manufacturing of glass. It is also used for oils, fats, and waxes (15%), foods and drinks (10%), textiles (5%), dyes and colors (5%), etc.

2.2.2.3 Petroleum Refinery

A petroleum refinery is an industrial processing plant where crude oil is processed and refined into more useful petroleum products, such as light distillates (e.g., liquefied petroleum gas [LPG], gasoline, naphtha), middle distillates (e.g., kerosene, diesel), heavy distillates and residuum (e.g., heavy fuel oil, lubricating oils, wax, and asphalt) [16].

Figure 2.13 is a photo of a typical refinery complex, which produces LPG, motor gasoline, diesel, jet fuel, low sulphur waxy residue, and fuel gas for both domestic and export markets. Around the world, on average, there is an increase of refinery facilities due to increases in chemical demand.

According to IEA statistics, the shares of the petrochemical and chemical industry (including feedstock) in total industrial final energy use by region in 2004 were: North America — 41%, Western Europe — 31%, China and India — 20%, Japan and Korea — 39%, Middle East — 20% and the Commonwealth of Independent States (CIS) — 24% (IEA statistics).

According to Bloomberg News [18], recently, refining capacity in Africa, China, Iraq, and India, respectively, grew by 8.9%, 12%, 6.8%, and 3.6%.

Also, according to Bloomberg News [?], there is an estimated 5–6 million bpd of new refinery capacities coming onstream from Asia and the Middle East, some of which is focused on supplying OECD product demand.

Generally, the strategic location of a refinery should consider the following important parameters:

1. *Site Safety.* The site should be safe from all possibilities, and has to be sufficiently far from residential areas and sensitive areas.

2. *Good infrastructure and facilities.* Infrastructure should be available for raw materials supply and temporary storage. Refineries that consume a large amount of steam and cooling water for heat rejection need to have an abundant source of water, such as adjacent to the sea.

Table 2.1 Energy use in the chemical and petrochemical industry (excluding electricity) in 2004 (IEA statistics)

	Amount	LHV	Feedstock Energy Needed	Fuel		Total Fuel + Feedstock
	Mt/yr	GJ/t	EJ/yr	GJ/t	EJ/yr	EJ/yr
Ethylene	103.3	47.2	4.9	13	1.3	6.2
Propylene	65.3	46.7	3.0	13	0.8	3.9
Butadiene	9.4	47.0	0.4	13	0.1	0.6
Butylene	20.3	47.0	1.0	10	0.2	1.2
Benzene	36.7	42.6	1.6	7	0.3	1.8
Toluene	18.4	42.6	0.8	7	0.1	0.9
Xylenes	33.7	41.3	1.4	7	0.2	1.6
Methanol	34.7	21.1	0.7	10	0.3	1.1
Ammonia	140.0	21	2.9	19	2.7	5.6
Carbon black	9.0	32.8	0.3	30	0.3	0.6
Soda ash	38.0	0.0	0.0	11	0.4	0.4
Olefins processing excl. polymerization	100.0	0.0	0.0	10	1.0	1.0
Polymerisation	50.0	0.0	0.0	5	0.3	0.3
Chlorine and Sodium Hydroxide	45.0	0.0	0.0	2	0.1	0.1
Total			**17.0**		**8.2**	**25.2**

3. *Transport and shipment requirement.* Good transportation is crucial to a successful refinery plant, and includes the easy shipment of products to markets, normally via big pipelines, tankers, and large carriers.

4. *Waste and environment considerations.* Facilities should be available for proper waste disposal and management.

5. *Energy availability.* The energy to operate the plant should be available and at affordable cost.

2.2.2.3.1 Petroleum Refinery and Energy Usage According to IEA (2007) [7], the chemical and petrochemical industry accounts for 30% of global industrial energy use and 16% of direct CO_2 emissions. Table 2.1 depicts the energy use in the chemical and petrochemical industry in 2004, indicating the high level of energy use in the chemical and petrochemical industries. Feedstock energy accounts for more than half of total energy use in the industry.

According to the study by Ernst and Christina (2005) [13], refineries spend typically 50% of cash operating costs (i.e., excluding capital costs and depreciation) on energy, making energy a major cost factor and also an important opportunity for cost reduction. Energy use is also a major source of emissions in the refinery industry making energy efficiency improvement an attractive opportunity to reduce emissions and operating costs. Voluntary government programs aim to assist industry to improve competitiveness through increased energy efficiency and reduced environmental impact.

Next, we shall now consider the cement industries — a non-metallic industry, and have an overview of the associated energy applications.

2.2.2.4 Cement and Concrete Industries

According to IEA statistics, the non-metallic mineral industry accounts for about 10% of total final industrial energy use. And, cement accounts for about 70–80% of the energy use in the non-metallic minerals sub-sector, consuming 8.2 exajoules (EJ) of energy a year, i.e., around 7% of total industrial fuel use.*

Global cement production increased from 594 million tonnes (Mt) in 1970 to 2,292 Mt in 2005, with the biggest growth occurring in developing countries — in particular China, which accounted for 46% of global cement production in 2005; while India, Thailand, Brazil, Indonesia, Iran, Egypt, Vietnam, and Saudi Arabia accounted for 15% (IEA 2007) [7].

---

*Cement accounts for almost 25% of total direct CO_2 emissions in industry, although only about 7% of total industrial fuel use.

Figure 2.14 (SEE COLOR INSERT)
Typical cement plant. The plant has an installed annual Portland cement production capacity of 1 million metric tonnes. (Courtesy of author.)

Cement and concrete are among the most important building materials of modern infrastructure applications, used together with iron and steel. Cement is mixed with water and materials (such as sand, gravel, and crushed stone) to make concrete, where it acts as the "glue" that holds a concrete mixture firmly. Its stone-like qualities and ability to be poured into different forms makes concrete an excellent construction material for buildings, bridges and other applications.

Portland cement contains about 60% lime, 25% silica, and 5% alumina; iron oxide and gypsum make up the rest of the materials. Cement is made from finely ground "clinker," which consists of the calcium silicate minerals formed when limestone and other materials are burned at very high temperatures in a kiln.

The main energy-using processes are raw material preparation, clinker making, and finish grinding (or cement making). Clinker production is the most energy-intensive stage in cement production, accounting for over 90% of total industry energy use. Clinker is produced through a controlled high-temperature burn in a kiln of a measured blend of limestone and lesser quantities of siliceous, aluminous, and ferrous materials. Mixing these ingredients and exposing them to intense heat causes chemical reactions that convert the partially molten raw materials into pellets called *clinker*. Cement plants grind clinker and add a variety of additives to produce cement, while integrated plants both manufacture clinker and grind it to make cement. The production process consists of three main steps: raw material mining and preparation, clinker production, and finish grinding (Worrell et al. 2000 [14]).

Manufacturing cement is an energy-intensive process. It requires 3 to 6 million BTUs (British thermal units) of energy and 1.7 tons of raw materials, mostly limestone, to make one ton of clinker. Figure 2.14 is a photo of a typical cement plant [15]).

Next, we will proceed to look at another very important non-metalic industry class i.e. textiles, pulp and paper industries.

2.2.2.5   Textiles, Pulp, and Paper Industries

2.2.2.5.1   Textiles Production and Energy Consumption   Textiles are materials made from fibers. The most common material is cloth. Synthetic fibers such as nylon, polyester, and acrylic fibers are widely used nowadays (see Figure 2.15). In general, energy in the textile plant is mostly used in the forms of thermal and electricity applications (Table 2.2) as follows:

1. Electricity, as a common power source for machinery, cooling, and temperature control systems, etc.; and

2. Fossil fuel (e.g., oil, liquefied petroleum gas (LPG), coal, and natural gas) as a fuel for boilers, which generate steam and heating systems for providing drying, dyeing, washing, and other processes.

Table 2.2 Typical manufacturing type and energy consumption for textile plant application

Manufacturing type	Thermal Energy (%)	Electricity (%)
Fiber	60–70	30–40
Textile	40–45	55–60
Clothing	60–65	35–40
Finishing	85–90	10–15

Figure 2.15 (SEE COLOR INSERT)
Cloth produced from the textile industry. Depending on cloth types and requirements, about 40–90 % of thermal energy and 10–60 % of electricity energy are needed for textile production (also see Table 2.2). (Courtesy of author.)

2.2.2.5.2 Pulping and Paper Making In brief, wood pulp, which is mainly made from softwoods, is suitable for making newspapers; chemical pulp is used to produce better quality papers, where chemicals such as sodium sulfate solution are used to digest wood chips for producing quality papers.

What is pulping and how is paper made? Pulping is the process whereby the fibers in wood are separated and treated to produce a wood pulp, i.e., a raw material for making paper. Wood pulp is made primarily from softwoods such as pine, either produced by grinding or chemically digested.

In the grinding process, ground wood pulp is made by shredding logs in a huge grinding machine, which results in coarse pulp suitable for making newsprint for newspaper printing purposes. Chemical pulp is made by digesting wood chips in a chemical solution, such as sodium sulphate, for producing better quality paper. This treatment separates the wood chips from their binder, called lignin. Most pulp is first mixed with water to form slurry. The slurry is pumped directly to an integrated paper or paperboard plant where it may be mixed with other pulps, recycled fiber, and fillers such as China clay before going to the paper machine.

All paper machines have three basic elements, namely wet end, press section, and drying section. From an applied energy aspect, a closed cycle bleached craft pulp mill may have 10–20 % higher steam and electricity demands than a conventional pulp mill due the need for extra energy to concentrate bleach plant effluents and perform waste water treatments.

2.2.2.5.3 Energy Use in the Pulp and Paper Industry Energy use in the pulp and paper industry involves these pulp production and paper production processes (IEA):

1. Chemical pulping

2. Mechanical pulping

3. Paper recycling

4. Paper production

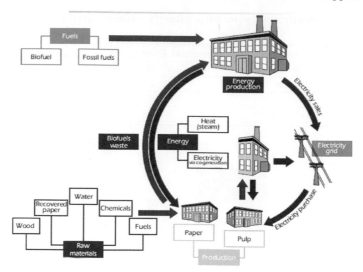

Figure 2.16
Uses and source of energy in the pulp and paper production process. (From Confederation of European Paper Industries (CEPI) [8] and (IEA 2007) [7].)

Nevertheless, according to the Confederation of European Paper Industries (CEPI) and EIA, unlike other energy intensive industries, the paper and pulp industry rely heavily on bioenergy, such that a significant portion of the heat and electricity required for pulp and paper applications is supplied internally via energy recycling from biomass waste/residues. (see Figure 2.16).

2.2.2.6   Metal Processing Industry

The metal working industries, processing metals for various applications, usually can be broadly divided into either *primary* metal processing (also called *upstream* metal processing) or secondary metal processing (also known as *downstream* metal processing). Primary metal processing heavily involves the manufacturing of machine components, machinery, instruments, and tools that are needed by other industries, sectors, and so on. Downstream metal processing steps include the fabrication of semi-finished products, such as tubes, strips, sheets, rods, wires and bars.
   Basic metal processing techniques include:

- Casting molten metals into a given shape, usually called foundry. Of primary importance is the casting of iron and steel. According to IEA [7], global production of cast iron is reported to be about 50 Mt, around 5% of total ferrous metal production. Cast iron contains 2–4% carbon and 1–3% silicon, where scrap iron or steel is often added.

- Smelting and refining of metal ores and scrap.

- Hammering or pressing metals into the shape of a die (hot or cold forging).

- Sintering (compressing and heating materials in powder form, including one or more metals).

- Shaping metals using lathe machinery.

- Welding and cutting sheet metal.

- Metal finishing.

A wide variety of techniques are available to finish metals, including grinding and polishing, abrasive blasting, and many surface finishing as well as coating techniques (such as electroplating, galvanizing, heat treatment, anodizing, powder coating, and so on).
   To produce refined metals, two metal recovery technologies are generally used:

1. Pyrometallurgical

2. Hydrometallurgical

Pyrometallurgical processes use heat to separate desired metals from other materials. These processes use differences between oxidation potentials, melting points, vapor pressures, densities, and miscibility of the ore components when melted.

Unlike in the pyrometallurgical process, in the hydrometallurgical technique, the desired metals are separated from other materials using techniques that capitalize on differences between constituent solubilities and/or electrochemical properties while in aqueous solutions. The high temperature required for the pyrometallurgical treatment of metals is obtained by burning fossil fuels or by using the exothermic reaction of the ore itself, e.g., flash smelting process.

The flash smelting process is an example of an energy-saving pyrometallurgical process in which iron and sulphur of the ore concentrate are oxidized. The exothermic reaction coupled with a heat recovery system can save energy for smelting.

The total energy use for the iron melting is about 5–10 GJ/t iron.

2.2.2.7 Mining Industry

About 50 billion metric tons per day of materials are removed from the Earth by mining, 70% of which are through the open-pit mines on the surface; the other 30% are extracted from underground mines. About 1342 Mt/year of iron ore is produced worldwide, with the top three producer, China, Brazil, and Australia, producing 60% of the total amount (Table 2.3).

The extraction processes of minerals, usually by *open pit mining* or *underground mining*, all demand high levels of energy.

2.2.2.7.1 Open Pit Mining
In an open-pit mining, minerals lie on or near the surface, and are extracted by various means, including *dragline conveyers* and *bucket-wheel excavators*. A large, powerful excavator can remove 9,000 m^3 or rock per hour.

When the rock is particularly hard in open pit mining, a high-energy *flame piercing* method is sometime used to cut the rock by literally melting it. Flame temperature as high as 3000° C can be obtained by combustion of fuel oil and oxygen.

The drager, in particular, the *bucket drager* used for mining ores, e.g., gold, in water, is found in big ore-processing plants, which can handle 15,000 metric tons of gravel per day. The mined materials are sieved and screened, in which finer materials are washed over the riffles, and the heavy ores settle out and are separated.

2.2.2.7.2 Underground Mining.
Generally, most of the available ores like coal are deposited underground. Therefore, boring shafts, construction of tunnels, and associated underground rail transport meet requirement for the miners and ores, all demanding both high cost and energy. A hoist or skip lifts the ore to the surface. In some mines, the ores are transported by conveyors. Ventilation equipment must be installed underground to keep the miners supplied with ample fresh air, and remove dust and unwanted gases such as methane.*

2.2.2.7.3 Other Mining Methods
Some minerals such as salt, potash, and magnesium are extracted from sea water by evaporation or electrolysis. Sulfur can be extracted by the Frasch process, in which hot water above the boiling point of sulfur is injected by pump to melt the underground sulfur deposit; subsequently, the high pressure air is used to force the water–sulfur mixture to the surface and the settling tanks, where the pure sulfur is obtained.

2.2.3 Energy Efficiency in Energy-Intensive Industry

2.2.3.0.4 What Is the Incentive and Why Do Energy-Intensive Industries Seek to Become More Energy Efficient?
As we all know that most companies or industries are of complex economic and social behavior in nature, a number of factors can simultaneously influence a specific company to adopt energy efficiency practices and applications. Figure 2.17 classifies the common reasons energy-intensive industries seek to become more energy efficient.

Brun and Gereffi (2011) [21] have categorized the reasons into three main areas, namely legal, financial, and social-driven, as follows:

---

*Methane gas has to be removed as it can become explosive when mixed with air and sparks.

Table 2.3 World iron ore mining distribution

Country	Mining (Mt/year)
Australia	231
Brazil	255
China	310
India	121
Russia	97
Ukraine	66
USA	55
Other	207
World	1,342

(From United States Geological Survey, 2005 [17].)

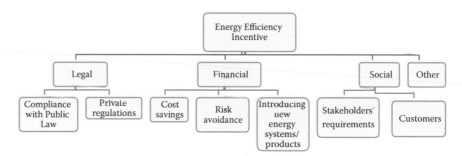

Figure 2.17
Energy efficiency incentive.

1. Legal — compliance with public laws and private regulations

2. Financial — cost savings; risk avoidance; brand names and new product development

3. Social — pressure from internal and external stakeholders

It was found that the greater the energy use, measured usually as a percent of operating costs or sales, the more incentive the firms have to reduce energy costs. Therefore, manufacturing companies in energy intensive industries like cement, chemicals, and petroleum refining will have strong financial incentives to invest in more efficient energy management systems and manufacturing processes or equipment that reduce their energy utilization.

Energy intensities* vary both among and within industries. The manufacturing of aluminum is known to be more energy-intensive than the manufacturing of steel. Within steel making itself there are several different technologies, each with its own minimum energy requirement. Similarly, several plants using the same steel making technology may have different energy intensities of production resulting from varying ages of plants, capacity utilization, and management techniques. Conservation measures are believed possible in all these determinants of production efficiency. In the United States, for example, it is estimated that such intensity conservation measures, with present technology, could result in 40 % energy savings, the bulk coming through the replacement of primary process facilities in chemical, steel, aluminum and paper industries. To date, however, we have seen that most industrial energy savings have been attributed to "housekeeping" measures. For example, in Europe about 50% of total industrial energy demand is generated by three highly energy-intensive industries–iron and steel, chemicals, and aluminum.

---

*We shall note at the outset that the energy intensity of industrial output (energy inputs required to produce unit of physical output) has significant implications for total energy demand and for conservation opportunities; not only is the structure of industrial output important, but so also is the choice of production process and the efficiency of production. The reductions in industry's energy intensity will result in significant energy savings. However, reductions do not mean energy conservation through physical reductions in output only; rather, given a certain output, improvements in production activities or processes will also result in less energy being consumed to produce the same output (than would have been the case had these improvements not taken place). (John E. Jankowski, Industrial Energy Demand and Conservation in Developing Countries, *Resources for the future*, Washington, D.C. Discussion Paper D.73A, Unpublished material, 110p).

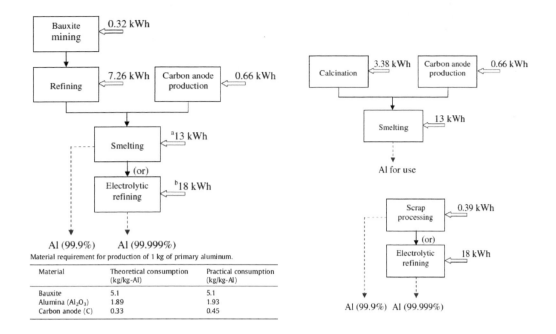

Figure 2.18

The simplified steps and energy consumption estimation. (*Left*): Aluminum primary production. (*Upper right*): Reaction product recycling. (*Lower right*): Secondary aluminum production. (From Wang et al., 2011 [12]; U.S. Department of Energy, 2007.)

As a result, the pattern of energy demand and conservation potential has been heavily influenced by the process technology of these three industries [11].

While the specific source and means of conservation potential varies within each manufacturing subsector, it is the process steam and direct heat applications that offer the greatest opportunities for energy savings. Together these two energy activities account for approximately 2/3 of total industrial energy inputs. The remaining 1/3 is accounted for by electricity, 26% — mostly for heating, lighting, and air conditioning, and by energy feedstocks about 9% of total industrial energy inputs [11].

2.2.4 *Case Study 1*: Comparing Aluminum Production Methods and Energy Consumption

The energy consumption and costs for different possible aluminum production processes have been estimated based on 1 kg of aluminum produced (U.S. Department of Energy [5]; Wang et al., 2011 [12]).

Aluminum production can be categorized based on three methods, that is, *primary aluminum production*; *aluminum production from recycled Al(OH)$_3$*, and *secondary production from recycled scrap*. Figure 2.18 shows the simplified steps for all the production methods together with the energy input requirements shown in kWh.

Primary aluminum is extracted from ore with typical steps described in Figure 2.18 (Left), which includes bauxite mining, bauxite refining, and electrochemical reduction of alumina in a smelter. Electrolytic refining is further required for producing aluminum of high purity (>99.99%). The energy content of hydrogen is represented by its lower heating value ($LHV_{H_2} = 119.93$ kJ/g).

The consumptions based on various methods can be estimated as follows [12]:

Primary aluminum production method:

Assuming a 100% hydrogen yield, the production of 1 kg of hydrogen corresponds to the consumption of 9 kg of aluminum. If primary aluminum is adopted, the overall energy efficiency (defined as the ratio of input energy to output energy) will be 17:4%, i.e.,

$$\left(\frac{119.93}{21.24 \times 3.6 \times 9} \right) \times 100\% = 17.4\,\%$$

Figure 2.19
SME definitions and number of employees. The limit of SME in EU is 250 employees; while in the US the limit is set to be 500. Obviously, any companies larger than this number are termed as big/large companies.

If the electricity price of US$0.0572/kWh is considered, production of 1 kg hydrogen for this situation will cost US$10.9.

Aluminum production from recycled Al(OH)$_3$:

If aluminum produced from recycled Al(OH)$_3$ is introduced, the overall energy efficiency can be increased to 21.7%, i.e.

$$\left(\frac{119.93}{17.04 \times 3.6 \times 9} \right) \times 100\% = 21.7\,\%$$

with a cost reduced to US$8.8/kg-H$_2$. (This price is still high compared to that of hydrogen produced from reforming and electrolysis.)

Reaction product recycling from scrap:

Now, the energy efficiency for use of secondary aluminum is as high as 949.1%, i.e.

$$\left(\frac{119.93}{0.39 \times 3.6 \times 9} \right) \times 100\% = 949.1\,\%$$

attributed to the low energy input required by scrap processing, which is only 1.8% of that needed by primary aluminum production! Use of secondary aluminum is an effective way to further reduce the hydrogen cost, which is estimated to be only US$0.2/kg- H$_2$.

2.3   Energy Applications in SME Energy Industries

2.3.1   What Are SME Energy Industries?

The term SME refers to "Small and Medium Enterprises," widely used in the European Union and in international organizations such as the United Nations. In the United States of America, the term "small and medium businesses" (SMBs) is commonly used instead. The industries with < 50 employees are often classified as "small," while those with < 250 employees "medium" [23]. On the other hand, the United States employs somewhat bigger figures in term of the number of employees: "small" businesses often refers to those with < 100 employees, while "medium-sized" businesses often refers to those with < 500 employees (Figure 2.19).

2.3.2   What Is the Difference between SMEs and Other Industries?

Unlike large companies in other industries, SME consist of smaller companies. For instance, in the automotive manufacturing sector, large companies manufacture multiple product lines, marketed under different brand names. Smaller SMEs involved in manufacturing consist of a single or few product lines. Large companies have advantages of *economy of scale*; smaller SEMs compete or focus on *specialized markets* or as *supporting companies* to the big industries.

2.3.3 What Are the Types of SME Energy-Related Industries?

Recognizing the importance of SMEs and their relation to energy and applications for boosting industrial development, around the world, many Governments have initiated a range of programs in diverse areas. A list of examples is given here:

1. Agriculture

2. Chemical and petrochemical

3. Construction

4. Electrical and Electronics

5. Food and Beverage

6. Green and renewable energy

7. Machinery and general engineering

8. Maintenance and reparing

9. Manufacturing

10. Metal processing and metal products

11. Pulp, paper, and printing

12. Plastic and rubber products

13. Textiles and leathers

14. Transport and equipment, including auto component industry

15. Wood and wood-based products.

These initiatives from the various governments are important in facilitating the growth of the SMEs, which in turn will support the energy-intensive industries. SMEs provide the bulk of employment and manufacturing capacity, in particular, in many developing countries like China and India.

In India, SMEs have significant shares in the metals, chemicals, food, and pulp and paper industries (GOI, 2005 [24]).

According to APEC, there are at least 39.8 million SMEs in China, accounting for 99% of the country's enterprises, 50% of asset value, 60% of turnover, 60% of exports, and 75% of employment (APEC, 2002) [25].

In the EU regions, many SMEs also deal with metal processing and metal products (IECSME, 2010).

2.3.4 Energy Efficiency and Applications in SME Energy Industry

Large enterprises usually have better provisions for comprehensive energy efficiency systems than the SMEs. For instance, in terms of energy efficiency in the European SMEs, it was reported that comprehensive systems for energy efficiency are much less in place in SMEs (4%) than in large enterprises (19%); the same applies for simple measures to save energy, which are used by 30% of SMEs but 46% of large enterprises. From the energy environment perspective, SMEs contribute roughly 64% of the industrial pollution in Europe. Up to 24% of SMEs actively engage in actions to reduce their environmental impact (mainly reduction of energy consumption); 0.4% of SMEs use a certified Environmental Management System (EMAS, ISO 14001 or other systems). Renewable energy sources will have to provide an average of 20% of all final energy consumption by the year 2020, with binding national targets.

Table 2.4 illustrates some SMEs' energy efficiency methodology for energy enhancement. Generally, four main areas in the SME sectors are important, i.e., technology, process, and space saving, as well as efficient energy management are required. From a technology perspective, use of efficient machineries is crucial. This includes efficient motors, heat exchangers and boilers. SMEs also could consider adopting new energy saving techniques for energy saving. A recent example of possible energy consumption saving in a melting process for aluminum casting application is given as a case study in the following subsection.

Table 2.4 Some SMEs' energy efficiency methodology

Area	Methodology
Technology	Use energy efficient machines (motor, heat exchanger, boiler)
	Installation of equipment for heat recycling, CHP, etc.
	Adopting new energy saving technology, e.g., casting.
	Improve insulation and reduce heat loss
	Installation of equipment for reducing leakage of electricity
Process	Recycling waste as energy resources (waste to energy)
	Using ambient heat and natural wind for drying
	Using new energy resource, e.g., LPG, to replace coal/diesel
Space	Using daylighting, LED, natural ventilation
Management	Applying efficient energy management

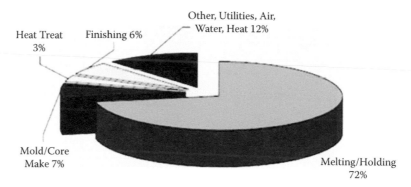

Figure 2.20
Typical casting tacit energy profile by process [2].

2.3.5 *Case Study 2*: Energy Consumption Saving Opportunity in Melting Process for Aluminum Casting

The energy application from melting and holding processes is a large part of the total energy used in the foundry. As shown in Figure 2.20, tacit energy* from the melting/holding process can amount to about 72% the process of the total energy usage [2].

Dai and Jolly (2011) [20] recently presented an industrial study in melting process by application of a new technique called *Constrained Rapid Induction Melting Single Shot Up-Casting* (CRIMSON). The method's main philosophy is that foundries, using an induction furnace, only need to melt the quantity of metal required to fill a single mould in a closed crucible rather than large batches that use unnecessary energy and create more rejects. The closed crucible, then, is transferred to a station and the melted metal is pushed up using a computer controlled anti-gravity filling method to fill the mould. Due to rapid melting, transfer, and filling in the new method, the holding time of melted metal is minimized, a drastic energy saving is achieved, and in the meantime the possibility of hydrogen absorption and formation of unwanted surface oxide film are reduced to a great extent.

Energy consumption estimation of melting process The energy required for heating A354 aluminum alloy to 760° C, without considering the chemical reaction, slag formation, and other heat losses, is as in Eq. 2.1:

$$Q = mC_{m1}(T_m - T_1) + Q_f + mC_{m2}(T_2 - T_m) \tag{2.1}$$

Where,
  Q is the heat that is needed to heat and melt the Al alloys;
  m is the mass of the Al alloy;

---

*Tacit energy is the term used to describe an energy value that equals the combination of onsite energy consumption, the process energy required to produce and transmit/transport the energy source, and feedstock energy.

C_{m1} is the specific heat for solid A354 Al alloys, 1.07 kJ/(kg°C);
C_{m2} is the specific heat for liquid A354 Al alloys, 1.05 kJ/(kg°C);
T_m is the melting temperature of A354 Al alloys, 596 °C;
T_1 is the environment temperature, 20 °C;
T_2 is the superheat temperature, 760 °C;
Q_f is the heat of fusion, 389 kJ/(kg°C).

Gas consumption calculation: In the study, gas is used to heat the Al alloys to its melting point 596 °C and melt the alloys. The heat required for heating and melting the Al alloys is calculated by Eq.2.2:

$$Q_1 = mC_{m1}(T_m - T_1) + Q_f \qquad (2.2)$$

$$Q_1 = 1.07\,(596 - 20) + 389 = 1005\,\text{kJ/kg}$$

Electricity consumption calculation: The electricity is used to overheat the melted Al alloys to 760 °C, the heat Q_2 for overheating the Al alloys can be calculated by,

$$Q_2 = mC_{m2}(T - T_{2m}) \qquad (2.3)$$

$$1.05(760 - 596) = 172\,\text{kJ/kg}$$

Total energy consumption: The total theoretical energy consumption is,

$$Q_{t1} = Q_1 + Q_2 = 1005 + 172 = 1177\,\text{kJ/kg} = 327\text{kWh/tonne}$$

It is estimated that the thermal efficiency is 7–19% for a crucible furnace using natural gas, and normal thermal efficiency (η_1) of 59–76% for an induction furnace using electricity [44].

Actual consumption of gas and electricity Actual consumption of LPG gas and electricity are $0.7 m^3/metrictons$ and 2800 kWh/metric tons, respectively. Energy density by mass of LPG is 49.6 MJ/kg.

Therefore, the total actual energy consumption, Q_{t2}, is 12,444 kWh/tonne=44.8 MJ/kg.

The thermal efficiency of using the LPG for melting the alloys:

$$\eta_1' = \frac{1005}{9644} = 10.4\,\%$$

The thermal efficiency of using electricity for melting the alloys:

$$\eta_2' = \frac{172}{2800} = 6.14\,\%$$

2.3.5.1 Calculation of Energy Consumption of New Casting Method

$$Q = mC_{m1}(T_m - T_1) + Q_f + mC_{m2}(T_2 - T_m) = 1145\,\text{kJ/kg}$$

See Table 2.5 for data.

The energy consumption measured during the melting was 1.98 GJ/metric tons (550 kWh/metric tons), therefore the thermal efficiency is,

$$\eta_c = \frac{1145}{1980} \times 100\% = 57.8\%$$

The difference in the thermal efficiency $\eta_2' = \frac{172}{2800} = 6.14\,\%$ of the holding furnace using electricity with the normal thermal efficiency ($\eta_1 = 59$ -76%) of an induction furnace was a much larger than expected. This suggests that there are great savings to be made in identifying why these poor efficiencies exist. The reason for the heat loss in an up-casting process facility can be attributed to radiation, conduction, and convection between the melt and the surrounding environment. When melting the same weight of the Al alloys, the old technique used 22.6 times more energy than the new casting facility.

The new technique replaced the large holding and melting furnaces by single shot melting, which results in efficiency improved by upwards of 40%, thus reducing production cost. It is estimated that 42.8 GJ/metric tons (11.9MWh/metric tons) can be saved for producing every metric tons of A354 casting alloys when using the new process.

Table 2.5 The actual measurement parameters

Parameter	Value
Weight of ingot	4 kg
Melting temperature	729 °C
Melting time	2 min
Injection time of U-caster	10 s
Holding time	20 s
Solidification time	23 s
Measured energy consumption for melting the ingot	7.92 MJ (2.2 kW h)

(From Dai and Jolly, 2011 [20].)

2.4 Energy Applications in Building Industries

In this section, we will describe the energy applications in building industries. In as far as applied energy is concerned, the building industry involves two kinds of energy applications, i.e., building construction application, and after construction, the occupant application. The latter consumes much of the energy use due to the energy consumption over a period of a much longer time. We will also discuss the energy efficiency in relation to the building energy consumption.

2.4.1 What Is the Building Industry and Its Applications?

Energy management in buildings is the control of energy use and cost while maintaining indoor environmental conditions to meet comfort and functional needs. Energy efficiency improvements need not sacrifice any functionality of the facility. Good energy management has the goal of reducing energy expenses to the lowest level possible without sacrificing comfort, productivity, or functionality. The energy manager should understand how energy is used in the building, to manage energy use and costs effectively. There are opportunities for savings by reducing the unit price of purchased energy, and by improving the efficiency and reducing the use of energy-consuming systems. Energy management procedures should be as simple, specific, and direct as possible. However, very large and complex facilities, such as hospital or university campuses, industrial complexes, or large office buildings, usually require a team effort and systematic processes [45].

Energy used in buildings and facilities composes a significant amount of the total energy used for all purposes, and thus affects energy resources. For instance, the American Society of Heating, Refrigerating and Air-conditioning Engineers (ASHRAE) recognizes the "effect of its technology on the environment and natural resources to protect the welfare of posterity." Many governmental agencies regulate energy conservation, often through the procedures to obtain building permits. Required efficiency values for building energy use strongly influence selection of Heating, Ventilating, Air-Conditioning, and Refrigeration (HVAC&R) systems and equipment and how they are applied ([27], Chapter 34).

In electric power industry worldwide, notable technologies for energy efficiency enhancement include efficient residual appliances. Heating, Ventilating, and Air-Conitioning (HVAC) equipment, light, motors and efficient industrial processes [15].

2.4.2 Energy Efficiency and Applications in Building Industries

2.4.2.1 Energy Consumed in Building Construction

Energy consumed in buildings depends on many factors. The main factors are the building type and size, building design and material use, building equipments' use, number of workers, and location of constructions, as remote areas often involve higher energy cost.

Consider a simple typical example:

A 3.5 m×3.5 m×3.14 m room construction, using common construction, would require about 1.8 ton of coal and 254 kWh of electricity, where the coal and electricity use include that which is not consumed directly during construction but are backward linkages in the construction process used in the manufacture of cement, bricks, tiles, and so forth. The assessment of the magnitude of energy consumption shows that a huge amount of energy is consumed in the housing sector alone [28].

Although smaller than household energy use, energy consumed in building-materials production is quite significant in energy budgets: in particular, for the materials industries, the energy costs for construction have been shown to account for over 20% of world fuel consumption [29].

Energy is consumed in housing construction mainly in three ways [28]:

1. In the procurement, manufacture, processing, and recycling of building materials

2. In transporting building materials to the building site

3. In on-site construction activities

A comparison of energy requirements of building materials is given in Table 2.6.

Table 2.6 Comparative energy requirements of building materials

Material	Primary energy requirement (GJ/ton)
High energy	
Steel	30-60
Lead, Zinc	25+
Glass	12-25
Cement	5-8
Plaster board	8-10
Medium energy	
Lime	3-5
Clay bricks and tiles	2-7
Gypsum plaster	1-4
Concrete:	
In-situ	0.8-1.5
Blocks	0.8-3.5
Precast	1.5-8
Sand lime bricks	0.8-1.2
Low energy	
Sand, aggregate	< 0.5

(From Tiwari, P. Energy efficiency and building construction in India,
Building and Environment 36 (2001) 1127–1135. With permission).

Figure 2.21 shows a typical modern building under construction. Unlike traditional buildings, modern buildings generally require heavier machinery, which includes lifting cranes, forklift trucks, and so forth, which usually run on internal combustion engines or electric motors.

2.4.2.2 On-Site Energy Resources

Most on-site energy for buildings in developed countries involves electricity and fossil fuels as primary on-site energy sources. Both fossil fuels and electricity can be described by their energy content (joules). This implies that energy forms are comparable and that an equivalence can be established. In reality, however, they are only comparable in energy terms when they are used to generate heat. Fossil fuels, for example, cannot directly drive motors or energize light bulbs. Conversely, electricity gives off heat as a byproduct regardless of whether it is used for running a motor or lighting a light bulb, and regardless of whether that heat is needed. Thus, electricity and fossil fuels have different characteristics, uses, and capabilities aside from any differences in their derivation.

Total energy demand in the commercial and residential sectors will grow at 1.4% and 0.8% per year, respectively. This results from increasing population and greater use of computers, telecommunications, and other office appliances, but it is offset by somewhat improved building and equipment efficiencies (ASHRAE, 2009, Chapter 34) [27].

Figure 2.21 (SEE COLOR INSERT)
A typical modern building under construction. Photo taken in March 2011, the building construction completed in September 2011. Depending on the building, the total quantity of energy consumed in a building during its lifetime is usually many times that consumed in its construction. (Courtesy of author.)

2.4.2.3 Near Future Energy Application Trend in Built Environments

Both widespread construction and conversion of buildings to very low energy consumption, and even ultra low or "zero" energy buildings, are part of the current and future scenario. For that purpose, the associated energy policy implications for buildings' efficiency standards and appliances are very big. A combination of built environment measures, heat pumps and air-conditioning, solar heating and cooling, as well as other highly efficient appliances and advanced lighting may reduce energy needs in buildings as well as shifting fuel use to renewables and low-carbon/green electricity (Holness, 2011 [31]).

2.4.2.3.1 Zero-Energy Building (ZEB) The Zero Energy Building is a complex energy application concept that implies "zero" energy usage for buildings. The term "zero" energy, "0" energy or net ZEB balance are often used interchangeably in various literature. Mathematically, we can write,

$$\sum energy\, applied = 0 \tag{2.4}$$

The approaches of applied energy can highlight different aspects of ZEB; in other words, the "zero" energy parameter can be in terms of one (or more) of the parameters in the following list:

- the final or delivered energy

- the end-use energy

- primary energy, e.g., Torcellini et al., 2006 [30]

- energy supplement with renewable, e.g., Iqbal, 2004 [31]

- CO_2 equivalent emission, e.g., Kilkis, 2007 [32]

- exergy or energy availability; e.g., Kilkis, 2007 [32]

- energy cost, e.g., Mertz et al., 2007 [33]

- life-cycle zero-energy building (LC-ZEB), e.g., Hernandez and Kenny, 2010 [36]

Some common issues in relation to the definitions of "zero-energy building" are given in Chapter 12.

2.5 Energy Applications in Transportation Industries

2.5.1 What Are the Transportation Industries?

In the past, human travel from place to place used non-machine power applications, i.e., by foot, horses, donkeys, and camels, as well as wind-driven ships, etc. Our current modern transport industries, however, essentially consist of machine-driven industries, which can be classified into 3 main categories:

1. Automotive industry

2. Air and Space transport industry

3. Sea transport industry

Rodrigue et al. (2009) [37] detailed the energy consumption of transportation application based on three strong modal variations as follows:

1. *Land transportation* accounts for the great majority of energy consumption. Road transportation alone consumes on average 85% of the total energy used by the transport sector in developed countries. This trend is not, however, uniform within the land transportation sector itself, as road transportation is almost the sole mode responsible for additional energy demands over the last 25 years. Despite a falling market share, rail transport, on the basis of 1 kg of oil equivalent, remains four times more efficient for passenger and twice as efficient for freight movement as road transport. Rail transport accounts for 6% of global transport's energy demand.

2. *Sea (Maritime) transportation* accounts for 90% of cross-border world trade as measured by volume. The nature of water transport and its economies of scale make it the most energy efficient mode since it uses only 7% of all the energy consumed by transport activities, a figure way below its contribution to the mobility of goods.

3. *Air transportation* plays an integral part in the globalization of transportation networks. The aviation industry accounts for 8% of the energy consumed by transportation. Air transport has high energy consumption levels, linked to high speeds. Fuel is the second most important budget for the air transport industry, accounting for 13–20% of total expenses. This accounts for about 1.2 million barrels per day. Technological innovations, such as more efficient engines and better aerodynamics, have led to a continuous improvement of the energy efficiency of each new generation of aircraft.

Figure 2.22 displays the energy consumption from each of the transport industries.

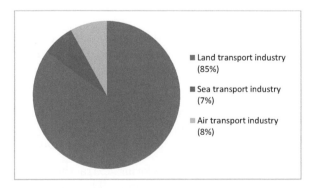

Figure 2.22
Energy use from transport industry. (Data from Rodrigue [37].)

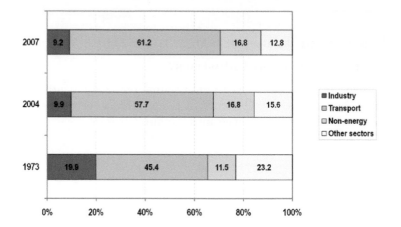

Figure 2.23
World Oil Energy Consumption by Sector, 1973–2007. (From International Energy Agency.)

2.5.2 Energy Consumption in Transport Industry

According to the International Energy Agency (IEA), transportation accounts for approximately 25% of world energy demand and more than 62% of all the oil used each year (Figure 2.23). The graphs also show that world oil consumption for transport applications increased from only 45% in 1973, to 58% in 2004, and up to 61% in 2007. This trend may largely be due to increased light vehicle usage, especially in developing countries.

The projection of energy usage by mode of transport [38] shows that light duty vehicles (essentially the private use of cars) is the sector with the highest energy usage, followed by freight trucks and air transport (Figure 2.24). The forecasts underlying these curves assume that world oil supplies will be sufficient to meet the increasing demand for petroleum-based fuels for transport and that the global economy will continue to grow.

2.5.2.1 Impacts of Transport on Energy Consumption

The impacts of transport on energy consumption are diverse, including those that are necessary for the provision of transport facilities [37]:

1. *Vehicle manufacture, maintenance, and disposal.* The energy spent for manufacturing and recycling vehicles is a direct function of vehicle complexity, material used, fleet size, and vehicle life cycle.

2. *Vehicle operation.* Mainly involves energy used to provide momentum to vehicles, namely as fuels, as well as for intermodal operations. The fuel markets for transportation activities are significant.

3. *Infrastructure construction and maintenance.* The building of roads, railways, bridges, tunnels, terminals, ports, and airports, and the provision of lighting and signaling equipment, require a substantial amount of energy. They have a direct relationship with vehicle operations since extensive networks are associated with large amounts of traffic.

4. *Administration of transport business.* The expenses involved in planning, developing, and managing transport infrastructures and operations involves time, capital, and skill that must be included in the total energy consumed by the transport sector. This is particularly the case for public transit.

5. *Energy production and trade.* The processes of exploring, extracting, refining, and distributing fuels or generating and transmitting energy also require power sources. The transformation of 100 units of primary energy in the form of crude oil produces only 85 units of energy in the form of gasoline. Any changes in transport energy demands influence the pattern and flows of the world's energy markets.

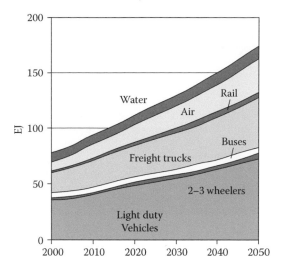

Figure 2.24
Projected energy usage, by transport sector, in ExaJoules (10^{18} Joule). (From IPCC [40].)

2.5.3 Examples of Automotive Transportation Industry and Common Usages

While the transportation industry is involved in all forms of road, sea, and air transportation, the automotive industry is the industry involved in road transportation of motor vehicles, but essentially also deals with the design, development, and manufacturing (includes assembly lines, automation, and robots) of vehicles, railroads and road constructions. Figure 2.25 is a photo of a typical car, a product from an automative industry.

2.5.4 Examples of Air Transportation Industry and Common Applications

The air transport industry provides transportation of passengers, cargo, mail and perishable goods, with the location including all operating airports and flying fields and terminal services, as well as air craft constructions.

Figure 2.25 (SEE COLOR INSERT)
Typical car — a product of the automotive industry. This 2.0 liter ~100 kW car can accommodate 8 people, with a fuel consumption rate of around 11 km per liter of petrol. (Courtesy of author.)

Figure 2.26
A typical passenger aircraft flying in the air. This aircraft is flying at an average speed range of 700–800 km/hour, which can consumes fuel at around 10–15 times more than a land vehicle's consumption. Fuel is an airline's second largest expense after labor cost. (Courtesy of author.)

2.5.4.1 Air Transport and Travel Industry

According to the Air Transportation Association [42], fuel is an airline's second largest expense after labor cost. Figure 2.26 is a photo showing a typical airplane flying across the sky.

2.5.5 Examples of Sea Transportation Industry and Common Applications

Sea is the most energy efficient transport, but fuel is an important component of a ship's operating costs. Sea or maritime transportation accounts for 90% of cross-border world trade as measured by volume. The nature of water transport and its economies of scale make it the most energy efficient mode, since it uses only 7% of all the energy consumed by transport activities, a figure way below its contribution to the mobility of goods [37].

Big tankers or large carriers are used to transport crude oil (see Figure 2.27), natural gas, and LNG across oceans, and continents.

Figure 2.28 shows a typical small maritime transport vehicle where a diesel-driven ferry is seen carrying vehicles and passengers, providing a means of crossing the river.

Figure 2.29 shows another river transport, but the lightweight boats have only a relatively small capacity. Each of the lightweight boats has of around a 5 kW diesel-engine-driven propellor designed for leisure which is limited to two to three passengers transportation. The manufacturing of boats with low capacity and the boats themselves such as those depicted in the photo are usually suited perfectly for SMEs industry applications.

Figure 2.27
A typical large crude oil carrier. (From miscellaneous.)

Figure 2.28 (SEE COLOR INSERT)
Ferry transport. This typical diesel-driven ferry can accommodate 24 vehicles per trip for river crossing transportation applications. The ferry employs four Cummins engines, each with norminal capacity of 261kW @ 1800 rpm. (*Left*): during loading; (*Right*): river transportation. (Courtesy of author.)

Figure 2.29
Small boat transport. (Courtesy of author.)

We shall also note that sea (marine) transport industry also includes ship building and locomotives.

2.5.6 Brief Overview of Climate Change and Greenhouse Emissions Mitigation from Transport Industry

The contribution of the transport systems, including road, air, and sea, are making the climate change through the emission of greenhouse (GHG) gases and other emissions.

The transport sector was estimated to contribute approximately 14% of the total greenhouse gas emissions in 2004. Such emissions include not only CO_2 but also CH_4, N_2O, and fugitive gases from refrigerated systems used in transportation, such as CFC-12 and CFC-134a. There are a range of measures and technologies available to reduce the emissions from the road transport sector. Lighter vehicles, lower average speeds, the use of more efficient engines, and the use of less greenhouse-intensive fuels (for example, natural gas in place of oil-based fuels) could all play a role (Love et al. 2010 [38]).

Aircraft contribute to climate change in two principal ways — through the emission of greenhouse gases such as CO2, NOx, and radiatively significant particles such as soot, and through the generation of contrails which in turn may have an impact on the global heat balance.

According to the International Maritime Organization (IMO) 2009 [39], international shipping in 2007 accounted for about 2.7% (870 million metric tons) of the global man-made CO2 emissions, and mid-range emission scenarios suggest that by 2050, in the absence of reduction policies, ship emissions may grow 150–250% (compared to 2007 emissions) as a result of growth in world trade. The IMO green house gases Study concludes that there is a significant potential for reduction of green house gases through technical and operational measures. The mitigation measures available are essentially two-fold: reducing the speed at which ships travel (slow steaming) and implementing new technologies into ship and engine design that make their overall operations more greenhouse-gas efficient.

2.5.7   Energy Efficiency and Applications in Transport Industry: Current and Future

2.5.7.1   Road Transportation Scenario

Energy and emissions in the road transport sector are saved largely through:

1. *Improvements in the efficiency of conventional vehicles;*

2. *Improvement through the increased use of fuel cell vehicles.*

3. *Improvement through the increased use of electric battery vehicles; and*

4. *Improvement through the increased use of hybrid vehicles.*

Hybrid vehicles are designed to consume low-carbon biofuels principally as a replacement for gasoline to fuel cars. Low-carbon biofuels are expected to play a significant role in the future, within the limits of sustainable crop production. While electric batteries and hydrogen fuel cells are the main alternatives for cars, it is difficult to judge at this stage which of these technologies — or which combination of them — will be the most competitive (see, e.g., [41]).

2.5.7.2   Sea (Marine) Transportation Industry Scenario

The IMO Marine Environment Protection Committee [39] adopted in July 2009 a package of technical and operational measures providing for a program on "Energy Efficient Design Index" for new ships, a "Ship Energy Management Plan" for new and existing ships, and an "Energy Efficient Operational Indicator" for existing ships.

The possible energy application scenarios are:

1. *Usage of diesel engines and energy efficiency.* Currently, approximately 86% of shipping is driven by diesel engines that have a 30- to 40-year lifetime, and so the rate at which greater engine efficiency could be deployed is relatively slow.

2. *Use of alternative fuels.* Options include the greater use of liquefied natural gas (LNG), which would have the co-benefit of energy efficiency improvement and reduced emissions of NOx and SOx that occur from current marine fuels.

We shall further discuss environment considerations in Chapter 4, in particular in Section 4.4 through Section 4.6.

2.6   Residential Energy Applications

For completeness, apart from the conventional industries, we will briefly review the energy applications from other energy application-related industries. One of the biggest such "energy industries" is the residential energy applications industry.

2.6.1   Residential Rnergy Applications

Nakagami (2011) [43] surveyed household energy consumption and related indicators in various western and Asian countries. In the western countries, household energy consumption shows a trend toward saturation, but in the

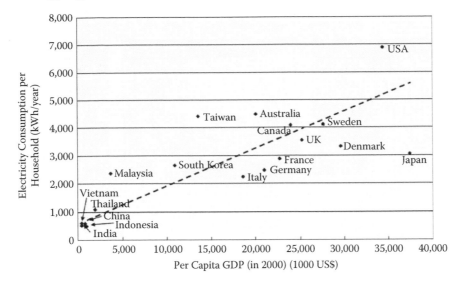

Figure 2.30
Correlation between electricity consumption (excluding cooling and water heating) and per capita GDP. (After Nakagami, 2011 [43]. With permission.)

Asian countries it is likely that household energy consumption will continue to rise. Figure 2.30 illustrates the relation between per capita GDP and electricity consumption, excluding that used for cooling and water heating. On the whole, as per capita GDP rises, so does electricity consumption.

Although its per capita GDP is about the same as Japan's, electricity consumption is highest in the United States of America, at more than twice that of Japan. When the breakdown of electricity end use is considered, much electricity is used for cooking in Canada, Sweden, and England. In countries such as Canada, Australia, and the United States of America, in addition to refrigerators, many households have freezers, with high electricity consumption.

Now, among Asian countries, there are many electric cooking appliances, such as the electric pot, and these consume much electricity. Although per capita GDP of South Korea is only about one fourth that of Japan, per household electricity consumption is at about the same level. Per capita GDP of Taiwan is a little higher than that of South Korea, but per capita household electricity consumption exceeds those of both South Korea and Japan. For the other Asian countries, it is likely that the saturation of household electrical appliances will increase. As the number of air conditioners in use grows, it is possible that electricity consumption will grow to exceed the levels of Japanese and Taiwanese households. In particular, the latent cooling demand in tropical regions is larger than that of Taiwan, and it is likely that increases in standards of living will have a major influence on electricity consumption.

2.7   *Application Projects, Assignments, Questions, and Problems*

Application Project

Linking knowledge with applied energy design and practice

1. Pay an industrial visit to any of the industries. For instance, a typical energy-intensive industry such as a cement industry. Divide students into groups of five persons. (A) Energy conversion system, (B) Energy resources and electrical applications, (C) Energy component and system efficiency, (D) Renewable energy and futuristic design. All groups have to come up with a single Overall Report (containing executive summary, introduction, main contents, discussion, summary, and so on), with overall presentations from all each groups, and to be attended by member of staffs and personnel(s) from industries.

Figure 2.31 (SEE COLOR INSERT)
A typical affordable-cost airboat constructed/assembled by students. The airboat or other energy systems can be designed, assembled, and constructed in-house. (After Abdullah et al. 2010 [26].)

2. Involve students with a team project of assembly of a typical energy system. For example, a typical car, or a typical airboat can be assembled from affordable components or even waste components/materials obtained (see example, Figure 2.31). An application project of this kind or similar can create not only fun but encourage team work spirit among students. It could also enhance the students' awareness and understanding of the importance of energy applications and related issues. Depending on projects, students can work in groups of four to five persons per group.

   (A)   Determine the following: (i) fuel consumption, (ii) energy efficiency, (iii) cost study, etc.

   (B)   Describe possible ways that improve the energy system you constructed.

Assignments

1. Each student is given a different company and the associated company data from any of the industries we have covered in class.

   (a) Analyze the company given to you in term of energy intensity, energy usage, and applications.

   (b) In your opinion, what is the current level of energy application of the company? Give an energy rating to the company (e.g., rating: 20%, 40%, 60%, 80%, 100%).

   (c) Assume that you are employed by the company as an energy adviser, and suggest ways that the company can be improved. Also, give an energy target, emission target (if any).

   Note: *Individual report: data analysis*

2. Each student is given 9 to 10 journals or other literature related to a specific area of industry (e.g., transport or building applications). Write a literature review based on the materials given. You literature review must include the following:

   (a) Abstract or executive summary

   (b) Introduction

   (b) Background study

   (c) Energy resources, supply, and applications

   (d) Energy survey analysis including energy intensity

   (e) Conclusion

   Note: *Individual report: literature review*

Questions and Problems

Multiple choice questions

1. The following are the methods practiced by oil and gas industries for improving energy production efficiency, *except*:

   (A) Finding new wells and improving current wells' operations

   (B) Horizontal well, directional, and extended-reach drilling

   (C) Improve in completions technology

   (D) Smart well technology and control

   (E) Improvement in coal drilling technology

   (F) Enhanced oil recovery (EOR)

2. The energy industry is a generic term for all of the industries involved in the production and sale of energy, including fuel extraction, energy generation, manufacturing, refining, and energy distribution. *Energy power industry* comprises the following main industries, *except*:

   (A) Petroleum and gas industry

   (B) Coal industry

   (C) Pulp and paper industry

   (D) Nuclear power industry

   (E) Alternative energy industry

3. The chemical and petrochemical industry accounts for more than 30% of the total industry energy use worldwide. Which of the following correctly describe(s) the chemical and petrochemical industry in relation to energy applications?

   I The raw materials for most plastics are hydrocarbons obtained from oil refining. To produce various products and different forms, plastics are shaped by moulding, extrusion, and laminating processes.

   II The energy intensive processes include electrolysis, steam cracking, etc.

   III Examples of energy-intensive petrochemicals are ammonia, ethylene, and propylene productions.

   IV Examples of energy-intensive, heavy chemicals are: ammonia, sulfuric acid, sodium hydroxide, and sodium calorate.

   (A) I only

   (B) I and II only

   (C) II and III only

   (D) III and IV only

   (E) All of the above

4. The transport sector was estimated to contribute approximately 14% of the total greenhouse gas emissions in 2004. Which of the following is/are the unwanted gas emissions due to transpiration applications?

   I CO_2

   II CH_4

   III N_2O

   IV CFC-12

   V CFC-134a.

(A) I only

(B) I and II only

(C) I, II, and III only

(D) I, II, III, and IV

(E) All of the above

5. Energy-intensive industry consumes large amounts of heat and electric energy to transform energy processes or materials for our everyday use. Energy intensive industry comprises the following main industries, *except*:

(A) Aluminum and iron industries

(B) Petroleum refining industry

(C) Pulp and paper industry

(D) Nuclear power industry

(E) Alternative energy industry

(F) Cement industry

6. The following statements correctly describe SME industries, *except*,

(A) SME industries consists of smaller companies.

(B) SMEs involved in manufacturing consist of a single or a few product lines compared to large companies.

(C) SEMs usually focus on specialized markets or as supporting companies to big industries.

(D) SME industries refers to the industries that can have up to 500 employees.

(E) SME industries are normally superior to large enterprises in providing comprehensive systems for energy efficiency.

7. Nakagami (2011) surveyed and compared household energy consumption and related indicators in various western and Asian countries. Figure 2.30 is the generic graph illustrating the relationship between per capita GDP and electricity consumption, excluding that used for cooling and water heating. Which of the following can be *correctly* inferred from the findings obtained?

I On the whole, as per capita GDP rises, so does electricity consumption.

II Although its per capita GDP is about the same as Japan's, electricity consumption is highest in the United States of America, at more than twice that of Japan.

III In countries such as Canada, Australia, and the United States of America, in addition to refrigerators, many households have freezers, with high electricity consumption.

IV Although per capita GDP of South Korea is only about one fourth that of Japan, per household electricity consumption is at about the same level.

V Per capita GDP of Taiwan is a little higher than that of South Korea, but per capita household electricity consumption exceeds those of both South Korea and Japan.

(A) I only

(B) I and II only

(C) I, II, and III only

(D) I, II, III, and IV

(E) All of the above

Theoretical questions

1. Define energy industry. Explain briefly the various categories of energy industry.

2. Explain the methods practiced by oil and gas industries for improving energy production.

3. Explain the methods practiced by mining industries for improving coal production.

4. Briefly discuss the energy use in a mining industry.

5. Define an SME industry and briefly explain its differences with big enterprise.

6. Give three motivation factors and discuss why energy-intensive industries seek to become more energy efficient.

7. Why are the world's aluminum and steel smelting industries located in close to specific areas or where are they located?

8. Name the two sources of aluminum production, and briefly explain what they are.

9. With the aid of a block diagram, explain how aluminum smelting is processed.

10. Name the three sources of aluminum production, and briefly explain what they are.

11. Name three of the world's most energy-intensive petrochemical and plastic commodities, and briefly explain what they are.

12. Give three application advantages of steel.

13. "The impacts of transport on energy consumption are diverse, including many that are necessary for the provision of transport facilities" (Rodrigue et al., 2009, [37]). List and briefly discuss any three of the impacts relating to the energy application of transport on energy consumption.

14. Discuss briefly how aircraft can contribute to climate change.

15. Briefly discuss the mitigation measures available for reduction of greenhouse gases (GHG) in maritime transport applications.

16. How can SME industries contribute to maritime transportation?

Homework problems

1. *Energy-intensive industry.* In a typical foundry operating at an environment temperature of 30°C, the melting process is conducted on a typical aluminum alloy for casting purposes. Determine the following:

   (a) The heat required for heating and melting the alloy to its melting point 596° C, by LPG gas.
   (b) The heat required to heat the alloy to 760° C, by electricity.
   (c) Total energy consumption.
   (d) The thermal efficiency of using LPG for melting.
   (e) The thermal efficiency of using electricity for melting.

   Data: the specific heat for solid A354 Al alloys, 1.07 kJ/(kg°C); the specific heat for liquid A354 Al alloys, 1.05 kJ/(kg°C); the melting temperature of A354 Al alloys, 596° C; the superheat temperature, 760° C; the heat of fusion, 389 kJ/(kg°C).

2. *Building industry — heating application.* In a typical hot-water heating system that uses hot water heating coil for heating application, from available manufacturing data, the manufacturer states that heating is capable of delivering 190 kW of heat, under the following given design parameters: Heat exchanger type = single pass cross-flow (tube row Nr =1); entering water temperature: 80° C, water mass flow rate: 4.20 kg/s, entering air temperature = 20 °C, air mass flow rate = 6.71 kg/s. Determine the following:

   (a) Hot-fluid capacity rate and cold-fluid capacity rate, in kW/K, respectively.

N_r	Side of C_{min}	Relation
1	Air	$\varepsilon = \dfrac{1}{C^*}[1 - e^{-C^*(1-e^{-NTU})}]$
	Tube fluid	$\varepsilon = 1 - e^{-(1-e^{-NTU\,C^*})/C^*}$
2	Air	$\varepsilon = \dfrac{1}{C^*}[1 - e^{-2KC^*}(1 + C^*K^2)]$
		$K = 1 - e^{-NTU/2}$
	Tube fluid	$\varepsilon = 1 - e^{-2K/C^*}\left(1 + \dfrac{K^2}{C^*}\right)$
		$K = 1 - e^{-NTU\,C^*/2}$
3	Air	$\varepsilon = \dfrac{1}{C^*}\left[1 - e^{-3KC^*}\left(1 + C^*K^2(3-K) + \dfrac{3(C^*)^2 K^4}{2}\right)\right]$
		$K = 1 - e^{-NTU/3}$
	Tube fluid	$\varepsilon = 1 - e^{-3K/C^*}\left(1 + \dfrac{K^2(3-K)}{C^*} + \dfrac{3K^4}{2(C^*)^2}\right)$
		$K = 1 - e^{-NTU\,C^*/3}$

Figure 2.32
The ε-NTU relations for a heat exchanger with a single pass cross-flow arrangement and different numbers of tube rows $N_r=1,2,3$.

(b) Capacity ratio, C^*.

(c) Effectiveness of the hot water system, ϵ.

(d) The *NTU* value.

(e) The heat transfer *UA*.

Note: The ε-NTU relations for a single pass cross-flow arrangement and different numbers of tube rows is given in Figure 2.32.

(Answer: 17.62 kW/K; 6.76 kW/K; *0.384; 0.468; 0.9041; 7.28 kW/K)*

3. *SME — Energy consumption and cost evaluation.* For a typical SME-based small factory, the annual consumption and related cost of various fuels used to power the associated energy activities are as given in Table 2.7. If the unit cost price is \$300/ton, \$570/ton, \$2/m^3, and \$0.3/ kWh for coal, gas oil, natural gas, and electricity, respectively, determine the following:

(a) Calculate the total annual energy consumption (in MJ).

(b) Determine the total annual cost spent on fuel for the factory.

(c) Calculate the cost of fuel per MJ used for each fuel,

(d) Based on (c) above, suggest how annual fuel consumption can be saved.

(Answer: 147,118 MJ; \$ 2,949; cost/MJ is 0.0107, 0.0125, 0.052 and 0.0833 \$/MJ for coal, gas oil, gas and electricity, respectively.)

Table 2.7 Annual energy consumption and cost of an SME-based factory

Fuel type	Quantity used	Heating value
Coal	2.5 ton	28,000 MJ/ton
Gas oil	1.2 ton	45,490 MJ/ton
Natural Gas	5000 m^3	38.5 MJ/ton
Electricity	3050 kWh	3.6 MJ/kWh

Answers to multiple choice questions:

1. (E) is the method for improving coal production.

2. (C) Pulp and paper industry is considered an *energy-intensive industry*.

3. (E) All of the above.

4. (E) All of the above.

5. (D)

6. (E)

7. (E)

Bibliography

[1] Hiraki, T., Akiyama, T., Exergetic life cycle assessment of new waste aluminum treatment system with co-production of pressurized hydrogen and aluminum hydroxide. *Int J Hydro Energy*, 2009; 34:153–61.

[2] Schifo, J.F., Radia, J.T. Theoretical/best practice energy use in metal casting operations. KERAMIDA Environmental, Inc., Indianapolis, May 2004.

[3] World Coal Institute, WCI case study, November 2008.

[4] http://www1.eere.energy.gov/industry/industries_technologies/index.html, accessed on June 23 2011.

[5] U.S. Department of Energy, U.S. energy requirements for aluminum production: historical perspective. Theoretical limits and current practices; 2007.

[6] U.S Energy Information Administration. International Energy Outlook 2009: World Energy and economic outlook; 2009, available online at: http://www.eia.doe.gov/oiaf/ieo/world.html, http://www.eia.doe.gov/oiaf/ieo/ highlights.html, http://www.eia.doe.gov/oiaf/ieo/industrial.html, http://www. eia.doe.gov/emeu/aer/eh/total.html, accessed August 31, 2011.

[7] International Energy Agency (IEA). Tracking industrial energy use and CO_2 emissions. Paris: IEA/OECD, 2007, Paris.

[8] CEPI (2006), Europe Global Champion in Paper Recycling: Paper Industries Meet Ambitious Target, Press Release July. Brussels, Belgium.

[9] Electrochemistry Encyclopedia, available at: http://electrochem.cwru.edu/encycl/. Accessed on June 24, 2011.

[10] International Iron and Steel Institute, "Statistics on Energy in the Steel Industry (1990 Update,)" Brussels (1990).

[11] Jankowski, J. E., Industrial Energy Demand and Conservation in Developing Countries, Resources for the future,Washington, D.C., Discussion Paper D.73A, Unpublished material, 110p.

[12] Wang, H., Leung, D. Y.C., Leung, M. K.H. Energy analysis of hydrogen and electricity production from aluminum-based processes, Applied Energy, 2011, doi:10.1016/j.apenergy.2011.02.018

[13] Worrell, E. and Galitsky, C., Energy efficiency improvement and cost saving opportunities for petroleum refineries, *Internationl Energy Studies*, 2005-02-15.

[14] Worrell, E., Martin, N., Price, L. Potentials for energy efficiency improvement in the US cement industry, *Energy* 25 (2000) 1189–1214.

[15] Levine, M. D., Koomey, J. G., Price, L., Geller, H., and Nadeli, S. Electricity end-use-efficiency: Experience with technologies, markets and policies throughout the World, *Energy*, Vol. 20, No. 1, pp. 37-61, 1995.

[16] Leffler, W.L. (1985). *Petroleum refining for the nontechnical person* (2nd Edition). PennWell Books.

[17] United States Geological Survey (USGS) (2005), Iron Ore. Available at: http://minerals.usgs.gov/minerals/ pubs/commodity/iron_ore/, accessed on, Sept 26, 2011.

[18] BP, BP Statistical Review of World Energy, 2011.

[19] Bloomberg News. "China, Japan May Cut Refining Capacity," Bloomberg News, January 26, 2011, via Factiva, © Bloomberg Ltd.

[20] Dai, X., Jolly, M. Potential energy savings by application of the novel CRIMSON aluminium casting process, *Applied Energy*, 2011, doi:10.1016/j.apenergy.2010.12.029.

[21] Brun, L. C., and Gereffi, G. 2011, The Multiple Patways to Industrial Energy Efficiency - A systems and value chain approach, Center on Globalization, Governance & Competitiveness, Duke University.

[22] Gielen, D., Taylor, P. Indicators for industrial energy efficiency in India, *Energy* 34 (2009) 962–969.

[23] (2003-05-06). Recommendation 2003/361/EC: SME Definition. Available at: http://ec.europa.eu/enterprise/ policies/sme/facts-figures-analysis/sme-definition/index_en.htm Retrieved April 28, 2011.

[24] GOI, 2005: Annual Report, 2004–2005 of the Ministry of Environment and Forest. Government of India, New Delhi.

[25] APEC, 2002: Profiles in SMEs and SME Issues, 1990-2000. Asia-Pacific Economic Cooperation, Singapore, World Scientific Publishing.

[26] Abdullah, M.O., Yek, P. N. Y., Hamdan, S., Junaidi, E., Kuek, P. An Airboat for Rural Riverine Transportation and Mangrove Marine Environment Applications, *International Journal of Research and Review in Applied Sciences*, (2)3, March, 2010, pp. 199–210.

[27] The ASHRAE Handbook 2009 Fundamentals, American Society of Heating, Refrigerating and Air-conditioning Inc. USA: Atlanta.

[28] Piyush Tiwari, Energy efficiency and building construction in India, Building and Environment 36 (2001) 1127–1135.

[29] PF. Chapman, The energy cost of materials. *Energy Policy*, 1974, 2(2).

[30] Torcellini, P., Pless, S., Deru, M., Crawley, D. Zero Energy Buildings: A Critical Look at the Definition, in: ACEEE Summer Stud, Pacific Grove, California, USA, 2006.

[31] Iqbal, M.T. A feasibility study of a zero energy home in Newfoundland, *Renewable Energy* 29 (2) (2004) 277–289.

[32] Kilkis, S. A New Metric for Net-zero Carbon Buildings, in: Energy Sustainability Conference, Long Beach, California, July 27–30, 2007. ASME.

[33] Mertz, G.A., Raffio, G.S., Kissock, K. Cost Optimization of Net-zero Energy House, in: Energy Sustainability, California, USA, July 27–30, 2007.

[34] Gordon V.R. Holness, On the path to Net Zero: How do we get there from here? *ASHRAE Journal*, June 2011, pp. 50–60.

[35] Metz, B., Davidson, O.R., Bosch, O.R., Dave R. and Meyer, L.A. Intergovernmental Panel on Climate Change (IPCC), Climate Change 2007: Mitigation of Climate Change. Contribution of Working Group III to the Fourth Assessment Report of the Intergovernmental Panel on Climate Change, Cambridge and New York: Cambridge University Press, 2007.

[36] Patxi Hernandez, Paul Kenny, From net energy to zero energy buildings: Defining life cycle zero energy buildings (LC-ZEB), Energy and Buildings, 42 (2010) 815–821.

[37] Rodrigue, J.-P., Comtois, C. and Slack, B. (2009), The geography of transport system, New York: Routledge Taylor & Francis Group, 352 pages.

[38] Love, G., Soares, A., Püempel, H. Climate Change, Climate Variability and Transportation, *Procedia Environmental Sciences* 1, (2010), 130–145.

[39] International Maritime Organization (IMO), Prevention of Air Pollution from Ships, Second IMO GHG Study, 2009. Available at http://www.transportenvironment.org/docs/mepc59_ghg_study.pdf.

[40] Metz, B., O.R. Davidson, P.R. Bosch, R. Dave and L.A. Meyer, Intergovernmental Panel on Climate Change (IPCC), Climate Change 2007: Mitigation of Climate Change. Contribution of Working Group III to the Fourth Assessment Report of the Intergovernmental Panel on Climate Change, Cambridge and New York, Cambridge University Press, 2007.

[41] GreenFacts, http://www.greenfacts.org/en/energy-technologies/l-2/3-emission-reductions-by-by-sector.htm#1, accessed August 13, 2011.

[42] Air Transportation Association, http://www.airlines.org/pages/home.aspx, accessed August 15, 2011.

[43] Hidetoshi Nakagami, International comparison of residential energy consumption, Jyukankyo Research Institute Inc., Japan, http://www.inive.org/members_area/medias/pdf/Inive/IAQVEC2007/Nakagami.pdf, accessed Sept 5, 2011.

[44] BSC. Incorporated, advanced melting technologies: energy saving concepts and opportunities for the metal casting industry, report. US Department of Energy; 2005.

[45] ASHRAE. 2011 ASHRAE® Handbook, HVAC applications, SI Edition. Atlanta, Georgia USA.

3

Energy Resources, Supply, and Demand

In this chapter, we will cover and examine one of the most fundamental aspects of energy application, in relation to the following elements:

- Energy resources;

- Energy supply; and

- Energy demand.

Also, we see the relationships between the elements as well as their importance to our energy applications — as a which influences our worldwide energy application scenarios.

3.1   Introduction to Energy Resources, Supply, and Demand

Our energy resources, energy supply, and energy demand can be represented by the optimistic, balanced plots shown in Figure 3.1. It is a general trend that, over the years, as total energy demand increases due to our daily energy requirements, total energy supply also increases in order to meet those demands.*

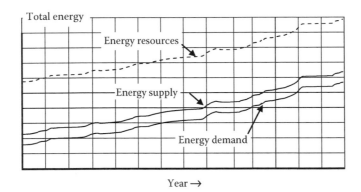

Figure 3.1
The interrelation of energy resources, energy supply, and energy demand based on optimistic, balanced total energy application (OBTE) scenarios.

---

*It is to be noted that the plot of total energy resources is also shown increasing year by year rather than as a fixed plateau; this does not violate the energy conservation laws as outlined in Chapter 1. The total amount of energy in the earth's system remains unchanged, that is, theoretically, it is the total net available energy reserves that decrease with time (see Section 3.2.2).

3.2   Energy Resources

3.2.1   What Is Energy Resources?

Energy resources is a term used to define ways of getting energy for our energy applications.

Our world's energy resources basically have close links with the sun, which travels marvelously and consistently around the globe, as a result of which the sun also contributes useful rays and the associated energy on the earth. Over the years, some of that amount of energy has been preserved in the Earth in the form of fossil fuels, while some other energy is continuously dispersed through energy conversion either directly (such as solar) or indirectly (such as wind, hydro, wave, and tidal energy).

3.2.1.1   Energy Resources Comparison

Energy source ratio can be used to indicate the amount of a typical energy source as compared to that of another energy source. For example,

Energy source ratio of fossil fuel to renewable energy,

$$= \frac{\text{Energy from fossil fuels}}{\text{Energy from renewable energy}} \qquad (3.1)$$

where, fossil fuels refer to oil, gas and coal.

Example 3.1   Figure 3.2 provides a comparison of worldwide reserve of various fuel sources in 2004. What is the energy source ratio of fossil fuel to renewable energy in that year?

Solution: Energy source ratio $= \frac{(38+23+26)}{(6+6+1)} = 6.7\#$

3.2.1.2   Reserves to Production [R/P] Ratio

The R/P ratio is an acronym for "reserve to production ratio."

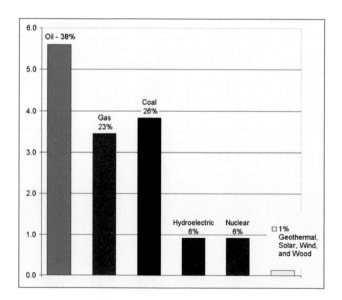

Figure 3.2
Worldwide energy sources, 2004. (From U.S. Department of Energy.)

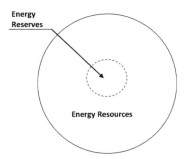

Figure 3.3

Energy resources and energy reserves for our energy applications. Note that the energy reserves are drawn in a dashed line as they are very much dependent on the current estimations.

Table 3.1 The top ten hard coal producers in 2008

Country	Million metric Ton	Percentage
China	2,716	50%
USA	993	18%
India	484	8.8%
Australia	332	5.6%
South Africa	251	4.6%
Russian Federation	246	4.5%
Indonesia	229	4.2%
Kazakhstan	100	1.8%
Poland	84	1.5%
Colombia	74	1.3%

(From SER, World Energy Council.)

3.2.2 How Much Energy Reserves from the Vast Amount of Energy Resources Do We Have?

Here we should define *energy reserves* as the energy available that had been evaluated and deemed commercially viable by energy authorities (for instance by BP, IEA, OPEC, World Energy Council, etc.). On the other hand, *energy resources* is a term we use to refer to the energy sources that are useful, and that will be possible or feasible for energy applications (see Figure 3.3).

Below is a the mini-summary of world energy reserves compiled from various available references mentioned above, in particular, from the World Energy Council and IEA:

Coal Reserves. World coal reserves amount to some 860 billion tons, of which 405 billion (47%) is classified as bituminous coal (including anthracite), 260 billion (30%) as sub-bituminous and 195 billion (23%) as lignite.

The countries with the largest recorded coal reserves are China, the United States of America, and the Russian Federation, with nearly 60% of global reserves between them, followed by Australia and India.

Now, in terms of coal producers, China, the United States and India are the top three, with total production amounting to 77% of the world's hard coal production. China alone produces 50% of the world's coal production. Table 3.1 shows the top ten coal producers as of 2008.

Oil reserves. First, we should note that most oil and natural gas liquid come together in the same oil wells, thus only oil is mentioned here for convenience purpose. We shall touch on natural gas reserves subsequently on the next page after oil reserves.

Global proved reserves* of crude oil and natural gas liquid (NGL) are reported to be approximately 1,239 billion barrels. In 2008, production was 82.1 million barrels per day.

---

*'Proved reserves' have been widely accepted as being a quantity that can be recovered economically from a known reservoir with reasonable certainty after McKelvey [1] who categorized resources both with respect to geological certainty and economic viability in 1972.

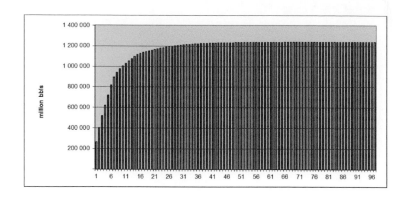

Figure 3.4
Cumulative reserves by country, plotted in order of decreasing increment. (From WEC SER. With permission of the World Energy Council, London, www.worldenergy.org.)

The distribution of reserves is such that most of the quantities are concentrated in the largest fields and found in the countries where these are located. Figure 3.4 illustrates this point well.

Production bears a different relationship to the reported proved reserves in different countries, as seen in Figure 3.5(a) About 66% of the global proved reserves are produced at a rate of about 1.2% per year (a reserves-to-production [R/P] ratio of about 85 years) from only six countries, while about 21% are produced at a rate of about 6% per year (an R/P ratio of about 17 years). The remaining 13% is produced by three countries at a rate of about 3.2% per year (at an R/P ratio of about 32 years). These are average values taken over several countries together. Variation can be marked from one country to another, particularly in the large group with low R/P ratios. The top six produced about 26 million b/d in 2008, the next three, about 14 million b/d, and the 88 countries in the rest of the world, about 42 million b/d. Figure 3.5(b) compares the average R/P ratios for the three groups of countries with the world average.

Natural gas reserve. At the end of 2008, 103 countries were identified as possessing proved reserves of natural gas, with an aggregate volume of approximately 186 trillion standard cubic meters. The world's largest reserves of natural gas are held by the Russian Federation, Iran, Qatar, Turkmenistan and Saudi Arabia.

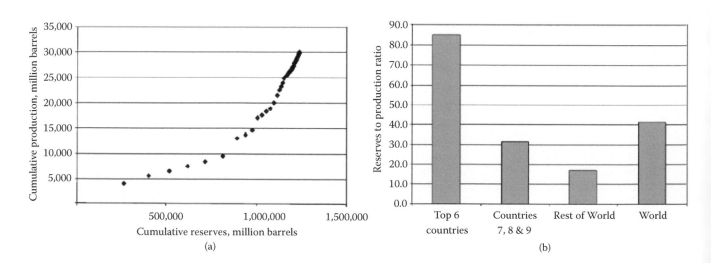

Figure 3.5
(a) Cumulative production versus cumulative reserves plotted by country in order of decreasing reserves increment. (b) Comparison of reserves to production ratios in 2008. (From WEC SER. With permission of the World Energy Council, London, www.worldenergy.org.)

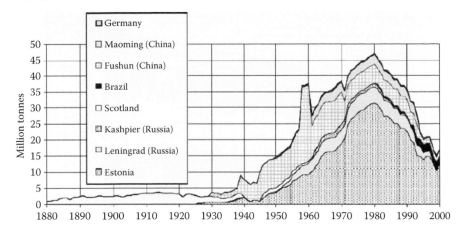

Figure 3.6
Oil shale mined from deposits in Brazil, China, Estonia, Germany, Russia, and Scotland, 1880–2000. (From USGS.)

Proved reserves of natural gas have been identified in every World Energy Council region, with the highest volumes in the Middle East (41%), Europe (including the whole of the Russian Federation) (27%), and Asia (15%). OPEC's proved reserves were some 93 trillion cubic meters at the end of 2008, equivalent to just over 50% of the world total. The corresponding total for the members of the CIS was just over 60 tcm, representing 33% of global reserves.

Oil shale. Total world resources of shale oil are estimated at 4.8 trillion barrels [23]. However, oil shale is not widely extracted currently, but limited to a few countries, notably Brazil, China, Estonia, Germany, Scotland, and Russia (see Figure 3.6). This is because petroleum-based crude oil is cheaper to produce than shale oil due to the additional costs of mining and extracting the energy from oil shale.

3.3 Energy Supply and Energy Demand

3.3.1 What Is *Energy Supply* and *Energy Demand?*

Energy supply is a term refering to production of fuels as well as other energy inputs from available resources. *Energy demand* is the energy input required to provide various energy applications to sustain our socioeconomic and energy activities.

3.3.2 Energy Supply and Demand from Our World Energy Resources

As in the previous section, 3.2.2, we shall also compile the energy supply/energy demand data available to provide a mini-summary, in particular, from World Energy Council and IEA, in terms of the main energy sources.

Figure 3.7 shows the world primary energy consumption* profile upto 2030 in terms of various energy resources.

Coal supply, demand, and consumption. Our world benefits from a vast supply of coal. The coal supply enables us to develop our economic structure, provide jobs worldwide, and alleviate poverty. From an energy application perspective, most significant contributions are for electricity generation in coal power plants, in particular:

1. Use for steel and aluminum production

2. Use in cement manufacturing

3. Use as a liquid fuel

---

*We shall note that "consumption" is a more familiar concept for most people, where it represents the total amount of energy used. The rate of energy consumption (or the rate of energy use) is often refer to energy demand.

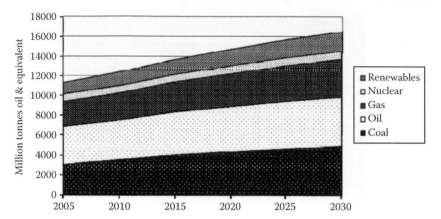

Figure 3.7
World primary energy consumption. (From IGU. With permission of the World Energy Council, London, www.worldenergy.org.)

Around 5.8 billion tons of hard coal and 953 million tons of brown coal were used worldwide in 2008. Since 2000, global coal consumption has grown faster than that of any other fuel — at 4.9% per year. The five largest coal users — China, the US, India, Japan, and Russia — account for around 72% of total global coal use.

The use of coal is expected to rise by over 60% by 2030, with developing countries responsible for around 97% of this increase. China and India alone will contribute 85% of the increase in demand for coal over this period. Most of this is in the power generation sector, with coal's share in global electricity generation set to increase from 41% to 44% by 2030, according to the International Energy Agency (IEA).

The biggest market for coal is Asia, which currently accounts for 56% of global coal consumption. China, and to a lesser extent India, are responsible for a significant proportion of this. Many countries do not have natural energy resources sufficient to cover their energy needs, and therefore need to import energy. Japan, Taiwan, China, and Republic of Korea, for example, import significant quantities of steam coal for electricity generation and coking coal for steel production. Countries possessing large, indigenous sources of coal will continue to use this affordable source of energy to raise electrification levels. In fact, the rapid electrification in South Africa, India and China would have been impossible without affordable coal.

Natural gas supply, demand, and consumption.

Figure 3.8 shows the natural gas supply and production by region. Total natural gas production in North America will increase from 722 bcm in 2005 to 900 bcm in 2030. Gas production in Latin America and the Caribbean almost doubles between now and 2030. The indigenous production of natural gas in Europe (except for Norway) is in decline and from 2004 the United Kingdom has been a net importer. Currently, half of the gas demand in Europe is

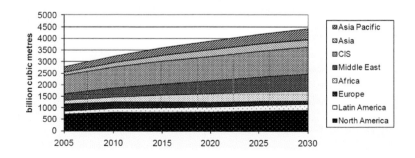

Figure 3.8
Natural gas supply, production by region. (From IGU. With permission of the World Energy Council, London, www.worldenergy.org.)

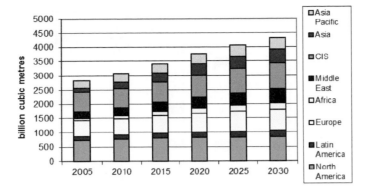

Figure 3.9
World natural gas demand by region. (From IGU. With permission of the World Energy Council, London, www.worldenergy.org.)

covered by domestic production. The other half is imported from Russia (25%), Africa (20%, mainly Algeria), and the Middle East (5%). Gas production in Africa will double between now and 2030, growing to 450 bcm/yr, with Algeria, Egypt, Nigeria, and Libya the main suppliers. Because of the huge gas reserves and substantial investment in the exploration and production sector, production of natural gas in the Middle East is increasing significantly. Gas production will increase from 290 bcm in 2005 to 740 bcm in 2030. The largest producer in the Middle East is Iran, which produced 132 bcm in 2007 and held its place as the fourth largest producer of natural gas in the world. The main gas-producing countries in the CIS are Russia, Kazakhstan, Turkmenistan, Uzbekistan, and Azerbaijan. The rates of economic growth, demand in the domestic and export gas markets, the level of oil and gas prices, as well as success in attracting investment in the development of new gas fields, will affect future production levels. Depending on these factors, production may vary within a range of 1,070 bcm to 1,280 bcm by 2030. The gas production in Asia is expected to grow to around 300 bcm in 2020 and then to stabilize at this level until 2030. China is the main supplier, followed by India and then Pakistan, Myanmar, and Bangladesh. In the Asia Pacific region, Indonesia is the main supplying country with substantial gas reserves, directly followed by Australia and Malaysia. Production has the potential to grow to 450 bcm/yr, which means almost twice as much as current levels.

Natural gas demand is projected to increase by 1.6% per year between 2007 and 2030 to a total of 4.4 tcm. The biggest consuming regions are North America and Commonwealth of Independent States (CIS) followed by Europe. The most dynamic regions are Asia (almost doubling from now to 2030), Africa, and the Middle East (Figure 3.9).

For the Residential and Commercial Sector, a moderate growth is expected from 0.7 tcm at present to well over 0.9 tcm in 2030. Although all regions show some growth in this sector, a significant increase is foreseen in Asia, mainly driven by the number of households to be provided with gas (see Figure 3.10).

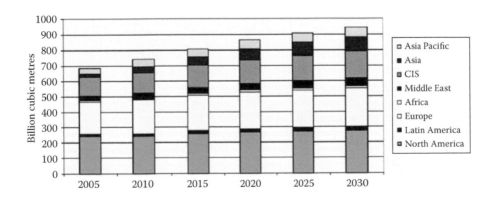

Figure 3.10
Natural gas demand, residential and commercial sector by region. (From IGU. With permission of the World Energy Council, London, www.worldenergy.org.)

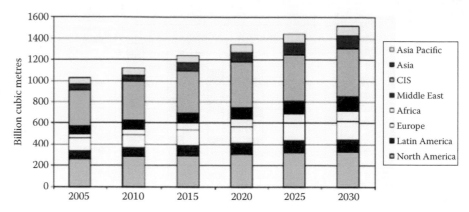

Figure 3.11
Natural gas demand, by region. (From IGU. With permission of the World Energy Council, London, www.worldenergy.org.)

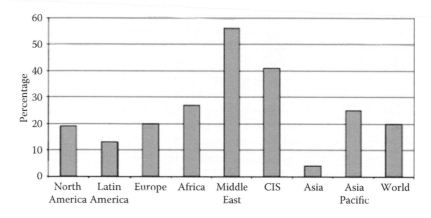

Figure 3.12
Regional share of, natural gas for power generation in 2006, by region. (From IGU. With permission of the World Energy Council, London, www.worldenergy.org.)

The Industrial Sector can benefit from the economic and environmental advantages of natural gas; low $CO2$ emissions and efficient combustibility enable gas to increase its market share. Industrial gas demand is expected to grow from 1 tcm to 1.5 tcm in 2030. Gas demand in the Commonwealth of Independent States (CIS) has the potential to grow strongly, although a major area of uncertainty during this period is the timing of gas consumers' reactions to the rising cost of gas. From a relative point of view, Asia has the highest growth figures. Industrial gas consumption will more than double by 2030, mainly driven by economic development in China and India. Combined heat and power (CHP) will probably expand in almost all regions (see Figure 3.11).

In the Power Sector: the increase in global gas demand during recent years was driven mainly by the power sector. The current global gas share in power generation applications is more than 20%, based on the electricity generated. There are, however, strong regional differences — in the Middle East this share is around 60%, while in Asia only 4% is generated by gas (coal being the principal fuel) (Figure 3.12).

In the Transport Sector, despite large potential, gas consumption for natural gas vehicles is expected to remain small, increasing from around 18 bcm currently to 60 bcm in 2030. Regionally this sector is currently most significant in Latin America, using about 8 bcm/yr. The main regions with growth in this sector are the Middle East and Asia (see Figure 3.13).

Liquefied natural gas (LNG) The expected global share of LNG is 400 bcm in 2015 and 750 bcm in 2030, corresponding to 17% of global gas demand. Qatar, in particular, is expected to play a major role as the largest supplier of LNG.

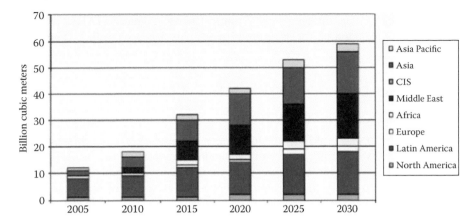

Figure 3.13
Natural gas demand.(From IGU. With permission of the World Energy Council, London, www.worldenergy.org.)

LNG receiving terminal usage patterns differs by region. In the Pacific area, where LNG is generally used as a base gas source without wide use of underground storage, seasonal demand fluctuations are absorbed by redundancy in LNG terminal capacities. In Europe and North America, with more underground storage facilities, higher utilization rates are achieved.

Uranium   Each year, the IAEA, BP, and World Energy Council, provide a range of projections on current and future nuclear electricity generation. In 2009 projection for 2030, the range of nuclear electricity generation varies between 3,711 TWh and 5,930 TWh with 2009 at about 2,560 TWh. The corresponding reactor uranium requirements would range between 105,000 tU and 140,000 tU by 2030. Figure 3.14 (a) depicts the top uranium producers as reported in 2008, while Figure 3.14 (b) shows the increased number of worldwide uranium resources.

According to IAEA, the current 6.3 mtU of identified uranium resources suffices to fuel the global 2008 reactor requirements for about 98 years — a reserves-to-production ratio much larger than for most commercially traded minerals and commodities, including oil and natural gas. Nevertheless, unlike the remnants of fossil fuels, spent nuclear fuel when it leaves the reactor still contains some 95% of its original energy content. Reprocessing and recycling of unspent uranium and the plutonium generated during its residence in the reactor can extend the availability of identified resources to several thousands of years, depending on reactor configuration and fuel cycle.

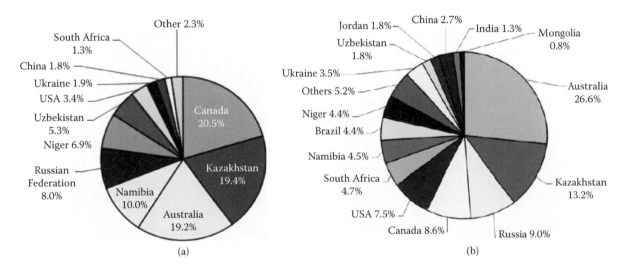

Figure 3.14
World Uranium supply: (a) Top uranium producers in 2008. (From WNA [15]). (b) World supply or distribution of identified uranium resources in 2010. (From World Energy Council 2010 and NEA/IAEA, 2010 [16].)

According to H-Holger Rogner of the International Atomic Energy Agency (IAEA) "Nuclear fuel resources are plentiful but they need the mobilization of above-ground investment funds to unlock their below-ground potentials" [16].

3.3.3   Energy Supply and Demand in Some Selected Countries

3.3.3.1   Asia Pacific Region

The Asia-Pacific region has the most rapidly growing energy demand in the world and will continue to have an increasing impact on world energy demand. Issues of particular importance to the Asia-Pacific region include the supply and demand balance of energy resources and the substitutability of fuels. In addition, national policies are often driven by price and supply security and by investment and development strategies at the national, regional, and international levels. Electricity growth in the Asia-Pacific region will be a major determinant in defining demand for coal. In the past, coal demand and supply were dominated by four nations: Australia, China, India, and Japan [5]. However, recently the nations are extended to include a few more countries, notably South Korea, Hong Kong, Taiwan, Singapore, and Malaysia.

Energy resources and energy consumption patterns The economies in the Asia Pacific region vary significantly in both energy resources and energy consumption patterns. Some, such as Australia, China, Indonesia, and Malaysia, have rich hydrocarbon endowments, whereas others, such as Japan, Singapore, South Korea, and Taiwan, have to import almost all of their oil.

Energy consumption patterns among countries in the region differ in several ways. For example, China's oil consumption accounts for less than 20% of its total energy consumption, whereas more than 55% of Japan's energy consumption is in the form of crude oil and oil products.

Oil demand Oil demand depends on the following primary factors [5]:

(i) Economic performance: The country's stage of economic development, growth rate of GDP, and foreign trade will affect energy intensity, end use patterns, and the affordability and availability of energy in the country.

(ii) International oil prices: International oil prices have direct impacts on oil demand and energy policies of oil importing countries.

(iii) Domestic oil pricing: The domestic product-pricing regime plays a major role in shaping domestic demand.

(iv) Substitution: The availability of indigenous energy resources provides substitution potential and increases the country's energy options.

The Asia-Pacific region is the only part of the world that has produced major new demand in the world oil market in recent years. Given the size of the region's population, 60% (about 4 billion people) of the total world population, the region has great potential to expand oil demand. The region's impressive economic growth in recent years is expected to continue during the next decade, and the region will therefore continue to have the world's highest growth rate of oil demand. The region's oil demand has already surpassed Western Europe's and will soon overtake North America's.

Natural gas demand The use of gas is relatively new in the Asia-Pacific region. As the governments and industries become familiar with established and new uses of gas, and as they prepare to make the heavy start-up investments in infrastructure and distribution, the consumption of gas is certain to increase rapidly. Natural gas has become the fuel of choice particularly among the region's electric power utilities, which place a premium on gas because of security of supply, minimal price volatility, and environment-friendly qualities. It is expected that gas use will expand, as additional discoveries are made, as the distribution system develops, and as a "gas utilization culture" emerges in the region.

Coal demand Dependence on coal to satisfy primary energy needs is higher in the Asia-Pacific region than in any other region of the world. About 46% of the region's commercial energy requirement is filled by coal, compared with 21% for the remainder of the world. The dominant factor in the high percentage of coal use in the region is China, the world's leading producer and consumer of coal. Coal is by far the region's most abundant fossil fuel resource, and it will remain the dominant primary fuel consumed.The rate of growth of coal consumption in the region is dependent not only on economic growth rates in the region but also on coal's ability to meet increasingly stringent environmental regulations, particularly those associated with coal use in generating electricity. Coal is a major contributor to air pollution, and it is facing challenges from cleaner alternatives: natural gas, nuclear energy, and hydropower. Coal is broadly classified as either thermal or metallurgical coal. Thermal coal dominates the market and is expected to account for more than 90% of the consumption growth during the outlook period. Thermal coal is used in electricity generation, industrial boilers, and cement manufacturing; it is also used in households for

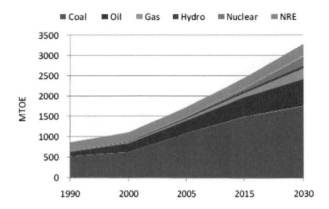

Figure 3.15
Primary energy supply of China. (From APERC analysis, 2009.)

cooking and heating. Metallurgical coal is used primarily for making steel. In recent years, technological advances have allowed partial substitution of lower quality coals, both metallurgical and thermal, as pulverized coal injection in steel making (Intarapravich et al. [5]).

Coal consumption Since most Asian economies have limited economic options for a shift away from coal, the region will account for 60–70 percent of world growth in coal consumption. China will continue to be the largest coal producer and consumer in the world.

3.3.3.1.1   China   *Energy resources and primary energy supply:*
China is rich in energy resources, particularly coal. In 2006, China was the largest producer and consumer of coal in the world, as well as the fifth largest producer, and second largest consumer, of oil.

China's total primary energy supply is projected to grow at an average annual rate of 2.6%. Among the fossil fuels, natural gas will grow at the fastest pace (7.7% per year), followed by oil (3.1%) and coal (1.9%). Nuclear energy is expected to play a key role in reducing China's CO_2 emissions; it has a projected growth of 11.9% over the period, while new renewable energy has a projected growth of about 1% (see Figure 3.15).

Energy demand and consumption:
China's demand is forecast to reach 9.87 million b/d in 2011, up from 9.34 million b/d in 2010. Recent growth in China's energy demand (see Figure 3.16) has mainly been driven by rapid growth in industry; in 2006, industry as a whole accounted for 43.8% of final energy consumption [7]. The main area of growth has been in heavy industry since 2001. Within this sector, energy use is dominated by smelting of ferrous metals like aluminum (29% of all industrial energy used in 2006), followed by non-metal mineral industries (21%) [6].

China's total energy consumption, according to the official data, decreased impressively during 1997–1998 and increased sharply during 2003–2007, which in turn resulted in energy intensity fluctuation. Furthermore, contrary to most of the earlier predictions, China's energy intensity increased in 2003–2005.

The transportation sector accounted for around 10.5% of final energy consumption in 2006; this increased at 9.1% annually from 2000 to 2005. This growth in demand is mainly driven by road transportation, which consumes 66.7% of this sector's energy use. Passenger vehicle numbers, including civil and private, have grown at an average annual rate of 24% from 2000 to 2007. While China is making significant investments in public transportation, including high-speed rail and urban mass transit rail systems, private vehicle ownership is still expected to rise sharply [7].

As pointed out by Liao and Wei in 2010 [8], much literatures has explained this "unusual phenomenon" from different perspectives. Most of this focused on industrial and economic structure shifting and technical changes, including Berrah et al. 2007 [9], Hofman and Labar [10], and Sinton and Fridley [11].

1. *Crude oil:* China has proven oil reserves of 2,100 million metric tons, which will last 5.6 years at 2008 demand levels [7]. China's oil demand increased in response to rising motorization, car ownership, and industrial development. In 2005, crude oil accounted for 19% or one fifth of the total energy demand. Much of the growth in domestic energy demand for crude oil is being met by imports. The expansion of domestic crude oil production and refinery capacity has not been sufficient to match the rapid increase in demand for diesel and gasoline[7].

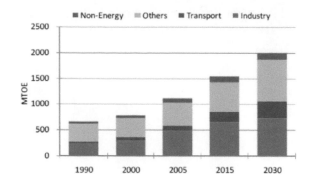

Figure 3.16
Energy demand of China. (From APERC analysis, 2009.)

2. *Natural gas:* China has proven natural gas reserves of 2,460 bcm, equivalent to 30 years' demand at 2008 levels [7].

3. *Coal*: According to recent estimates, China has recoverable coal reserves of some 114.5 billion tonnes, enough to last 44 years at 2007 demand levels [7]. Most of the China's existing electricity generation is coal fired, with coal accounting for 79% of electricity production in 2005.

4. *Hydroelectricity*: China is endowed with 400 GW of hydroelectric generation potential, more than any other economy in the world (2007 installed hydro capacity was 148 GW).

5. *Wind energy*: China has high potential wind-based generation of 1000 GW, including 750 GW offshore and 250 GW shore-based. [7].

6. *Solar energy*: China has been speeding up installation of solar photovoltaic power generation: in early 2007 the installed capacity was 20 MWp, which had risen to 200 MWp at the end of that year. China produced 1088 MWp of solar cells in 2007, exceeding production rates in Japan (920 MWp) and Europe (1062 MWp) [7].

3.3.3.1.2 India "India's energy demand will grow significantly. However, it is short of hydrocarbon and other energy resources. Though it has potential for hydro electricity and other renewable energies, it is limited in potential supply compared to India's need for electricity. Therefore, India has to develop other alternative resources, in particular, with nuclear and solar energy. A part from that energy efficiency an conservation are important top" [20].

According to Parikh et al. 2009 [21], who examined demand scenario from the supply perspectives ranging from coal, hydrocarbon, nuclear, and hydrogen, to hydro and other renewables, none of these renewables are substantial and India will have to rely on imports.

3.3.3.1.3 Japan

1. *Crude oil*: According to the 2008 OPEC Annual Statistical Bulletin, output of crude oil averaged 2.57 million b/d in 2008, of which the bulk was exported, almost all to Japan and other Asia/Pacific destinations.

2. *Natural gas and coal*: With respect to demand and supply, Japan does not have natural energy resources such as natural gas and coal sufficient to cover domestic energy needs, and therefore needs to import energy. Japan imports significant quantities of natural gas and steam coal for electricity generation, and coking coal for steel production.

3. *Nuclear energy*: According to IAEA data, there were 55 operable nuclear reactors at the end of 2008, with an aggregate generating capacity of 49,315 MWe gross, 47,278 MWe net. Within this total there were 28 BWRs (24,764 MWe gross, 23,908 MWe net), 23 PWRs (19,366 MWe gross, 18,420 MWe net), and four ABWRs (5,185 MWe gross, 4,950 MWe net). At the beginning of 2010, total net nuclear generating capacity was 46,823 MWe in 54 reactors, which provided about 29% of Japan's net generation of electricity during the year. The Japanese

WEC Member Committee expects that by the end of 2020 there will be 62 nuclear reactors in operation, with a total gross capacity of 60,197 MWe (approximately 57,700 MWe net).

4. *Solar Energy:* Tthe Japanese market increased by nearly 12% over 2007 and represented 16% of the group of 19. By year end a cumulative total of 2,144 MWp had been installed, some 84% higher than that of the next largest country, the United States of America. Grid-connected capacity at 2,053 MWp represented 96% of the total.

5. *Wind energy:* By end-2007 total installed power stood at 1,538 MW. During 2008, 342 MW were added bringing the total by year-end to 1,880 MW and with annual output of 2,919 GWh, from 1,508 turbines.

6. *Geothermal:* Japan is one of the world leaders in terms of generation of electricity from geothermal. Total capacity reached 535 MWe in 2006. The position was unchanged at the end of 2008. The existing 18 plants are located on the southern island of Kyushu, in northern Honshu, at Mori on Hokkaido, and on the island of Hachijo, some 300 km south of Tokyo. By far the most important utilization of geothermal hot water in Japan is for direct use. It can be classified into three categories: the thermal use of hot water, geo-heat pumps, and hot springs for bathing. Based on the consideration that there are more than 25,000 hot springs throughout the country, a figure of nearly 1,700 MWt expressed in terms of fuel alternative energy was thought to represent this use in 2006. This estimate accounts for some 80% of total direct use. When recreational hot-spring bathing is excluded, the estimated 2006 total installed direct use capacity was 400 MWt. Of this total, snow melting and air conditioning accounted for 38%; hot water supply and swimming pools, 31%; space heating, 19%; greenhouse heating, 9%; fish breeding 2%; and industrial and other uses, negligible. Other energy conversion in Japan includes ocean thermal energy conversion (OTEC).

3.3.3.1.4 Malaysia Malaysia is one of Southeast Asia's successful economies — GDP has grown steadily at an average annual rate of 6.1% since 2000, as can be clearly seen in Figure 3.17. According to APERC, its economic success has been principally based on manufacturing and resource extraction, although there are ongoing initiatives to expand services and higher-value-added activities.

According to APEC [7], three possible outlooks or scenarios of Malaysian energy demand and supply are as follows:

- Malaysia's primary energy supply is projected to grow at 2.8 percent a year, from 65.9 Mtoe in 2005 to 130.5 Mtoe in 2030. The growth is driven mainly by the demand for coal and gas in the electricity generation sector and oil products in the transport sector.

- The rising demand for coal in the electricity generation sector and for oil in the transport and industry sectors will make Malaysia a net energy importer by 2030. Net imports of energy will reach 17.9 Mtoe in 2030; in 2005 the economy had net exports of 31.1 Mtoe.

- Malaysia is seeking to ensure its long-term energy security and minimize negative environmental impacts, by intensifying its energy-efficiency initiatives and enhancing development of viable alternative energy sources, such as solar and biofuels.

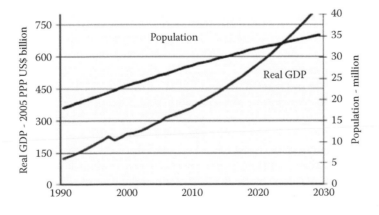

Figure 3.17
GDP and population of Malaysia. (From APERC analysis, 2009 [7].)

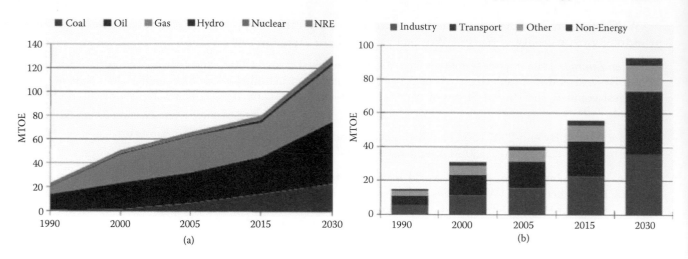

Figure 3.18
(a) Malaysian primary energy supply and (b) Malaysian final energy demand. (From Asia Pacific Energy Research Centre, APEC Energy Demand and Supply Outlook, 4th Edition, Economy Review, 2009. With permission.)

Energy resources and supply

Figure 3.18 shows the primary energy supply of Malaysia. Oil is one of the economy's main sources of energy; in 2006, the total oil supply was 26 Mtoe or 39% of the total energy supply.

Malaysia's primary energy supply is projected to grow at an average annual rate of 3%, reaching 130.5 Mtoe in 2030. Oil is expected to account for 40%, natural gas 36%, coal 18%, hydro 2%, and renewables 4% [7].

Energy demand

The final energy demand (Figure 3.18b) for the industrial sector is expected to grow at an average annual rate of 3.4%, reaching about 35.9 Mtoe by 2030. Electricity and oil are projected to be the fastest growing energy sources in the industrial sector; demand for electricity will grow at an average annual rate of 5.1%, while oil grows at 3.4%. The transport sector accounts for the largest portion of Malaysia's final energy demand, with a share of 37.3 Mtoe. The sector's energy demand is projected to grow at an average annual rate of 3.6% [7].

3.3.3.2   Europe Region

3.3.3.2.1   Denmark   According to Kaya and Yalcintas (2010) [13], Denmark possesses excellent energy application characteristics, in that it:

(i) is a world leader in the use of renewable energy (especially developing wind energy technology);
(ii) has one of the highest GDP's in the European Union (EU);
(iii) has the lowest energy intensity in the EU;
(iv) has an electricity usage per capita similar to those of California and Hawaii (6000–8000 kWh); and
(v) has an active energy policy to increase energy efficiency while maintaining high economic growth.

Denmark's GDP increased by 85% while energy usage increased only 8% for the period 1991–2005. In other words, Denmark produced more than many countries for every kWh of consumed electricity.

Figure 3.19 shows the electricity intensity index of the United States, Denmark, California, and Hawaii with year 1991 represented by 100% in the graph. We can note that Denmark consistently has the lowest electricity intensity over time, which reflects its constantly increasing GDP. On the other hand, Hawaii's electricity intensity increases due to inefficient electricity use. More importantly, Figure 3.19 also indicates that the energy usage per capita does not decrease the economic growth, which reflects the findings of the APS report. According to Kaya and Yalcintas, contrary to a common misconception, experiences of both Denmark and California show that an increase in energy efficiency stimulates economic growth if GDP is an indicator of economic growth. Thus, energy policy makers, engineers, scientists and entrepreneurs should place equal, if not more, emphasis on decreasing energy use by adapting policies and technologies to improve energy efficiency (Kaya and Yalcintas, 2010 [13]).

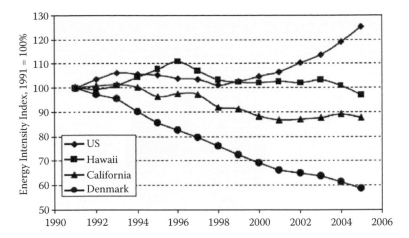

Figure 3.19
Comparison of electric intensities of Hawaii, US average, California and Denmark. (From Kaya and Yalcintas [13].)

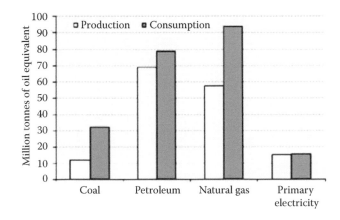

Figure 3.20
Production and consumption of UK primary fuels, 2010. (From Department of Energy & Climate Change, United Kingdom [14].)

3.3.3.2.2 United Kingdom (UK)   Figure 3.20 illustrates the figures for the production and consumption of individual primary fuels in 2010. In 2010, aggregate primary fuel consumption was not met by indigenous production; this continues the trend since 2004 when the UK became a net importer of fuel. However, as explained in subsequent chapters, the UK has traded fuels such as oil and gas regardless of whether it has been a net exporter or importer. In 2010 the UK imported more coal, crude oil, electricity, and gas than it exported; however, the UK remained a net exporter of petroleum products. There was particularly large growth in gas imports in 2010, particularly of Liquefied Natural Gas (LNG), following the expansion of import facilities in 2009, and an increase in gas exports.

   Now in terms of consumption, the energy consumption by the UK's main industrial groups in 2010 is as given in Figure 3.21. In 2010, about 27.5 million tons of oil equivalent were consumed by the main industrial groups. The largest consuming groups were chemicals (16.1 %), metal products, machinery and equipment (12.5%), food, beverages, and tobacco (11.6%), iron and steel and non-ferrous metals (8.0 %), and paper, printing and publishing (8.7%).

   The United Kingdom has an impressively great variety of renewables; see the Sankey diagram in Figure 3.22 showing the energy flows of renewables' supply and demand. Landfill gas, energy from gas, wind, wave, and tidal are the main supply of renewables used primarily for electricity production; most liquid biofuels are imported and supplied for transportation applications.

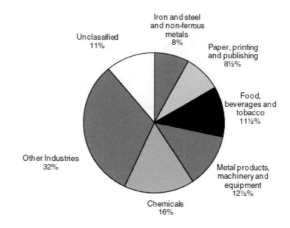

Figure 3.21
Energy consumption by UK main industrial groups, 2010. The graph is for a total final energy consumption by industry of 27.5 million tons of oil equivalent. (From UK Department of Energy & Climate Change [14].)

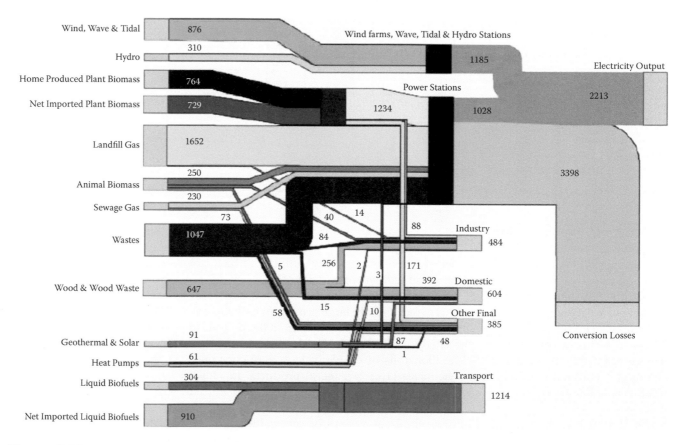

Figure 3.22
Renewable energy flowchart 2010 demand and supply in UK. (From UK Department of Energy & Climate Change [14].)

3.3.3.3 Middle East Region

3.3.3.3.1 Iran *Energy resources and supply in Iran*

Energy resources in Iran consist of the third largest oil reserves and the second largest natural gas reserves in the world. According to estimates in 2006, Iran produced about 5% of total global crude oil production. They produced 4.2 million barrels per day (670,000 m3/d) of total liquids and 3.8 million of those barrels were crude oil.

By the end of 2009, Iranian oil R/P ratio was 89.4 years, which is the world's highest. By 2009, Iran had 52 active rigs and 1,853 producing oil wells (IEA, 2009).

Energy demand and consumption in Iran

The energy consumption in the country is extraordinarily higher than international standards. Natural gas and oil consumption both account for about half of Iran's domestic energy consumption. With its heavy dependence on oil and gas revenues, Iran continues to explore for new sources of natural gas and oil.

Power generation capacity of Iranian thermal power plants reached 173 terawatt hours in 2007, accounting for 17.9% of power production in the Middle East and African region. Natural gas was the main form of energy in Iran in 2007, comprising over 55% of energy needs, while oil and hydroelectricity accounted for 42% and 2% respectively. Demand has also been supported by rapid increases in car production in recent years, and petrol consumption is estimated to have been around 1,800,000 barrels per day (286,000 m3/d) in 2007 (before rationing), of which about one-third is imported. These imports are proving expensive, costing the government about US$4bn in the first nine months of 2007/08. Nearly 40% of refined oil consumed by Iran is imported from India (Javadi, 2011) [4].

3.3.3.3.2 Jordan *Energy resources and supply*

Unlike other Middle Eastern countries, Jordan imports most of its energy needs as a result of lack of energy resources available. Its lack of conventional fossil energy resources, such as oil and gas, creates a burden on its economy due to the relatively high cost of imported oil, thus causing slow economic growth fairly critical for social development. Jordan has limited energy resources such as oil shale deposits, tar sands, a few low geothermal sources and biogas. Its oil shale, heavy oil, and tar sands are not of high quality. In addition, their exploration, development, and utilization are rather difficult, and currently not economically feasible; thus they do not contribute any significant power for now (Department of Statistics Amman, 2010 [12]).

Energy demand in Jordan

The energy consumption grew at an average of 14% per year and the energy bill averaged at 13% of GNP), and consumed most of the foreign exchange earned by exports of all Jordanian commodities in the last three decades. In 1996 the country imported 60,000 barrels per day of crude oil. Figure 3.23 shows the projected demand for crude oil in Jordan up to the year 2010. The transportation sector consumes 41% of the total consumption of primary energy, the industrial sector 21%, and households 20%. Figure 3.24 presents the electricity consumption of Jordan up to the year 2010. As a result, Jordan has to focus on renewable energy.

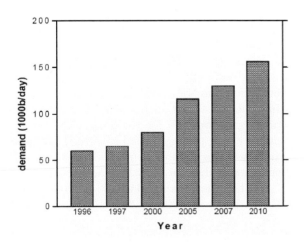

Figure 3.23
Crude oil demand for Jordan 1996-2010.

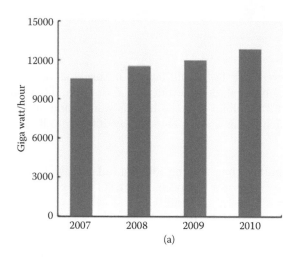

Indicator	2007	2008	2009	2010
Industrial	2,917	3,128	3,005.5	3,258.4
Household	4,017	4,459	4,888.3	5,219.7
Commercial	1,758	1,925	1,979.7	2,183.6
Water Pumping	1,592	1,713	1,772.5	1,866.9
Street Illumination	269	284	310.2	314.6
Total	10,553	11,509	11,956.3	12,843.2
Electricity Subscribers	1,262	1,352	1,426	1,498

(a)   (b)

Figure 3.24
Electric consumption for Jordan, 2007–2010. (a) Total electric consumption. (b) Electricity consumption (GW/hour) by purpose. (From Department of Statistics, Jordan [12].)

Renewable energy

A large sector of Jordanian electrical load can greatly benefit from solar energy, due to high irradiance and long sunny days in Jordan, whether these loads are connected to a grid or stand alone. The existing and planned projects for small and large wind farms are important for establishing a good alternative for existing conventional sources of energy [2].

In 1993, the share provided by solar water heaters was between 1.7% and 1.8%, photovoltaic systems gave 0.0016%, hydro power provided 0.06%, and wind power contributed 0.007%.

3.3.3.3.3   Kuwait   Kuwait is one of the main oil and gas producing countries.

In Kuwait, the consumption of oil — the country's main source of energy — is increasing year by year. In addition to the harsh climate and the rapid economic and construction growth in the country, there are further aspects of energy inefficiency, causing imbalance of the energy supply and demand scenario.

Energy supply and demand in Kuwait

As reported by Alotaibi [3], while 10% of the produced energy was being consumed locally in Kuwait in 1980, this percentage increased to 20% in 2005 and is expected to reach 40% by 2015. If this situation continues, the country will be forced to increase production or reduce exportation, both of which options will cause serious problems to the country in meeting future energy demands due to its dependence solely on oil as a source of energy and income.

The mean consumption of electricity per capita in Kuwait rose from 4912 kWh/person in 1978 to 8321 kWh/person in 1988; it then rose from 12,461 kWh/person to 13,142 kWh/person from 1998 to 2008. The overall consumed quantities of energy in 2005 were equivalent to 193 million barrels (approximately 530 thousand barrels daily), i.e., an increase of 95% compared to 1995. This fuel is mainly used for power plants, the oil sector, transportation, and households. Among the above four major users of fuel, the combined power and desalination plants (CPDP) are the dominant suppliers of electricity generation and water desalination, which accounted for approximately 54% of the total consumed energy in 2005, while the remaining energy was distributed among the other sectors: 28% for the oil sector, 17% for the transportation sector, and 1% for the household sector (Ministry of Electricity and Water, Kuwait, 2008). The installed capacity of power plants to produce electricity, according to the Ministry of Electricity and Water, has been increasing to handle the expanding demand. For example, the installed electricity and desalted water capacities increased more than three-fold, from 2128 MW and 78 MIG/d (million gallons per day) in 1978 to 7398 MW and 254 MIG/d in 1988, and they then reached 11,642 MW and 423.1 MIG/d in 2008. The annual rate of increase of installed capacities for the period from 1992 to 2008 is 3.2% and 4.1% for electricity and desalted water, respectively (see Figure 3.25).

Renewable energy in Kuwait: Focusing on the power plant sector, Alotaibi (2011) suggested efficient electricity production as well as the importance of alternative resources applications such as nuclear, solar, and wind energy to meet the future energy demand of Kuwait.

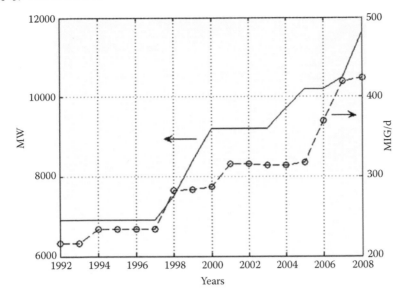

Figure 3.25
Installed capacity of electricity and desalination water plants from 1992 to 2008. (After Alotaibi, 2011 [3].)

3.3.3.4 North America Region

3.3.3.4.1 United States of America (USA)

Energy Resources, supply-demand At present, petroleum accounts for 37% of total energy supply for energy consumption in the United States. Natural gas provides 25% and coal provides 21% of US energy sources. Another 9% comes from nuclear power plants. Renewable energy sources account for 8% of consumption. Most of the renewables are derived from the hydropower and biomass applications such as wood, municipal waste, and agricultural crops (see Figure 3.26).

From the demand viewpoint, according to Energy Information Administration, from energy application sectors, transportation consumes 28% of the energy supply (94% of which is from petroleum, primarily for powering internal combustion engines of vehicles), and the industrial sector accounts for 20% or one fifth of the total primary energy supply. Residential and commercial applications uses 11%; and electrical power generation from grid amount 41%. The electricity power plants in the United States are mostly driven by coal (48%), nuclear (22%), and natural gas (18%). Some other important information in relation to United States fossil fuel energy supply and demand are summarized below (Energy Information Administration; The National Academies [24]):

1. *Crude oil*: Currently, the United States imports almost two third of its oil from a few countries. According to EIA, U.S. production of oil will remain approximately constant through 2030, while imports are projected to rise gradually to about 70% of consumption.

2. *Natural gas*: The annual volume of natural gas consumption is projected to rise from 21.8 trillion cubic feet (TCF) in 2006 to about 23.4 TCF in 2030. Although the United States imports less than 3% of its natural gas from outside North America, it is forecast that imports will increase in the next few decades, from 0.5 TCF per year in 2006 to 2.9 TCF per year in 2030.

3. *Coal*: U.S. coal reserves are plentiful, and estimated at about 270 billion tons. Demand is projected to increase by 30% between 2010 and 2030, propelled by rising use of electricity and possibly the expanded use of still-developing technology that converts coal to liquid fuel.

4. *Nuclear Fuel*: The United States has plenty of uranium fuel. According to U.S. government estimates, output from nuclear power plants is expected to increase only 18% by 2030.

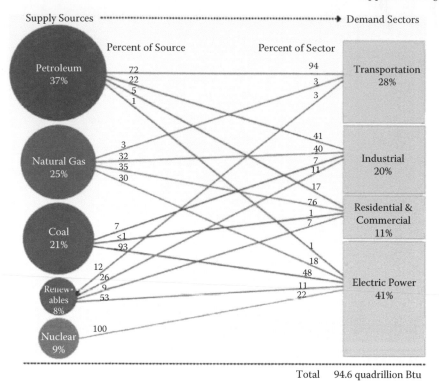

Figure 3.26
Energy sources and demand for the United States in 2009. (From Energy Information Administration, 2009 [17].)

3.3.4 How to Cope with Our Current and Future World Energy Supply and Demand Problems

As we have learned, the energy resources, and supply and demand problems differ from region to region and from country to country. Nevertheless, the problems can generally be tackled by a few important steps or measures:

1. *Finding and developing conventional and new sources of fossil fuels, i.e., oil, gas, and coal.*

2. *Using energy wisely, efficiently, and in an environmentally-friendly way.*

3. *Using renewables and investigating the next generation of energy sources.*

The details of the steps are as follows:

3.3.4.1   *1. Finding and developing conventional and new sources of fossil fuels.*

From the previous discussions, we have learned that at the current rate of fossil fuel production and the associated consumption rates, it is likely that we will have oil and gas for only about 60–70 years and coal for about 100–150 years. Nuclear fuel, on the other hand, may have longer supply potential, in the range of 500–700 years, due to the ability to reuse the spent fuels. Under these finite and urgent fuel requirements, alternatives/renewables and other energy methods have to be developed somewhat rapidly, together with energy efficiency improvements.

Nevertheless, many believe that even if the use of renewables doubles or triples over the next 25 years, the world is likely still to depend on fossil fuels and nuclear power for at least 50% –70% of its energy needs (see, e.g., APEC [7], BP outlook [19], Chevron [18]).

Therefore, oil producing companies around the world are trying to keep up with new searches for oil and gas reservoirs, whilst enhancing oil production by incorporating various technologies, such as "deep" drilling, horizontal drilling, enhanced oil recovery (EOR), and so forth. As a typical recent example in 2010, Chevron has invested $21.6 billion in exploration and production to develop projects for more oil and gas: new technologies are helping the oil and gas industries to see more clearly beneath the Earth's surface. These advances allow the energy and power industries to maximize the production of existing oil and natural gas fields (through, e.g., enhanced oil recovery, horizontal drilling, etc.) and to drill deeper than ever before to locate and recover resources that in the past were considered too difficult to develop. At the Tahiti Field in the deepwater U.S. Gulf of Mexico, Chevron successfully drilled the Tahiti's deepest producing well, which is more than 8,140 m, a record for the Gulf of Mexico, the oil production of which began in May 2009 [18].

Other fossil fuels such as shale oils are currently being extracted only in a few countries, such as Brazil, China, Estonia, Germany, and Israel. However, with the continuing decline of petroleum supplies, accompanied by increasing costs of petroleum-based products, oil shale presents opportunities for supplying some of the fossil energy needs of the world in the years ahead [23]. The same is true for bitumen, heavy oil, and other secondary fossil fuels.

We will further cover enhanced oil recovery (EOR) aspects, including horizontal drilling technology, in Chapter 5, Section 5.2, Energy Recovery.

3.3.4.2 *2. Using energy wisely, efficiently, and in an environmentally friendly way.*

From a philosophical point of view, energy saving (or wastage minimization) and energy efficiency improvement are the two cheapest and most plentiful forms of "new" or "free" energy we have. Basically, energy recovery (e.g., use of combined heat and power/cogeneration), energy savings, wastage minimization, energy improvements, and environmentally friendly goals (e.g., carbon reductions) are all interrelated.

All industries, government agencies, businesses, and residential applications have to encourage the adoption of energy efficiency, reduce energy wastage, and cut down on electric usage for our energy applications. Some of these endeavors are as follows:

- improve power plant and combustion engine efficiency

- eliminate pipeline and storage leakages

- instal more efficient heat exchangers, upgrading steam traps, boilers, etc.

- use of efficient motors, pumps, etc.

- green coal productions, etc.

3.3.4.3 *3. Developing Alternative Energy and other energy resources.*

The depletion of fossil fuels and growing energy demand around the globe are becoming more challenging day by day. Therefore, renewables such as biofuels (e.g., the second generation biofuels that will not subvert the food supply), geothermal energy, hydro power, solar energy, wind energy, and so forth will provide new sources of useful fuels for both power and clean energy.

It is vitally important that we all face the challenges and embrace the energy improvement opportunities. This is to ensure that energy extraction, energy production, and energy applications are done in an environmentally safe manner for sustaining economic growth and improving the quality of life of people around the world. We will cover alternative energy/renewables in Chapter 7 through Chapter 11.

3.4 Energy Flow Visualization and Sankey Diagram

Our daily energy applications are extremely complex and dynamic in nature, generally dealing with energy input, energy output, and the interrelation between energy input and output.

In order to understand the details of energy flow and energy utilization, apart from tabulation of the energy flows, we could use an *energy flow visualization* diagram or *Sankey diagram* to represent the energy flows of our energy systems.

Generally, our energy systems can be divided into two broad categories, i.e., the *macro* level and the *micro* level. In fact, all the micro levels add up to become the global macro level.

$$\sum World\,Macro\,energy\,system = \sum_{n=1}^{n=i} micro\,energy\,systems \tag{3.2}$$

Macro level refers to our world energy system just described in previous sub-sections, i.e., world energy resources, energy supply, and energy demand, while the micro energy systems refers to other energy systems. These micro energy systems can be as small as the energy flows in a typical battery.

3.4.1   What Is a Sankey Diagram and Energy Flow Visualization?

The increasing cost of energy has caused the energy policy makers and industries to examine ways of reducing energy consumption in energy systems and energy processes. Often, energy balances are used to study the various stages of a typical process, or over the whole process and even extending over the total energy system. And, a Sankey diagram is one of the most important flow diagram visualization tools. From an applied energy standpoint, the Sankey diagram is a very useful tool to represent an entire energy system (such as and energy demand and supply scenario) and energy systems in terms of component systems (such as a battery, a boiler, furnaces, power plants, etc.). This diagram represents visually various energy inputs, outputs, losses, and energy stored, so that overall energy improvements can be done in accordance with priority or preferences.

3.4.2   Applications Benefits of Sankey Diagram and Flow Visualization Analysis

The application benefits of a Sankey diagram are:

1. The Sankey diagram represents visually various energy inputs, outputs, losses, and energy stored, so that overall energy improvements can be done in accordance with priority or preference.

2. The Sankey diagram is a visual aid that can illustrate which energy flows represent useful energy and what flows are responsible for waste or emissions.

3. The Sankey diagram can help in locating dominant contributions of an energy component to an overall flow of energy.

4. Apart from energy, a Sankey diagram can be used to visualize material balance, cost transfers, or exergy between processes.

3.4.3   Basics of Sankey Diagram

Captain Henry Matthew Sankey (1853–1925), an Irish engineer who worked on steam engine improvements, was the first to show that the transfer of energy can be represented by a flow diagram. Here are some of the simple basics of a Sankey diagram:

1. In any energy systems, the mass and energy going into the system must balance with the mass and energy coming out from the system.

$$\text{Energy in} = \text{Energy consumed} + \text{Energy out} + \text{Energy stored} \tag{3.3}$$

2. Width of flow arrows is proportional to flow quantity.

3.4.4   Examples of Sankey Diagram Applications

3.4.4.1   Energy Application 1. Building industry — Go green to save the total energy loss.

Figure 3.27 is a Sankey diagram, together with three-dimensional energy systems, showing the total energy loss — essentially all the losses from efficiency losses in conventional power plants to the losses from our building. The diagram illustrates that "more than 90% of the energy extracted from the ground is wasted before it becomes useful work." The article calls for green buildings towards "zero energy building (ZEB)" where energy is produced onsite, encourages renewables, and minimizes all losses. The losses are presented by Sankey arrows bent down sharply, with the amount of losses proportionate to the size of the arrows. The 3D images of the equipment are placed on the diagram to further help in the visualization of the process steps where energy is lost.

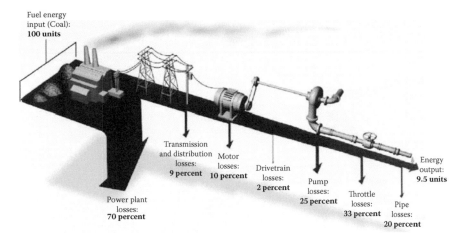

Fuel energy
input (Coal):
100 units

Transmission
and distribution Motor
losses: losses:
9 percent 10 percent

Drivetrain
losses:
2 percent

Pump
losses:
25 percent

Throttle
losses:
33 percent

Energy
output:
9.5 units

Pipe
losses:
20 percent

Power plant
losses:
70 percent

Figure 3.27
Total energy loss — from efficiency losses in conventional power plants to building losses. (Courtesy of WBCSD [22].)

3.4.4.2   Energy Application 2. General: Energy usage for United States in 2006 and 2009.

Figure 3.28 is a typical Sankey Diagram that depicts the flow of energy, measured in quadrillion (1 million billion) BTUs, across the energy system of the United States for 2006 and 2009, respectively, based on data from the Energy Information Administration (EIA) of the U.S. Department of Energy. The chart illustrates the connections between primary energy resources (fossil, nuclear, and renewables), shown at the far left, and end-use sectors categorized into residential, commercial, industrial, and transportation. Electricity powers the sectors to varying degrees and is positioned closer to the middle of the chart to display its inputs and outputs. Note that hydro, wind, and solar electricity inputs are expressed using fossil-fuel plants heat rate to more easily account for differences between the conversion efficiency of renewables and the fuel utilization for combustion- and nuclear-driven systems. This enables hydro, wind, and solar to be counted on a similar basis as coal, natural gas, and oil. For this reason, the sum of the inputs for electricity differs slightly from the displayed total electricity output.

In general, the Sankey diagrams are very informative, comprehensive and useful. Many energy supply-demand relations and direct observations can be made, such as those described below:

For the 2006 diagram: Distributed electricity represents only retail electricity sales and does not include small amounts of electricity imports or self-generation. Energy flows for non-thermal sources (i.e., hydro, wind, and solar) represent electricity generated from those sources. Electricity generation, transmission, and distribution losses include fuel and thermal energy inputs for electric generation and an estimated 9% transmission and distribution loss, as well as electricity consumed at power plants. Total lost energy includes these losses as well as losses based on estimates of end-use efficiency, including 80% efficiency for residential, commercial, and industrial sectors, 20% efficiency for light-duty vehicles, and 25% efficiency for aircraft [23, 24]). From the chart, in the Year 2006, the main energy resources are petroleum (39.57%), coal (22.53%) and natural gas (22.47%). From the energy use perspective, we can calculate that transportation applications account for 28.32% (Car [(17.16%) + freight (7.77%) + aviation (3.39%)] = 28.32%). And, industrial used 23.49% of the total energy. Useful energy is shown to be 42.51%.

Now, for the 2009 diagram, it can be seen that the two main energy resources are petroleum and natural gas: petroleum supply (35.27%), natural gas (23.37%); while coal supply drop to only 19.76%. From the energy use perspective, we can clearly see that transportation applications account for 26.98% while industrial used 21.78% of the total energy. Rejected energy is amounting to 54.64%; therefore, useful energy in 2009 improved to 45.36%.

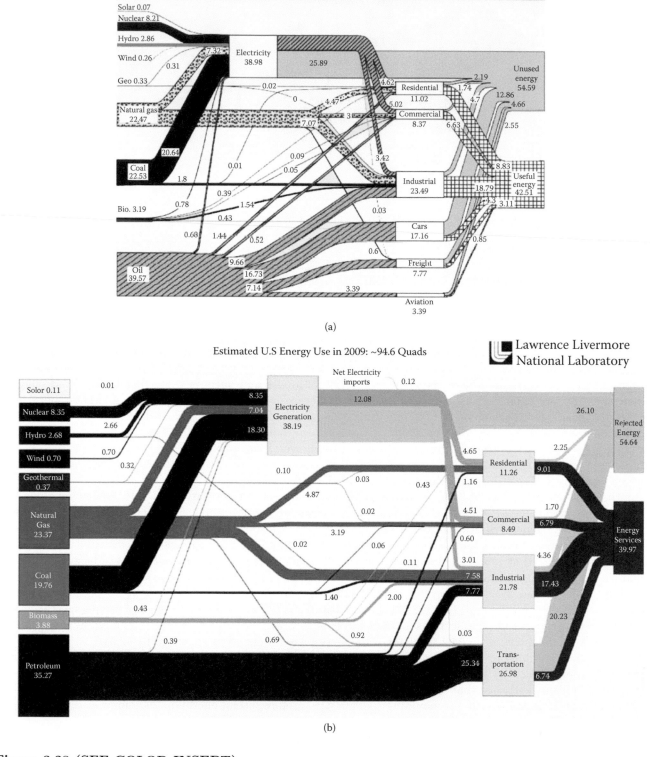

Figure 3.28 (SEE COLOR INSERT)
Estimated energy usage in (a) 2006 and (b) 2009. [From LLNL 2008 data based on DOE/EIA-0384 (2006) and LLNL 2010 data based on DOE/EIA-0384 (2009), respectively. Credit is given to the Lawrence Livermore National Laboratory and the Department of Energy, under whose auspices the work was performed.]

3.5 *Application Projects, Assignments, Questions, and Problems*

Application Project

Linking knowledge with applied energy design and practice

1. Based on the application example shown in Secttion 3.4.4.2 and Figure 3.28, obtain relevant data from the national energy policy related bodies such as department of energy, etc., the student group* is going to:

   (a) Construct a new Sankey diagram to show the latest energy demand and supply scenarios in your country.

   (b) Construct a new Sankey diagram to show the latest energy demand and supply scenarios in your country.

   (c) Compare your Sankey diagram with that of Figure 3.28.

   (d) Write a report describing the present energy scenarios.

Assignments

1. Compare and discuss the energy demand and supply for any two countries selected from different regions. [for example, compare Jordan (Middle eastern region) and Belgium (Europe region)]:

   (a) Middle eastern region

   (b) Asia Pacific Region

   (c) Europe Region

   (d) North America Region

2. Construct a Sankey diagram and discuss the energy applications based on cited journals for *any one* of the following energy schemes/systems:

   (a) A cement plant

   (b) A hydro-power plant

   (c) A typical building

   (d) A computer system

   Also, list and discuss ways that can improve energy applications for the energy schemes/systems you have chosen.

3. *Data analysis (Individual report/assignment)*. Analyze the Sankey diagram as depicted in Figure 3.22, which indicates the renewable energy flowchart in terms of demand and supply in the UK in 2010,

   (a) What is the ratio of electricity output per total energy input of all renewables?

   (b) Determine the percentage contribution of each of the following categories of renewables in supplying the primary energy:

       (i) Landfill gas

       (ii) Wastes

       (iii) Wind, wave, and tidal

   (c) What is the ratio of total energy conversion losses compared to total energy supply from renewables?

   (d) List and discuss briefly how the total energy conversion losses can possibly be minimized.

---

*Group project. Suggestion: three students per group.

Questions and Problems

Multiple choice questions

1. The oil demand for a nation depends on the following factors:

   I Economic performance

   II International oil prices

   III Domestic oil pricing

   IV Substitution and the availability of indigenous energy resources

   (A) I and II only

   (B) I and IV only

   (C) II and III only

   (D) None of the above

   (E) All the above

2. To meet future energy demand, which of the following are the technologies suitable for sustainable electricity production in Kuwait?

   I Efficient electricity production

   II Wind power generation

   III Solar power generation

   IV Nuclear power generation

   (A) I and II only

   (B) I and IV only

   (C) II and III only

   (D) None of the above

   (E) All the above

3. The following statements correctly describe the energy resources, supply, and demand in India, *except*:

   (A) India's energy demand will grow significantly.

   (B) India is short of hydrocarbon resources.

   (C) India has potential for hydro electricity and other renewable energies.

   (D) India's main options lie with nuclear, solar, energy efficiency, and conservation.

   (E) India will not have to rely on imports since there are substantial coal, hydrocarbon, nuclear, hydrogen, hydro, and other renewables etc.

4. Which of the following statements correctly describe the energy demand in the Asia Pacific region?

   (A) The Asia-Pacific region has the most rapidly growing energy demand in the world.

   (B) Dependence on coal to satisfy primary energy needs is higher in the Asia-Pacific region than in any other region of the world.

   (C) International oil prices have no impact on oil demand of oil importing countries in the region.

   (D) The Asia-Pacific region has produced major new demand in the world oil market in recent years.

   (E) The region's oil demand has already surpassed Western Europe's.

5. From a philosophical point of view, energy saving and energy efficiency improvement are the two cheapest and most plentiful forms of "new" or "free" energy we have. Which of the following statements is *not* related to the adoption of energy efficiency in both residential and industrial applications?

   (A) Improve power plants and combustion engines' efficiencies

   (B) Eliminate pipeline and storage leakages

   (C) Install more efficient heat exchangers and upgrad steam traps

   (D) Use efficient motors, pumps, etc.

   (E) Increase coal productions

6. The applications benefits of Sankey diagrams and flow visualization analysis include:

   I The Sankey diagram visually represents various energy inputs, outputs, losses, and energy stored, so that overall energy improvement can be done in accordance to with priority or preference.

   II The Sankey diagram is a visual aid that can illustrate which energy flows represent useful energy and what flows are responsible for waste or emissions.

   III The Sankey diagram can help in locating dominant contributions of an energy component to an overall flow of energy.

   IV Apart from energy, the Sankey diagram can be used to visualize material balance, cost transfers, or exergy between processes.

   (A) I and II only

   (B) I and IV only

   (C) II and III only

   (D) III and IV only

   (E) All the above

7. Which of the following statements correctly describe energy reserves and energy resources?

   (I) Energy reserves can be defined as the available energy that had been evaluated and deemed commercially viable by standard energy authorities.

   (II) Energy reserves can be defined as the available energy that had been evaluated and deemed not commercially viable by standard energy authorities.

   (III) Energy resources is a term we use to refer to the energy sources that are useful, and that will be possible/feasible for energy applications.

   (IV) Energy resources is a term we use to refer to the energy sources that are useful, and that may not be recovered for energy applications.

   (V) Energy reserves or proved reserves is a quantity that can be recovered economically from a known reservoir with reasonable certainty.

   (A) I and II only

   (B) I and III only

   (C) II and III only

   (D) I, III and V only

   (E) III and IV only

   (F) All the above

8. "Nuclear fuel resources are plentiful but they need the mobilization of above-ground investment funds to unlock their below-ground potentials." H-Holger Rogner, International Atomic Energy Agency (IAEA). Which of the following views is in support of the above statement?

(I) According to IAEA, the current 6.3 mtU of identified uranium resources suffices to fuel the global 2008 reactor requirements for about 98 years — a reserves-to-production ratio much larger than for most commercially traded minerals and commodities, including oil and natural gas.

(II) Unlike the remnants of fossil fuels, spent nuclear fuel when it leaves the reactor still contains some 95% of its original energy content.

(III) Reprocessing and recycling of unspent uranium and the plutonium generated during its residence in the reactor can extend the availability of Identified Resources to several thousands of years, depending on reactor configuration and fuel cycle.

(IV) There are no anticipated increases in nuclear power generation worldwide due to the increasing nuclear safety issue and unreliable of the nuclear source for energy applications.

(A) I and II only

(B) I and III only

(C) III and IV only

(D) I, II, and III only

(E) All the above

Theoretical questions

1. Define energy resources, energy supply, and energy demand.

2. Discuss the current energy supply and demand scenarios in a typical country in the Asia Pacific region and Middle East, respectively. Compare the two countries you have chosen in terms of current energy resources, energy supply, and demand.

3. What are the possible future energy problems to be faced in the oil producing countries, including Middle eastern countries? Briefly discuss and suggest how to cope with future energy requirements.

4. Currently, what are the main uses of coal? Discuss briefly the current world coal supply, demand, and consumption. Also, what will be the most likely future demand of coal, say in 2030?

5. Discuss briefly the current natural gas supply, demand, and consumption for the following regions:

   (a) Asia

   (b) Europe

   (c) North America

6. Why does Japan have to import most of its fuel from other countries? Describe the current energy supply and demand of Japan.

7. Discuss the role of nuclear energy for power production applications in Japan.

8. "Nuclear fuel resources are plentiful but they need the mobilization of above-ground investment funds to unlock their below-ground potentials." Statement by H-Holger Rogner, International Atomic Energy Agency (IAEA). Based on the above testament, elaborate and discuss the interrelation of nuclear energy resources, supply, and demand.

9. Unlike other Middle Eastern countries, why does Jordan have to focus on renewable energy and not fossil fuels?

10. Discuss in detail how to cope with our current and future world energy supply and demand problems.

11. Explain how the oil industries are finding more fossil fuels to meet energy supply.

12. What are the application benefits of a Sankey diagram?

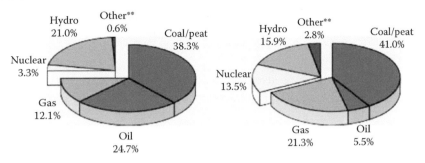

**Other includes geothermal, solar, wind, combustible renewables and waste, and heat.

Figure 3.29
Fuel share of electric generation for 1973 and 2008. (*Left*): 1973 total, 6,116 TWh; (*Right*): 2008 total, 20,181 TWh. (Courtesy of IEA.)

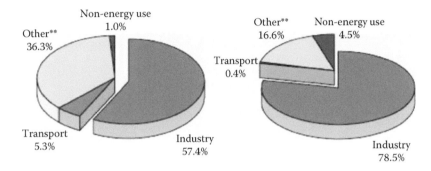

Figure 3.30
Comparison of 1973 and 2008 shares of world coal consumption. (*Left*): Year 1973, total energy use=621 Mtoe. (*Right*): For 2008, total energy use = 823 Mtoe. (Courtesy of IEA.)

Homework problems

1. *Energy resources comparison.* Figure 3.29 provides a comparison of worldwide electricity generation from various fuel sources in 1973 and 2008. Determine the following:

   (a) The energy source ratio of fossil fuel to renewable energy for 1973 and 2008.

   (b) Compare and discuss (a) above with respect to total fuel share.

   (*Answer: 3.5; 3.6; total fuel share =21.6%; 18.7%*)

2. *Energy consumption comparison.* Figure 3.30 provides a comparison of 1973 and 2008 shares of world coal consumption. Determine the following:

   (a) Compare coal use in 1973 and 2008 for industrial applications.

   (b) Explain the possible reasons for the increased use of coal as the energy resource and/or supply for industrial applications.

   (c) What are the possible implications of the increased use of coal?

3. *Energy visualization and Sankey diagram.* Based on the Sankey diagram as depicted in Figure 3.22, which indicates the renewable energy flowchart in terms of demand and supply in the United Kingdom in 2010:

   (a) Determine the percentage of electricity output per total energy input of all renewables.

   (b) Determine the percentage contribution of each of the following categories of renewables in supplying the primary energy:

    (i) Landfill gas

    (ii) Wastes

    (iii) Wind, wave, and tidal

(c) What is the ratio of total energy conversion losses compared to total energy supply from renewables?

(d) List and discuss briefly how the total energy conversion losses can possibly be minimized.

Answers to multiple choice questions:

1. (E) All of the above.

2. (E) All of the above.

3. (E) According to K. Parikh et al. (2009), who examined demand scenarios from the supply perspectives ranging from coal, hydrocarbon, nuclear, hydrogen, to hydro and other renewable, they deduced that none of these are substantial and India will have to rely on imports.

4. (C) International oil prices have direct impact on oil demand and energy policies of oil importing countries.

5. (E) Increased coal production does not improve efficiency; however, use of "green" coal production would improve power plant/machines' efficiency and the environment.

6. (E) All of the above

7. (D) I, III, and V only

8. (D) I, II, and III only

Bibliography

[1] McKelvey, V.E., 1972. Mineral Resource Estimates and Public Policy: *American Scientist*, V.60, No.1, p.32-40.

[2] Al zou'bi, M., Renewable Energy Potential and Characteristics in Jordan, *JJMIE*, Volume 4, Number 1, Jan. 2010, Pages 45–48.

[3] Alotaibi, S., Energy consumption in Kuwait: Prospects and future approaches, *Energy Policy, 39 (2011) 637–643*.

[4] Javadi, M. S., Ghanavati, F., Iran's Electricity Renewable Resource Planning, *Journal of American Science*, 2011;7(5).

[5] Intarapravich, D., Johnson, C. J., Li, B., Long, S., Pezeshki, S., Prawiraatmadja, W., Tang F. C. , Wo, K., Asia-Pacific Energy Supply and Demand to 2010, *Energy*, Vol. 21, No. 11, pp. 101–1039, 1996.

[6] IEA (2008) Energy balances of Non-OECD countries (2008 edition). International Energy Agency. Paris, France. http://www.iea.org/

[7] Asia Pacific Energy Research Centre (APERC), APEC Energy Demand and Supply Outlook, 4th Edition, Economy Review, 2009.

[8] Liao, H., Wei, Y.-M., China's energy consumption: A perspective from Divisia aggregation approach, *Energy* 35 (2010) 28–34.

[9] Berrah N., Feng, F., Priddle, R., Wang, L. *Energy in China: the closing window of opportunity*. Washington, DC: World Bank; 2007.

[10] Hofman, B., Labar, K. Structural change and energy use: evidence from China's provinces. World Bank; 2007.

[11] Sinton, J.E., Fridley, D.G., What goes up: recent trends in China's energy consumption. *Energy Policy* 2000; 28(10):671–87.

[12] Department of Statistics, Amman, *Annual Report on Environmental Statistics*, 2010, Amman, Jordan.

[13] Kaya, A., Yalcintas, M., Energy consumption trends in Hawaii, *Energy* 35 (2010) 1363–1367.

[14] Department of Energy & Climate Change, *Digest of United Kingdom Energy Statistics* 2011, United Kingdom, 2011.

[15] World Nuclear Association (WNA), 2009. World Uranium Mining. www.world-nuclear.org/info/inf23.html. Accessed July 20, 2011.

[16] NEA/IAEA, OECD Nuclear Energy Agency and International Atomic Energy Agency, 2010. "Uranium 2009: Resources, Production and Demand," a joint report by the OECD Nuclear Energy Agency and the International Atomic Energy Agency, OECD, Paris, France.

[17] Energy Information Administration, *Annual Energy Review*, 2009.

[18] http://www.chevron.com/globalissues/energysupplydemand./ Accessed Sept 15, 2011.

[19] BP Energy Outlook 2030, London, January 2011.

[20] Parikh, J., Parikh, K. India's energy needs and low carbon options, *Energy* 36 (2011) 3650-3658.

[21] Parikh, K.S., Karandikar, V., Rana, A., Dani, P.. Projecting India's energy requirements for policy formulation. *Energy* 2009;34(8), Elsevier.

[22] WBCSD, June 26, 2006, Making tomorrow's buildings more energy efficient, *Business & Sustainable Development News*.

[23] World Energy Outlook. 2010. International Energy Agency (IEA). Also online at http://www.iea.org/weo/2010.asp

[24] The National Academies, US, available at http://www.nationalacademies.org/ accessed August 24, 2011.

4

Energy Management and Analysis

In this chapter, we will consider energy management and the associated energy analysis usually encountered or employed in energy applications. We will first introduce a few energy management tools and the basic terms of management study, such as energy audits, and energy life-cycle analysis. We will also discuss the environmental aspects of energy applications, which include energy pollutants, energy safety, control, and the impacts associated with energy applications. Of primary importance also is the discussion of various energy policies and planning for our future energy sustainability and to promote clean energy.

4.1 Energy Audits

4.1.1 What Are Energy Audits?

Energy audits refers to a formal energy account in terms of energy consumption and the associated costs of an energy system, over a certain period, usually on a yearly basis. The audit or account is broken down accordintly into different sections, which are also called energy components. Utility bills or energy application data are collected and shown for the particular period to allow the auditor to evaluate the energy supply pattern, energy demand rate structures, and energy usage profiles.

4.1.2 Basic of Energy Audits

Similar to financial audits, the energy account or energy audit is broken down into different energy application sections.

An example of energy audit for building applications is given in Table 4.1. In this example, the audit allows one to see clearly the various parameters (for instance, product output, fuel consumption, power consumption, CO_2 emission, etc.) for each and every product component (such as cement, brick, glass, etc.). The building materials industry in this typical example (China) account for about half of global production. The industry used 6.6 EJ of fossil fuels and 0.8 EJ of electricity in 2006, which equals about 6% of all manufacturing industry final energy use and 23% of CO_2 emissions. Cement accounts for half of this energy use and three-quarters of CO_2 emissions. The industry can reduce its energy use by 29% and CO_2 emissions by 10%, if the government's policy targets for introduction of new technologies by 2010 are met.

Specific Energy Consumption *Specific energy consumption* is the energy consumed by the energy system per unit output, e.g., in the case of a building, it can be the energy consumption per unit area of the building.

Energy target *Energy target* is the energy goal that an entity (normally a government) is going to achieve in the future, normally 5 or 10 years ahead of the policy made. In the typical example above (see Table 4.1), the energy targets are twofold, i.e., to achieve the goals of energy saving and CO_2 reductions concurrently, and, at a reasonable amount for each and every one of the energy components concerned.

Energy bench mark *Energy bench mark* is the energy "control" or "base line" energy case based on past achievement experiences, such as those given by energy regulatory boards.

Table 4.1 Energy Audit: Energy use, CO_2 emissions and short-term reduction potentials in the chinese building materials industry, 2006

Products	Output in 2006	Fuel Consumption	Power Consumption	Direct CO_2 Emissions	BAT for Near and Medium Term	Energy Saving Target (in 2010)	CO_2 Emissions Reduction Target
		PJ/yr	*TWh*	*Mt/yr*		*PJ/yr*	*Mt/yr*
Cement	1 235 Mt	3 047	123.5	986	NSP	615	105
Clay brick and tile	520 billion standard bricks	1 319	15.9	148	Tunnel kilns	565	29
Building ceramics	5 024 million m²	967	24.2	23	Roller kilns	440	1
Lime	162 Mt	791	8.9	190	Maerz kiln	164	4
Flat glass	22.75 Mt	264	10.0	7	Float process	47	0
Glass fibres	1.41 Mt	88	0.0	2	Direct melt process	18	0
Sanitary wares	131 million pieces	41	0.7	1	Large batch kiln	8	0
Others	–	147	40.0	30	Energy saving technologies	76	3
All ceramic building materials	–	6 645	223.2	1 387	–	1 934	142

Note: BAT = Best available technology; NAP = New Suspension Preheater.
(Courtesy of Cu and Wang [1].)

4.1.3 Benefits and Applications of Energy Audits

The application advantages of energy audits are:

1. We can use energy audits as a tool to analyze energy systems to improve system efficiency, minimize costs, and maximize production.

2. Energy audits also allow us to make comparison studies between various energy schemes for corrective actions planning (see Example 4.1). The action plans may involve selecting energy resources, improving maintenance standards, operation process controls, leakage and losses identification, etc.

Example 4.1 Typical example of energy audit for comparison studies

Table 4.2 illustrates a simple example of how an energy audit or energy account can be used to compare a few energy schemes to power the appliances of buildings based on energy resources applications, in terms of percentage (%). The bench mark "zero house" is also given to indicate the energy level to achieve a typical "zero energy" reference building standard.

4.2 Energy Use and Fuel Consumption Study

In this section, we will first consider a theoretical idea on the relationship of primary energy and secondary energy usage with respect to the overall system efficiency. Thereafter, we will proceed to look at the fuel consumption aspect of energy application.

Table 4.2 A typical example of energy audit to compare energy application distribution of a few buildings based on the energy resources available

Energy resources	Building #1	Building #2	Building #3	"Zero energy" building
Electricity (%)	60%	30%	20%	-
LPG Gas (%)	30%	60%	10%	-
Diesel oil (%)	10%	-	-	-
Renewable energy (%)	-	10%	70%	100%
Energy saving components:				
LED saver	No	No	Yes	Yes
Day lighting	No	Yes	Yes	Yes
Other	No	No	Yes	Yes

4.2.1 Primary Energy — Secondary Energy Use Relationship and Overall Efficiency

With regards to energy use and fuel consumption in energy systems, we have to differentiate between primary and secondary energy consumption. Considering all the losses, it is appropriate that we look at the energy conversion chain from energy production right to the final energy consumer.

4.2.1.1 Example 4.2

Consider a simple table fan as shown in Figure 4.1.

(a) Determine the overall efficiency of primary energy utilization for the fan. Assume the efficiencies for various energy conversion processes are as follows:

- Crude oil (in refinery process) to fuel oil (e.g., diesel/petrol/other fuel oils): 92 %
- Fuel oil to electricity (in thermal power plant): 40 %
- Electricity transmission and distributions: 98%
- Conversion of electrical energy into mechanical energy (by fan): 70%

(b) What is the overall efficiency of primary energy utilization if the thermal power plant converts to higher efficiency natural-gas-fired power with efficiency of, say 55%)?

Figure 4.1
A typical table fan. This 500 W fan is designed for offices and residential applications. (Courtesy of author.)

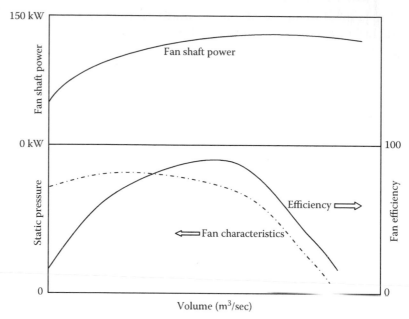

Figure 4.2
Fan efficiency and characteristics curves as a function of volume flow.

(c) Now, what is the overall efficiency of primary energy utilization if the thermal power plant switch to a higher efficient natural gas-fired power plant of 55% efficiency, and with further improved fan efficiency of 90%?

Solution:

From Chapter 1, Section 1.6, we learned that the overall efficiency $\eta_{overall}$ for the primary energy source is the product of all the individual conversion efficiencies.

Case (a) Overall efficiency $\eta_{overall} = 0.92 \times 0.70 \times 0.98 \times (0.40) = 0.25$ or 25%.#

Case (b) Overall efficiency $\eta_{overall} = 0.92 \times 0.70 \times 0.98 \times (0.55) = 0.32$ or 32%.# (an energy saving of 7%).

Case (c) Overall efficiency $\eta_{overall} = 0.92 \times (0.90) \times 0.98 \times (0.55) = 0.45$ or 45 %.# (a possible energy saving of 20%!).

We can note that over the years, most energy efficiencies from refinery processes, and electricity transmission and distribution will not have changed much; however, *thermal* power systems have undergone significant progress due to better utilization of heat and heat recycling technology. For instance, in the 1950s, the efficiency of thermal power plants was only around 18%; in 1980s the modern power plants had improved to an efficiency range of 35–37%. Today, our modern combined gas-fired power plants were reported to be able to reach an efficiency of possibly up to 58% or more.* In other words, due to continuous technology advancement, thermal power plants can reduce substantial primary energy demand!

Now, if every element in the energy conversion chain is further improved in efficiency, the overall efficiency will increase remarkably: Considering the electrical-mechanical conversion in the above application, the major associated energy loss is heat loss, which occurs in the impeller, bearing, belting, motor, and variable speed drive, i.e., $\sum loss = \sum loss_{impellear} + \sum loss_{bearing} + \sum loss_{belting\,bearing}$ loss. Such system loss can be reduced by optimization of the operating condition (see Figure 4.2), thus maximizing the electrical-mechanical conversion efficiency, which in turn contributes to the overall efficiency as given in Case (b).

The fan system component losses can be minimized by, for example, operating the system within optimization range so much so that the system efficiency approaches higher component efficiency (can reach 80–85%), thus resulting in higher overall efficiency as in Case (c).

---

*For example, the Sendai Thermal Power Station, Tohoku Electric Power Co., Inc., Japan. (Capacity: 446-MW, combined-cycle plant; efficiency: 58%) as reported in *Power magazine*, Angela Neville, "Sendai Plant Boosts Efficiency and Cuts Emissions," *Power*, April 1, 2011.

4.2.2 Fuel Consumption Study

Energy system design improvements and the use of higher efficiency components could have substantially reduced total energy consumption. Take, for instance, how the use of high efficiency fans replacing low efficiency fans could reduce total energy use and provide energy savings for a building (see Example 4.2.2.3).

4.2.2.1 *Fuel Consumption Rate*

Fuel consumption rate is the ratio of work done to the fuel use.

$$\text{Fuel consumption rate} = \frac{\text{Work done}}{\text{Total fuel use}} \qquad (4.1)$$

Example: fuel consumption rate for a typical car, is the number of miles traveled to the number of gasoline burned; the S1 unit is expressed in km/liter.

4.2.2.2 *Specific Fuel Consumption*

Specific fuel consumption is the ratio of the fuel use to its output in specific units.

$$\text{Specific fuel consumption} = \frac{\text{Total fuel use}}{\text{Energy output per specific unit}} \qquad (4.2)$$

Specific fuel consumption is a widely used measure of energy system performance for, e.g., an engine performance, the SI units of which is in kg per hour per kilowatt [(kg/h)/kW].

4.2.2.3 Example 4.3

An operator of a commercial building has decided to change to the use of a higher efficiency fan (Model A) replacing the old fan (Model D). The data available from the fan manufacturer is given in Table 4.3.

(a) Determine how much savings per year if the fan operates 16 hours per day, and the electric rate is $0.1 per kW-hour. The drive loss is estimated at 0.45 hp with minimum efficiency of 91.7%.

(b) How can the operator further improve the fan operation and for further saving?

Solution:

(a) Smaller fans are typically less efficient than larger fans for a given fan type.

Model A: Total mechanical power is

$$P_{mechanical} = fan\,shaft\,bhp + hp\,drive\,loss = 6.84 + 0.45 = 7.29\,hp$$

Total power conversion required by the fan motor is

$$P_{fan\,motor} = 7.29\,hp \times \frac{0.746\,\frac{kW}{hp}}{0.917} = 5.93\,kW$$

Total annual cost

$$Annual\,Cost = 5.93\,kW \times 16\,\frac{hours}{day} \times 365\,\frac{days}{year} \times \frac{\$0.10}{kw-hour} = \$3463$$

Table 4.3 Fan specifications with ducted outlet, 11.07 m³/sec and required pressure

Model	Fan size (in.)	Fan speed (rpm)	Fan power (bhp*)	Fan class	FEG**
A	24	1579	6.84	I	85
B	27	1289	6.24	I	85
C	30	1033	5.73	I	85
D	33	887	5.67	I	85

* bhp = brake house power
** FEG = fan efficiency grade

Model B: Total mechanical power is

$$P_{mechanical} = fan\,shaft\,bhp + hp\,drive\,loss = (5.67) + 0.45 = 6.12\,hp$$

Total power conversion required by the fan motor is

$$P_{fan\,motor} = 6.12\,hp \times \frac{0.746\,\frac{kW}{hp}}{0.917} = 4.98\,kW$$

Total annual cost

$$Annual\,Cost = 4.98\,kW \times 16\,\frac{hours}{day} \times 365\,\frac{days}{year} \times \frac{\$0.10}{kw-hour} = \$2908$$

Therefore, annual savings = \$3463 - \$2908 = \$555\#

(b) Use of variable frequency drive (VFD) where VFDs allow fans to respond to system requirements and run at reduced speeds.

4.2.2.4 Econometrics modeling study on fuel consumption

Over the years, there were many econometrics modeling studies on world fuel consumption.

Greening et al. 2007 [2] emphasized that "industrial energy consumption is one of the hardest for end-uses to analyze, model and forecast." Their main contribution is the development of econometrics models for worldwide fossil fuel consumption, based on demand side variables, i.e., fuel price, production, net imports, and reserves.

Shafiee and Topal (2008) [3] use econometrics techniques to find out the effect of variables on fossil fuel consumption and to demonstrate the trend of fossil fuel consumption, based on the linear model:

$$C_t = \alpha + \beta_1 PRO_t + \beta_2 P_t + \beta_3 NI_t + \beta_4 RES_t + \varepsilon \tag{4.3}$$

where,

C_t = world fossil fuel consumption
PRO_t = U.S. fossil fuel production
P_t = world real fossil fuel price
NI_t = U.S. fossil fuel net import
RES_t = world fossil fuel proven reserves

The research found that while the consumption of fossil fuels worldwide has increased, trends in U.S. production and net imports have been dependent on the type of fossil fuels. Most of the U.S. coal and gas production has been for domestic use, which is why it does not have a strong influence on worldwide fossil fuel prices. Moreover, the reserves of fossil fuels have not shown any diminution during the last couple of decades and predictions that they were about to run out are not substantiated.

4.3 Energy Life-Cycle Analysis

4.3.1 What Is Energy Life-Cycle Analysis?

Energy life-cycle analysis or assessment (LCA) is a quantitative method for assessing the energy input and outputs, as well as energy impacts (for example, energy savings and environmental impacts) of energy products or energy component systems from cradle to grave, i.e., from creation to disposal, including energy production and recovery. It is an approach in which all energy inputs to a product are accounted for — either as direct energy inputs during manufacture or other energy inputs required to produce components, materials, or services needed for the entire processes.

4.3.2 Basis and Type of Energy Life-Cycle Analysis

According to the ISO 14040:2006 [4] and ISO 14044:2006 [5] standards, a Life Cycle Assessment is carried out in four phases or main components:

1. *Goal and scope.* The ISO standards require that the goal and scope of an LCA be clearly defined and consistent with the intended application.

2. *Life cycle inventory* (LCI). The LCI analysis involves creating an inventory of flows from and to a product (energy) system. Inventory flows include inputs of water, energy, and raw materials, and releases to air, land, and water.

3. *Life cycle impact assessment* (LCIA). It is aimed at evaluating the significance of potential environmental impacts based on the LCI flow results, which consist of the following elements:

   (a)   Selection of impact categories, category indicators, and characterization models;

   (b)   Classification stage, where the inventory parameters are sorted and assigned to specific impact categories; and

   (c)   Impact measurement, where the categorized LCI flows are characterized, using one of many possible LCIA methodologies, into common equivalence units that are then summed to provide an overall impact category total.

4. *LCA interpretation.* LCA Interpretation is a systematic technique to identify, quantify, check, and evaluate information from the results of LCI and/or LCIA assessment, and the results of which are summarized. The outcome of the interpretation phase is a set of conclusions and recommendations for the study.

In the following subsections, we summarize three examples of case studies conducted on life-cycle of energy applications.

4.3.2.1 Case Study 1: Life-Cycle Study of Soy Bean Diesel Oil

Pradhan et al. 2009 [6]had conducted a life-cycle assessment (LCA) to quantify and compare the environmental and energy flows associated both with biodiesel and petroleum-based diesel. The fossil energy ratio (FER) is used to measure the energy balance of biodiesel defined as the ratio of the energy output of the final biofuel product to the fossil energy required to produce the biofuel.

Energy requirements for producing soybeans were estimated for both direct energy, such as diesel fuel and gasoline, and indirect energy, such as fertilizers and pesticides. Pradhan et al. found that diesel fuel use required the most energy on the farm, followed by fertilizers and herbicides. Next, the energy required to transport soybeans from the farm to processing plants was estimated based on information from the Greenhouse Gases, Regulated Emissions, and Energy Use in Transportation (GREET) Model. It requires about 6,393 British Thermal Units (Btu) to transport 1 bushel of soybeans to a processing facility.

The model used in the study was designed to represent a processing facility that combines a soybean processing plant with a biodiesel conversion unit producing 9.8 million gallons of biodiesel, 151,515 tons of soybean meal, 9,000 tons of soybean hulls, and 4,380 tons of crude glycerin. The soybean crusher uses energy in the form of electricity to power motors and provide lighting. Natural gas and process steam are used to provide heat for drying. Hexane is used for oil extraction. The total amount of energy required for removing the soybean oil is about 23,000 Btu per gallon of biodiesel. The soybean oil is converted into biodiesel using a process called transesterification, which is done by reacting the oil with an alcohol and a catalyst in large reactors. This reaction also results in the production of crude glycerin, which is a valuable coproduct. The conversion of the soybean oil into biodiesel and the treatment of the glycerin requires almost 19,000 Btu per gallon of biodiesel. Energy is also required to ship the biodiesel from the processing plant to marketing outlets. Using the GREET model, it was determined that on average it requires about 1,000 Btu to ship a gallon of biodiesel to its final destination. Combining the energy input estimates from the four subsystems completed the base case life-cycle assessment for biodiesel. After adjusting the energy inputs by energy efficiency factors and allocating energy by coproducts, the total energy required to produce a gallon of biodiesel was 25,696 Btu.

The study by Pradhan et al. (see Figure 4.3) showed that biodiesel conversion used the most energy, accounting for about 60% of the total energy required in the life-cycle inventory. Soybean agriculture accounted for 18% of the total

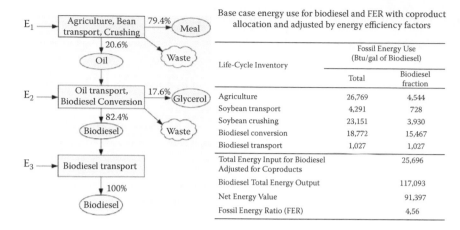

Figure 4.3
Typical results from the life-cycle analysis. (After Pradhan et al. 2009 [6].)

energy requirements, followed by soybean crushing, which required almost 15% of the total energy. The net energy value (i.e., biodiesel energy output, minus fossil energy input) was about 91,000 Btu per gallon. The estimated FER of biodiesel was 4.56. The life study approach used is detailed below:

The secondary inputs added were farm machinery, building materials for a crushing plant, and building materials for a biodiesel conversion plant. When the input energy for both agricultural machinery and building material are added to the inventory, FER declines to 4.40, still considerably higher than the 3.2 FER reported elsewhere. The final step in the life cycle analysis was to examine the effect of rising soybean yields on the FER of biodiesel. The analysis found that the FER of soybean biodiesel is expected to reach 4.69 when projected soybean yield reaches 45 bushels per acre in 2015. This is about a 3% increase compared to the 2002 FER estimate. This result suggests that the FER of biodiesel will continue to improve over time. In addition to higher yields, improvements can be expected to occur in other areas of the life cycle as the agricultural sector, along with the biodiesel industry, continues to make energy efficiency gains in order to lower production costs.

How to estimate the fossil energy ratio (FER) is defined in Eq. 4.4,

$$FER = \frac{Renewable\ fuel\ energy\ output}{Fossil\ energy\ input} \tag{4.4}$$

$$Energy\ input\ allocation\ for\ biodiesel = E1f1 + E2f2 + E3 \tag{4.5}$$

where,

$E1$ is energy input for agriculture, soybean transport, and soybean crushing;

$f1$ is the mass fraction of soybean oil used to produce biodiesel;

$E2$ is the energy used during transesterification and the transport of the soybean oil;

$f2$ is mass fraction of the transesterified oil used to produce biodiesel;

$E3$ is energy input for biodiesel transport.

Combining the energy input estimates from the four subsystems completes the base case life-cycle assessment for biodiesel. The phases are often interdependent in that the results of one phase will inform how other phases are completed.

All estimates of electricity generation were based on weighted average of all sources of power used in the United States, including coal, natural gas, nuclear, and hydroelectric. Electricity use only includes electricity generated from fossil sources, which on a natural average equals 70%. After adjusting the inputs by energy efficiencies and allocating energy by coproducts, the total energy required to produce a gallon of biodiesel is 25,696 Btu.

4.3.2.2 Case Study 2: Life-Cycle Study of a Coal Power Plant [26]

Worldwide emissions of greenhouse gasses (GHGs) due to human activities are increasing and with it the concentration of GHGs in the atmosphere, resulting in climatic change. One of the options to mitigate GHG emissions is the implementation of CO_2 capture and storage (CCS) in the electricity supply system. The basic idea is that CO_2 is captured from power plants and sequestrated underground for many thousands of years.

Table 4.4 Life cycle inventory results for key air pollutants, emissions to water, resource consumption, and production of waste and by-products [26]

Substance	Unit (per kWh)	Case 1	Case 2	Case 3
Emissions to atmosphere				
CO_2	g	1050	805	200
NO_x	g	1.94	1.03	1.39
SO_2	g	1.41	0.71	0.84
Methane	g	1.47	1.13	1.51
HF	mg	11.98	1.38	0.64
HCl	mg	14.10	7.68	3.90
Hg	μg	22.01	6.77	9.66
Particulate matter <10 μm	mg	97.83	67.33	84.92
Particulate matter >10 μm	g	1.51	1.11	1.46
MEA	mg	2.63×10^{-4}	1.99×10^{-4}	12.25
NH_3	mg	63.73	47.03	248.48
PAH[a]	μg	46.39	35.52	48.02
NMVOC[b]	mg	119.35	91.06	127.23
Emissions to water				
Hg	μg	3.75	3.22	4.53
PAH[a]	μg	7.22	5.49	8.70
Nitrate	mg	28.86	26.84	67.97
Resources				
Coal direct	g	441	338	444
Coal total	g	447	343	451
NH_3	g	0.75	0.39	2.13
MEA	g	2.60×10^{-7}	1.97×10^{-7}	2.04
NaOH	g	0.12	0.11	0.39
Limestone	g	7.73	5.64	7.51
Quicklime	g	1.06	0.01	0.03
Wastes and by-products				
Gypsum	g	-1.39×10^{-5}	9.08	11.91
Reclaimer bottoms	g	–	–	2.10
Total waste	g	140.42	107.74	146.31

[a] Polycyclic aromatic hydrocarbons.
[b] Non-methane volatile organic compounds.

During the operation of the power plant a vast amount of resources are consumed, such as coal, limestone, ammonia, chemicals, and water for cooling and for the steam cycle.

In order to assess the environmental impacts of three pulverized coal fired electricity supply chains with and without carbon capture and storage (CCS), Koornneef et al. 2008 [26] had conducted a life cycle assessment on a cradle-to-grave basis. The life cycle assessment methodology distinguishes three cases of comparison studies as follows:

Case 1: the reference case and representing the average sub-critical pulverized coal fired power plant operating in the Netherlands in the year 2000.

Case 2: a state-of-the-art ultra-supercritical pulverized coal fired power plant as proposed by several companies to be installed in the coming years (2011–2013) in the Netherlands. This power plant can be considered the best available technology at present.

Case 3: a state-of-the-art coal fired power plant, equal to Case 2, equipped with a post-combustion capture facility based on chemical absorption of CO_2 with monoethanolamine (MEA). CO_2 capture with the use of MEA is already widely applied in the chemical industry, though it is still in the pilot and demonstration phase for application in coal fired power plants. The technology is assumed to be available in the near future. The CCS chain further comprises compression, transport, and underground storage of the CO_2.

The chain with CCS comprises post-combustion CO_2 capture with monoethanolamine, compression, transport by pipeline, and storage in a geological reservoir. The two reference chains represent subcritical and state-of-the-art ultra-supercritical pulverized coal fired electricity generation. A detailed greenhouse gas (GHG) balance for the three chains had been established, and disclosed environmental trade-offs and co-benefits due to CO_2 capture, transport and storage.

Their results (Table 4.4) show that due to CCS, the GHG emissions per kWh are reduced substantially to 243 g/kWh. This is a reduction of 78% and 71% compared to the sub-critical and state-of-the-art power plant, respectively. The removal of CO_2 is partially offset by increased GHG emissions in upstream and downstream processes, to a small extent (0.7 g/kWh) caused by the CCS infrastructure. An environmental co-benefit is expected, following from the deeper reduction of hydrogen fluoride and hydrogen chloride emissions. Most notable environmental trade-offs are the increase in human toxicity, ozone layer depletion, and fresh water ecotoxicity potential for which the CCS chain is outperformed by both other chains. The state-of-the-art power plant without CCS also shows a better score for the eutrophication, acidification, and photochemical oxidation potential despite the deeper reduction of SOx and NOx in the CCS power plant.

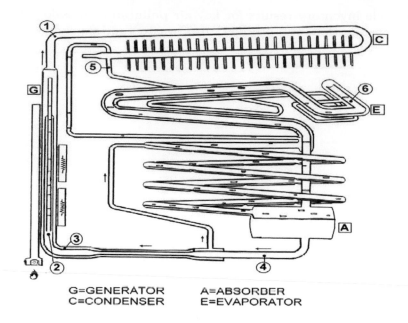

G=GENERATOR   A=ABSORBER
C=CONDENSER   E=EVAPORATOR

Figure 4.4
Schematic diagram of a typical vapor absorption refrigerator [24]; model: TAR/EV.

4.3.2.3   Case Study 3: Life-Cycle Analysis of an Adsorption Air-Conditioner Powered by Three Different Energy Resources [25]

In this life-cycle analysis study, aH_2O–NH_3–H_2 absorption cooling machine is considered (see Figure 4.4). The cooling machine uses ammonia as refrigerant, water as absorbent, and hydrogen as equalizer for pressure.

In the current study, the analysis conducted is categorized into two parts, i.e., without time effect, and with time effect.

4.3.2.3.1   Life-Cycle Analysis without Time Effect Consideration

The economical analysis (without time effect) carried out includes the capital cost, life cycle cost, cash flow diagram, accounting rate of return, payback period, and the graph of annual saving as functions of time. The brief explanations of parameters and equations used in this study are summarized as follows:

(a) Capital cost, C

$$C = C_m \tag{4.6}$$

where,

$C=$ capital cost;
$C_m=$ cost of machine.

Capital cost refers to the capital needed to purchase equipments, including an electric heater or fuel burner. The assumption made is that the transport fee and insulation fee are included in the cost of the equipment.

(b) Life cycle cost

$$C_{LS}=C+C_F \tag{4.7}$$

where,

$C_{LS}=$ life cycle cost;
$C_F=$ cost of fuel.

Life cycle cost is the sum of the capital cost plus the fuel cost. The maintenance and associated replacement cost are negligible (estimated to be less than 3%), as the system is intended for residential application in this study.

(c) Annual operating cost

$$C_{OP}=C_F + C_M \tag{4.8}$$

where,

C_{OP} = annual operating cost.

The maintenance fee is neglected in this investigation, because this study is for residential usage, where maintenance service can be neglected. Methods (a) through (c) above deal with the cost in relation to the system powered by the individual energy resources. The following methods, outlined from (d) through (h), are used to compare the cost parameters in relation to various energy sources, i.e., optional case (alternative energy sources) with the original case (electric energy).

(d) Accounting rate of return, ARR

For the ARR method, the annual fuel cost of the absorption refrigerator powered by LPG or photovoltaic, F is first calculated. Fe is the fuel cost of the electrical grid system, which is Watts multiplied with the cost. Then, the Net Annual Saving, NAT is calculated by Eq. 4.9 as follows:

$$NAT = F - Fe - Depreciation \tag{4.9}$$

where,

NAT = net annual saving;
F = fuel cost;
Fe = fuel cost of the electrical grid system.

The Net Annual Saving refers to the real saving value, minus the decrease in resale price after usage. The depreciation is neglected in the current study as a small scale residential machine does not require resale after using. Eq. 4.9 calculates the average net annual saving in n year.

$$ANAT = \frac{\sum_n NAT}{n} \tag{4.10}$$

where,

$ANAT$ = average net annual saving.
Finally, ARR is calculated as,

$$ARR = \frac{ANAT}{C} \tag{4.11}$$

where,

ARR = accounting rate of return.

(e) Payback method

The payback is important to find out the number of years required to recover all the capital cost. That is, mathematically, payback period is described by Eq. 4.12, an iterative procedure until the running total, RT is zero.

$$RT = C - NAT \tag{4.12}$$

where,

RT = running tools.

(f) Net saving cost diagram

Net saving cost diagram is used to show the net annual saving, based on Eq. 4.9, of each year in a specific period.

(g) Annual saving vs. time graph

In this method, the annual saving cost with time is plotted in line graph format for analysis.

4.3.2.3.2 Life-Cycle Analysis with Time Effect Consideration (h) Internal rate of return analysis, IRR
For the IRR method, Net Present Value, NPV is first determined. All the cash flow is converted to an equivalent base. From Keown [22], single payment present worth,

$$P = \left[\frac{1}{(1+i)^n} \right] S \tag{4.13}$$

where,

P = present value;
S = future value.

Uniform series compound present worth, P is then calculated by Eq. 4.14 as follows [22]:

$$P = R\left[\frac{(1+i)^n - 1}{i(i+1)^n}\right] \tag{4.14}$$

where,

R=uniform future value.

IRR is used to find the discount rate when NPV is equal to zero. A graph is plotted to find the discount rate when NPV is equal to zero.

(i) Cash flow diagram with inflation

According to Thumann and Mehta [23], inflation is the rate of increase in the average price of goods. The Real value parameter is determined by

$$Real\,value = \frac{S}{(1 + RRI)^N} \tag{4.15}$$

where,

RRI= real rate of interest.

If we combine the effect of the real rate of interest RRT, and the overall rate of inflation I_R, the uniform face value. R, is,

$$R = (1 + I_R)(1 + RRI) - 1 \tag{4.16}$$

So, the equation for inflation becomes:

$$Real\,value = \frac{S}{(1 + R)^N} \tag{4.17}$$

A computer program has been developed based on the above equations, the flow chart of which is given in Figure 4.5.

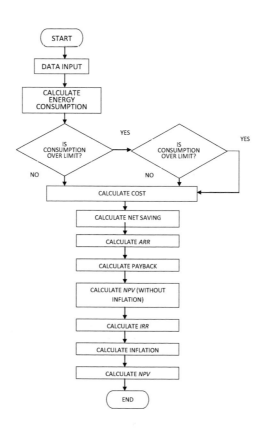

Figure 4.5

Flowchart of the energy life cycle analysis simulation program. (After Abdullah and Tang, 2010 [25].)

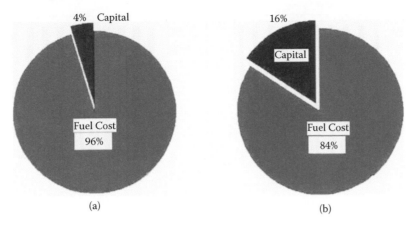

Figure 4.6
The 10 year life cycle cost: (a) electrical heater; (b) LPG burner. (After Abdullah and Tang, 2010.)

4.3.2.3.3   Energy Economic Analysis* and Life-Cycle Study for the Absorption Application by Different Energy Sources
From the capital cost analysis, it is clearly shown that the electrical heater is the cheapest energy alternative, with capital cost of only $US2857. Then, it is followed by the LPG burner, which costs around $US176. Due to the higher manufacturing cost and technology, the photovoltaic system is the most expensive alternative which costs about $US2836.

From 10-year life cycle cost analysis, it depicts the estimated costs of owning and operating an electric/fuel system to heat an absorption refrigerator system. As can be clearly seen from Figure 4.6(a), most of the expense of the electrical system is in the fuel cost. In the 10-year life cycle cost, the total amount of cost used is $US128. The initial cost of the electrical heater system is only around 4% of the total cost incurred over the life of the system. It means around 96% of cost expenses is in the fuel cost consumption. The amount for fuel cost is RM4168.01. The maintenance fee is neglected here because residential usage seldom arranges maintenance for the device.

Then, from Figure 4.6(b), the capital cost of the LPG burner system is 16% of the total cost. It is higher than the electrical heater system. But its overall expense is lower than that of the electrical heater system, which is RM3802.66 and its fuel cost for 10 years is just $US942. The overall cost thus is cheaper and more cost-effective than the electrical heater system in the long run.

As for the photovoltaic system, it has the most attractive benefit of having zero fuel cost. But it has the highest capital cost which, in this case, is $US2836. So, the result of its 10-year life cycle is totally its capital cost.

From the annual operating cost diagram (Figure 4.7a), it shows that the photovoltaic system produces almost zero fuel cost. The electrical heater consumes at a higher fuel cost than the LPG burner system. For electrical supply, every year it needs to pay a fuel cost of around $US123. For the LPG, the fuel cost per year is $US94. Hence, the LPG is around 23% cheaper than the electrical cost.

The above mentioned methods are used to analyze the system separately. The following methods are comparisons made between the optional (LPG/photovoltaic) and the original (electric) project, such as accounting rate of return, payback period, the net saving cost diagram, and internal rate of return.

The accounting rate of return result shows that the LPG burner is more profitable than the photovoltaic system because it gives a higher ARR. Nevertheless, if one studies carefully, the photovoltaic system is instead more profitable. This is because the capital cost for the photovoltaic system is higher, and so the users could actually save $US123 each year if the photovoltaic system is adopted. For the LPG burner, on the other hand, only $US28 is saved. Therefore, ARR alone is not suitable to compare the systems, which have different capital costs. The payback period shows the benefits of the LPG burner in terms of capital cost. It just requires 6.215 years for payback compared to the photovoltaic heater system, which needs a period of 23.135 years. In other words, it is 272% longer than the LPG burner system.

When the net savings cost diagram is analyzed, the photovoltaic is better in net saving. This is because, each year, it could save fuel cost > 300% of LPG burner fuel cost.

---

*Note that all prices are quoted in malaysian currency (RM) in the original journal article; here we adopt the United States currency, $US, such that the conversion rate of 1$ US = RM3.40 (approximate) is used herewith.

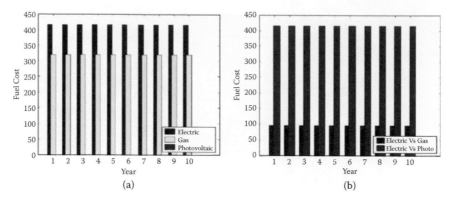

Figure 4.7
Fuel cost comparison: (a) Annual Operating Cost. It is to be noted that the graph of photovoltaic is not shown because the fuel cost and the maintenance cost is assumed to be zero. (b) Net Saving Cost. (After Abdullah and Tang, 2010 [25].)

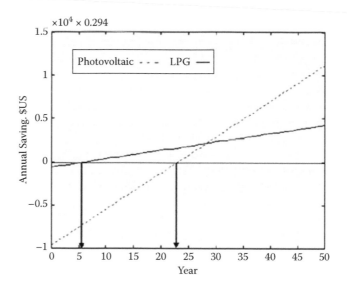

Figure 4.8
Annual savings versus years.

Figure 4.8 depicts the plots of the annual savings versus time for photovoltaic and LPG, and for a period of 50 years. The first information that could be gathered here is the payback period. The plots show that the LPG burner system and the photovoltaic system have a payback period of approximately 6.2 years and 23.1 years, respectively. The high payback period for the photovoltaic system is mainly due to its extremely high system capital cost.

The earning time of the projects is one of the most important pieces of economic information. As can be seen from the figure, therefore, the LPG burner has gained profits faster and higher than the photovoltaic system. But this situation is going to change after 27 years. From the twenty seventh year, the annual cost savings of the photovoltaic system is expected to be higher than that of the LPG burner system. The difference of values becomes more pronounced thereafter. The difference can well be calculated from the area above the x-axis, which amounts to $US1866 within 50 years. Based on the IRR method, it strongly proves that the LPG burner is the preferred choice; it needs a discount rate of 9.4675 to make Net Present Value equal to zero in 10 years period. The photovoltaic, however, cannot achieve any positive discount rate.

The methods mentioned previously were calculated without inflation. The following, however, are the analyses that consider the inflation factor. Figure 4.9 shows the photovoltaic gives a cheaper fuel cost (zero fuel cost) compared to LPG and electric from grid. Next, the fuel cost comparison between the electrical and LPG can be deduced for a period of 10 years. During the initial 4 years, the LPG fuel gas is cheaper than the electrical fuel cost. It is

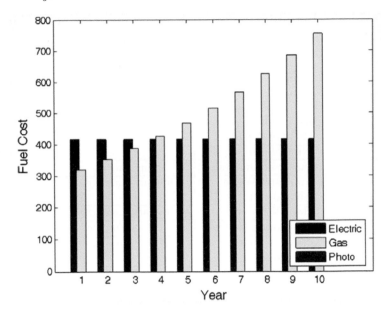

Figure 4.9
Annual operating cost (with inflation). It is to be noted that the graph of photovoltaic is not shown because the fuel cost and the maintenance cost is assumed to be zero.

Table 4.5 Result summary of life cycle analysis

Energy source	Electrical	LPG	Photovoltaic
Capital cost	$US57	$US176	$US2836
10-year life cycle cost	4% (capital cost) 96% (operating cost)	16%(capital Cost) 84%(operating cost)	100% (capital cost) 0% (operating cost)
Annual operating cost diagram	$US123	$US94	$US0
ARR	-	16.089	4.322
Payback	-	6.215	23.135
IRR	-	9.467	-51.182
Net saving cost	-	$US28	$US123
Annual operating cost (with inflation)	Highest within the first 3 year period	Highest after 4th year	Zero

because the electrical price is constant, and does not inflate over 10 years, as far as the present study and location is concerned. However, the LPG fuel cost has inflated every year so that after a 4-year period, the fuel of the LPG is more expensive than the electrical fuel cost.

The summary of the overall result is shown in Table 4.5. In summary, the electricity from grid is the best alternative for short term consideration of less than 10 years. This is because it has low capital cost and constant fuel cost. LPG is the second alternative of choice. The main disadvantage of the LPG burner is the fuel cost, which may increase every year. Photovoltaic is the good alternative only when the long-term project is planned such as over a 10-year period.

4.4   Energy, Environment, and Health

From the ecology energy point of view, our Earth was and is in harmonious and energy-balanced condition. From the total sunlight arriving on Earth, most of it (around 45-47%) bounces back into space. About 30% of this energy

Table 4.6 Selected pollutants and their environmental and health effects

	Pollutants	Environmental or health effects	Remarks/ example references
Coal and oil Power plants	Carbon Dioxide (air)	Global warming	[17]
	Nitrous Oxides (air)	Smog	
	Particulate or Dust (air)	Lung cancer	
	Thermal pollution (liquid)	Kill marine life	
	TSP (air/liquid)	Allergic disease	[14]
	Heavy metals (air/liquid)		[21]
Nuclear power plants	Particulate or Dust (air)	Lung cancer	[18]
	Radioactive Waste (liquid/solid)	Difficult in disposal	
	Radiation (air)	Radiation, cancer	[18]
	Heat Emissions (water)	Kill marine life	
Oil spills	Oil contaminants (liquid)	Kill marine life	
	Nitric oxides (NO)	Acid rain	[20]
	Sulfur dioxide (SO_2)	Acid rain	[20]
Vehicles and transport (esp. exhaust)	CO (air)	Angina, Brain damage	
	NO_x, HCs, SO_2 (air)	Acid rain, cancer	[19]
	TSP, lead (solids)	Allergic, cancer	[19]

is radiated back in the Earth's atmosphere as heat; around 20% of this energy enters the hydro-cycle; around 3% of the energy is absorbed by waves and wind as heat; and less than 1% of the energy is used in photosynthesis.

The adverse effects of excessive or uncontrolled energy conversion and utilization include those related to emissions from thermal power plants (especially coal and oil-powered plants), nuclear power plants, strip mining, water pollution, oil spills, and acid rainfall.

We all know that carbon dioxide traps infra-red rays from the sun; but excess of carbon dioxide in our Earth's atmosphere results in unwanted heat buildup and has caused global warming; see, for instance, [17]. Nitrogen dioxide adds to the unwanted acidity of rainfall and affects human health [20]. Airways and lung function can be affected by nitrogen dioxide, particularly in asthma sufferers. Almost all nitrogen oxides are emitted from the combustion of fossil fuels. Sulfur dioxide and total suspended solids (TSP) are believed to cause some health problems such as allergies, respiration disorders and other airway related-diseases, in particular among children; see, e.g., Kramer et al. 1999 [14]. Other pollutants such as heavy metals may also cause serious environmental problems and have to be carefully controlled.

Table 4.6 lists some pollutants together with their environmental and health effects.

Oil spill and oil contaminants.   Oil spillage from oil fields, if left untreated, will be detrimental to our marine ecology including marine life. For example, during the April 2010 massive oil well "blowout" and subsequent oil spill in the Gulf ocean, thousands of gallons of oil spewing out into the Gulf of Mexico and surrounding causing worry about the marine life, environment, and ecology.

Apart from oil well blow outs, oil spills can also come from oil rig installations during drilling or from the transportation of oil from an oil field to shore. However, in general, oil spillages are low relative to total oil production.

Figure 4.10 shows a photograph of a massive oil spill and the associated long oil slick.

Acid rain   Nitric oxide (NO) and sulfur dioxide (SO_2) are the precursors for acid rainfall. The following oxidation process occur:

$$NO + SO_2 + O_2 \longrightarrow NO_2 + SO_3 \tag{4.18}$$

The process subsequently forms acids of nitrogen and sulfur. The sulfur also forms sulfate aerosol. One of the effects is that the sulfate aerosol reflects sunlight — causing some parts of the world to be cooler due to insufficient greenhouse effects.

Figure 4.10 (SEE COLOR INSERT)
An example of a massive oil spill. As seen in the photo, a hundreds of kilometers long oil slick propagated from the deep water oil well drilling rig, Gulf of Mexico. (From NASA.)

Global warming and greenhouse effect   The term "Global warming" refers to the rise of our Earth's temperature. It is a result of our human activities, those that increase concentrations of greenhouse gases in the atmosphere, especially burning of fossil fuels by power plants and internal combustion engines of vehicles, and large scale deforestation as well as other "high carbon" energy applications.

For example, chimneys from power plants and factories (see Figure 4.11) belch out CO_2 and other pollutants to the air. Together with the hot air mixtures the CO_2 gases are carried high up into the atmosphere, as a result of which gases build up, trapping heat, leading to higher global temperatures.

Thermal pollution   The term "thermal pollution" refers to unwanted discharge of thermal energy into waters. Thermal power stations discharge large amounts of energy into water since water is a cost-effective medium for heat rejection from industrial condensers of power plants. For instance, the thermal plant as described in Figure 4.12 rejects heat using pumped sea water from the nearby sea. However, if this heated water from condensers is discharged into smaller water bodies such as lakes or rivers, it will cause a temperature rise of the water. As a result, the ability of water to hold dissolved gasses will be reduced with temperature rise. Aquatic life in the lake or river

Figure 4.11 (SEE COLOR INSERT)
Pollution from a typical SME-related factory. (Courtesy of author.)

(a)   (b)

Figure 4.12 (SEE COLOR INSERT)
A thermal power plant using sea water for effective water rejection. This power plant has two cooling ponds whose water is continuously recycled from the nearby South China Sea. (a) The power plant. (Courtesy of Sejingkat Power Corporation.) (b) One of the cooling water ponds. (Courtesy of author.)

will die at above 35°C when the dissolved oxygen will be too low for the aquatic creatures and plants.

To estimate the amount of thermal energy released to the environment from a thermal power plant, a thermal emission factor called the *thermal discharge index* (TDI) is usually used.

$$Thermal\ discharge\ index(TDI) = \frac{Q_{out,thermal}}{Q_{out,electric}} \tag{4.19}$$

where,

$Q_{out,thermal}$ = Thermal power output discharged to environment, $kW_{thermal}$ or $MW_{thermal}$
$Q_{out,electric}$ = Electrical power output from power plant, $kW_{electrical}$ or $MW_{electrical}$

Nuclear radioactive radiation   The sources of nuclear radiation are mainly from uranium mining and processing, nuclear waste disposal, nuclear accidents, and nuclear weapons tests (the latter two limited to those living within the reactor's vicinity). For instance, a study of cancer mortality among workers in the nuclear industries has shown that there are possibilities of effects of radiation on mortality rates, induced cancer, and other related effects on human health (Cardis et al., 1995, 2007 [7][18]).

What are the general industrial practices to reduce thermal pollution?   Our industry normally removes the waste heat from the condensers by transfering the water source to specific reservoirs for heat rejection, to the atmosphere, and then it is fed back to the sensitive water source. The specific reservoirs are, for instance, a separate lake, cooling pond, and big cooling towers. Usually heat rejection at sea is the most preferred and cost-effective method.

Greenhouse effect and global climate change   The greenhouse effect is a process in the atmosphere that traps heat in the Earth's atmosphere. It is the result of physical interaction between components of the atmosphere, i.e., greenhouse gases and incoming sunlight. The main greenhouse gases are:$CO_2, SO_2, CH_4, N_2O, CFCs, HCFC$, etc.

Carbon dioxide differs from other power-plant emissions like sulfur dioxide (SO_2) because it is not the local impact of CO_2 emissions, but the impacts arising from the accumulation of CO_2 in the environment, called "CO_2 mobile pools," that is a concern and needs to be controlled. The excess accumulation of CO_2 in the environment can cause global climate change as well as other environmental problems [36].

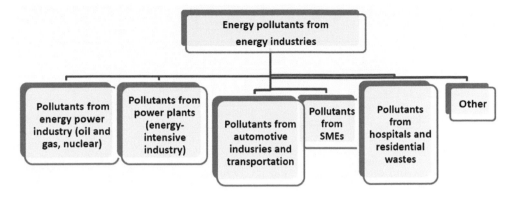

Figure 4.13
Classification of energy pollutants from energy industries.

4.5 Energy Pollutants, Safety and Controls

4.5.1 What Are the Energy Pollutants?

Energy pollutants are used or waste materials that pollute our environments, i.e., air, water, or soil, and which may cause unwanted effects. Almost all energy production and its associated applications involve some forms of environmental pollution. Different energy sources and applications pollute in a different way, and to a different degree. For instance, pollutants such as nitrous oxides help to create smog and haze, and make it difficult for elderly people and people with lung problems to breathe. Global warming is caused by the tendency for some gases, like carbon dioxide, to trap heat in the earth's atmosphere. All the pollutants around the world, collectively, result in a gradual increase in our Earth's temperatures, causing changes in weather patterns and melting of the icebergs as well as rising ocean levels. Figure 4.13 is a somewhat comprehensive classification of the main energy pollutants from our energy industries.

4.5.1.1 Pollutants from Energy Power Industry (Oil and Gas, Nuclear)

Pollutants from the oil and gas industries are those pollutants from drilling rigs and drilling mud, waste oil, or leakage from pipeline and flue gas. Pollutants and also contributed from the major oil spills. Pollutants from nuclear industry are usually in the form of radiation and radioactive waste.

4.5.1.2 Pollutants from Power Plants (Energy-Intensive Industry)

Depending on the type and resources of power plants, pollutants from power plants include nitrous oxides, carbon dioxide, solid particulates, etc. (Figure 4.14).

Consider energy generation by burning coal in coal-fired power plants. The combustion of coal is a major source of CO_2 emissions from the flue gas generated.

Example 4.5 During combustion of coal, flue gas is produced in the presence of air. A typical coal with calorific value of 5 kW/kg has an ash content of 7.2% by weight. For complete combustion of 1 kg of a typical coal, 8 kg of air is required for the combustion process. What is the quantity of flue gas generated by burning 500 kg coal?

Solution: Quantity of flue gas (per kg of coal) = quantity of air for combustion (per kg of coal) + quantity of fuel - quantity of ash (per kg of coal)

$$= 8\,\text{kg} + 1\,\text{kg} - (0.072)\,\text{kg} = 8.928\,\text{kg}$$

Therefore,

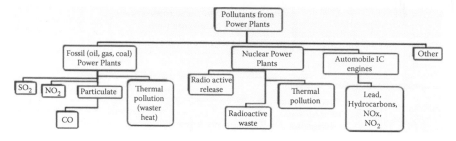

Figure 4.14
Pollutants from power plants and internal combustion (IC) engines.

Quantity of flue gas by burning 500 kg of coal = 500×8.928 = 4,464 kg #.

The amount of flue gas components and pollutants such as CO_2, oxides of nitrogen, sulfur dioxide, particulate matter, and heavy metals can be analyzed accordingly.

In response to increasingly stringent environmental regulations, "clean coal" or "low carbon" technologies are being developed to reduce harmful emissions as well as to improve the efficiency of these power plants.

4.5.1.3 Pollutants from SMEs

Depending on the type of SMEs are pollutants from SMEs usually less than those pollutants contributed from energy-intensive industries. Smoke, fumes and fly-ash are the notable pollutants from SMEs.

4.5.1.4 Pollutants from Automotive/Transportation Industries

Transportation industries, especially road vehicles applications, are the second major source of pollutants. The use of fossil fuel-powered internal combustion (IC) engines in the vehicles emit CO, NO_x, HCs, SO_2, and other toxic substances such as total suspended particles (TSP) and lead (Figure 4.14) Apart from air pollution, transport industries also contribute to noise pollution, in particular, affecting noise-sensitive urban areas such as hospitals.

Four main factors leading to serious pollution:

1. Increase of road vehicles.

2. Poor quality of vehicle use with low combustion efficiency and poor exhaust system.

3. Low-quality of fuel use.

4. High concentration of cars in large cities and specific areas.

How to minimize pollutants from transport industries

1. Improve fuel efficiency in vehicle use.

2. Use of better fuel quality including suitable fuel catalyst, etc.

3. Use of environmentally-friendly fuels such as biofuels.

4. Continuous improvement of environment control, emission standards, and other environmental measures.

5. Better energy management and town planning, especially for big cities with high concentration of populations.

6. Use of hybrid cars or hydrogen/battery driven cars.

4.5.1.5 Pollutants from Industrial and Residential Wastes

4.5.1.5.1 Wastewater Treatment Efficiency Evaluation

In typical wastewater treatment, usually the physical, chemical, and biological characteristics and properties of waste water before and after treatments (such as dissolved oxygen, biochemical oxygen demand, chemical oxygen demand, total organic carbon, total suspended solids, salinity, pH, conductivity, turbidity, etc.) can be analyzed in detail in accordance with standard methods, e.g., the Standard Methods for the Examination of Water and Wastewater (1999) [13], etc. The overall treatment capacity and treatment efficiency of the system can be calculated based upon the influent concentrations and the effluent requirements.

$$Efficiency = [(C_{in} - C_{out})/C_{in}] \times 100 \tag{4.20}$$

where,

C_{in}= Influent concentration (typically mg/L);
C_{out} = Effluent concentration (typically mg/L); and
Efficiency is expressed as a percentage (%).

The treatment capacity over time, for example in a biochemical treatment processes, is usually modeled as a first-order equation:

$$C_t/C_0 = e^{-kt} \tag{4.21}$$

where,

C_t = Concentration at time; t (typically in mg/L);
C_0 = Initial concentration at time = 0 (typically in mg/L);
k = reaction rate constant (typically in $days^{-1}$); and
t = time (typically in days).

4.6 Impact of Energy Use on Our Society and the Environment

Generally, the impact of energy use is mainly due to our society's energy scenarios, which can be broadly classified into two categories:

1. *Energy insufficiency or "energy poverty."* According to the International Energy Agency (IEA, 2010 [15]), today, there are 1.4 billion (20%) people around the world that lack access to electricity. Some 85% of them are in rural areas. The number of people relying on the traditional use of biomass is projected to rise from 2.7 billion today to 2.8 billion in 2030. Using World Health Organization (WHO) estimates, linked to the projection of biomass use and the associated smoke, it is estimated that household air pollution from the use of traditional unregulated biomass in inefficient stoves would lead to over 1.5 million premature deaths per year, over 4000 per day, in 2030, greater than estimates for premature deaths from malaria, tuberculosis, or HIV/AIDS. Here, we will not discuss the details of this category due to the length limitation of this book, but we leave it to the readers to access the vast amount of information available.

2. *Imbalanced energy sufficiency.* This general category applies to around 80% or so of the world population.

Figure 4.15 is a simplified diagram showing the classification of the impact of energy applications. The impact can be classified into environment and social impact. The growing of intensive energy usage, in particular large scale industrial energy applications and related activities, has contributed huge impact on our society and environment. For example, the problems associated with fossil fuel extraction, transport, production, and applications all associated with carbon and harmful emissions have now grown to cause various serious global issues such as climate change, acid rain, and the greenhouse effect. Such problems have now also evolved to become major worldwide political issues and the subject of international debates and regulations.

The impact covers the social, economic and political aspects at all levels — local, national and international. Energy wastage from fossil fuel applications, in particular, coal and oil fed power plants, affects the environment in numerous ways: the carbon emissions left in the air ruin the Earth's atmosphere, which has an even larger impact on the Earth as well as our society.

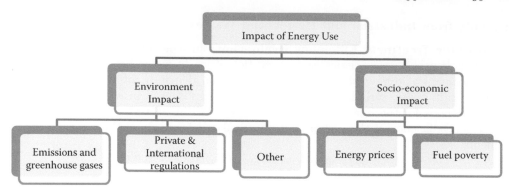

Figure 4.15
Impact of energy use.

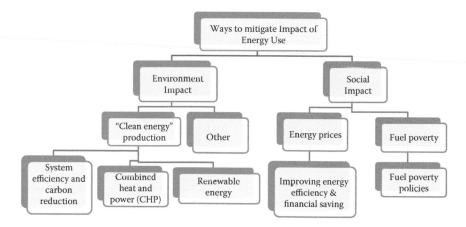

Figure 4.16
Ways to mitigate the impact of energy use.

According to the Department of Energy and Climate Change UK[8], climate change is already happening and its impact could, in the longer term, be substantial. Its impact could include many millions more people exposed to hunger, water stress (water shortage) and flooding. Additionally, low-lying areas, wetlands, and small islands will be exposed to risks from rising sea levels, especially in South East Asia. There will be irreversible losses of biodiversity.

4.6.1   Ways to Mitigate Impact of Energy Use

Figure 4.16 illustrates various strategies that can be implemented to mitigate the impact of energy use. Clean energy production includes:

1. System efficiency and the associated carbon reduction;

2. Combined heat and power (CHP) systems;

3. Use of nuclear energy;* and

4. Renewable energy systems.

---

*From an applied energy perspective, in the meantime, nuclear energy, as a high density energy provider, is suitable to provide sufficient and clean energy required worldwide at least before 2030, and gradually to be replaced by pure renewable energy in the future — perhaps after 2030!

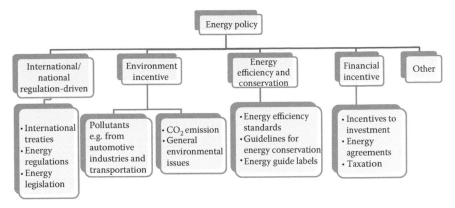

Figure 4.17
Classification of energy policy, attributes and some associated programs.

4.7 Energy Policy, Planning, and Statistics (Energy Target, Production and Consumption)

4.7.1 Energy Policy

Energy Policy is basically the energy goal or target. According to Jacobs (2009) [9], energy policy is the manner by which a given entity (often governmental) has decided to address issues of energy development including energy production, distribution, and consumption. The attributes of energy policy may include legislation, international treaties, incentives to investment, agreements, guidelines for energy conservation, taxation, energy efficiency standards, and energy guide labels.*

Energy policy can be viewed as a tool for developing a *strategic plan* for a specific period, for instance, a typical long-term strategic plan can cover a period of 5 to 10 years, or more, with a primary target to increase energy intensity,† improve energy efficiency, reduce greenhouse gas emissions, and promote energy conservation and energy sustainability.

The various types of energy policies and associated programs that have been used or planned in countries worldwide can be broadly summarized into five categories (Figure 4.17).

Energy policies are used widely in the industrial sector to meet specific energy use or energy efficiency targets. Industrial energy policy can be viewed as a tool for developing a long-term strategic plan, covering a period of 5–10 years, for increasing industrial energy efficiency and reducing greenhouse gas emissions. This policy engages not only the engineers and management at industrial facilities, but also includes government, industry associations, financial institutions, and others (e.g., Abdelaziz et al., 2011 [10]; Price and Worrell, 2000, [11].)

4.7.1.1 Financial Incentive or Fiscal Policy

Financial incentive or *Fiscal policies* include imposition of taxes, tax rebates, investment tax credits, and establishing investment bank lending criteria for promotion of energy efficiency. Taxation policies are a mandatory means for influencing the introduction of energy efficiency and environment criteria, example, through the use of tax rebates or investment tax credits. Investment bank lending criteria can be established to give higher priority for funding projects that improve energy efficiency [11].

4.7.1.2 Energy Regulations

Regulations and protocols are mandatory tools for implementing policies, e.g. the Kyoto Protocol.

---

*Energy guide labels generally estimate how much electrical energy the appliances (e.g., refrigerators, washers, furnaces) use, compares energy use consumption of similar products, and estimates annual operating cost.
†Energy intensity is generally defined as the energy consumption divided by the economic output; and energy efficiency is the reciprocal of energy intensity.

Table 4.7 Selected main energy policies in China from 2007–2010

Year	Policies	Policy makers
2007	Eleventh five-year plan for energy development	NDRC
	Medium & long term plan for renewable energy	
2008	Regulations on energy conservation of public institutions	NDRC
2009	Law on promoting the circular economy	NPC
2010	On strengthening energy management contract and promoting energy saving service industry	NDRC

NDRC - National development and reformed commission

NPC - National People's Congress

4.7.1.3 Energy Standards

Energy standards are applied to particular energy systems (such as power plants standards) or energy equipment used (such as motors, boilers, etc.).

4.7.1.4 Example of Energy Policy Implemented

China energy policy. According to Zhang et al., 2011 [12], the Medium and Long-Term Plan for Energy Conservation is a milestone in China's main energy policy, followed by financial incentives and market mechanisms. The recent main China policies are given in Table 4.7.

The policy of Eleventh Five-Year Plan for Energy Development, 2007, has a goal of reducing energy intensity by 20% between 2006 and 2010. At the end of 2009, 15.6% reduction in energy intensity as of 2005 has been reported.

Korean energy policy. One main objective of the Korean government's energy policy was to provide low-cost energy supplies to encourage and sustain economic development. This policy was successful, as proved by unprecedented economic growth. But the low energy prices also discouraged investment in energy efficiency technologies, hindering the government's efforts to improve energy efficiency. In December 1979, the Korean government began to implement comprehensive energy efficiency programs based on the 1979 Rational Energy Utilization Act. With its subsequent amendments, this law provides the legal basis for enforcing the government's current energy efficiency policy. The Korean government's energy efficiency efforts primarily consist of two programs:

- Energy efficient lighting program. In the lighting "or" 'e' label program, the Korea Electric Power Corporation (KEPCO) provides incentives for eligible e labels to the manufacturers of energy efficient lighting devices. Any customer who buys KEPCO-designated energy efficient measures, whether the purchase is for replacement or new installations, is eligible for incentives. From 1994 to 2001, the lighting program saved 337 MW, and it hopes to save 565 MW by 2015.

- Inverters for improving motor efficiency. The motor efficiency program focuses on inverters that improve the power conversion efficiency of the motor. Customers who save 10 kW or more by installing inverters on specified motors receive incentives. The motor program started in 2001 and saved 2 MW in 2001, 5 MW in 2002, and 6 MW in 2003. It hopes to save 1004 MW by 2015.

Some of the energy policies selected are given in Table 4.8.

4.7.2 Energy Planning and Statistics

For the best utilization of energy sustainability and energy technology for future applications, over the years there have been numerous energy plans and statistics available worldwide. Some of the planning and statistics are given below:

Table 4.8 Selected energy policies in various countries

Country	Policies
Australia	Energy smart business program; Greenhouse challenge
Canada	Industrial program for energy conservation
Denmark	Agreements on industrial energy efficiency
France	Voluntary agreement on CO_2 reduction
Germany	Declaration of German industry on global warming prevention
Japan	Basic Act on Energy Policy
Malaysia	National renewable energy policy
Netherlands	Long term agreements on energy efficiency
Sweden	Eco-energy
UK	Climate change levy, energy efficiency best practice program

4.7.2.1 Energy Planning and Statistics for the United States [27], [35]

According to McKay [27] and Greenblatt [35] article on Google's 2008 plan* for a 40% defossilization of the United States of America [35], the main features of the plan are three fold: efficiency measures, electrification of transport, and electricity production from renewables.

The reductions in energy use and emissions plan includes:

- Fossil fuel-based electricity generation by 88%

- Vehicle oil consumption by 44%

- Dependence on imported oil (currently 10 million barrels per day) by 37%

- Electricity-sector CO_2 emissions by 95%

- Personal vehicle sector CO_2 emissions by 44%

- U.S. CO_2 emissions overall by 49% (from 41%)

As to the electricity applications, the electricity production plan includes:

- 10.6 kWh/d/p of wind power

- 2.7 kWh/d/p of solar photovoltaic

- 1.9 kWh/d/p of concentrating solar power

- 1.7 kWh/d/p of biomass

- 5.8 kWh/d/p of geothermal power by 2030

That's a total of 23 kWh/d/p of new renewables. The United States of America also assumes a small increase in nuclear power from 7.2 kWh/d/p to 8.3 kWh/d/p, and no change in hydroelectricity. Natural gas would continue to be used, contributing 4 kWh/d/p ([27], p. 239).

4.7.2.2 Energy Planning and Statistics for Japan (adapted from [28])

Recently, a 66-page document of Japan's new basic energy plan (BEP) was approved by the cabinet of then-Prime Minister Yukio Hatoyama and released to the public on June 18, 2011.

The BEP lays out seven goals called "basic viewpoints" of Japanese energy policy:
- enhancing overall energy security
- strengthening policy to counter global warming
- achieving economic growth, with energy as a core driver

---

*The Clean Energy 2030 plan proposed in 2008 is a strategic path to between the United States off Fossil fuels, in particular coal and oil, for electricity generation by 2030.

- ensuring the safety of the energy supply
- ensuring the efficient functioning of energy markets
- restructuring the energy industry
- gaining public understanding

To achieve the above goals, the BEP establishes five ambitious targets for 2030:

Target 1: The first target is to double Japan's "energy self-sufficiency ratio" (currently 18%) to about 40% and its "self-developed fossil fuel supply ratio" (currently 26%) to about 50%; and, as a result, to raise its "energy independence ratio" (currently 38%) to about 70%. The latter figure is currently the average among the members of the Organization for Economic Cooperation and Development (OECD).

The energy independence (EI) ratio is the percentage of Japan's primary energy supply, which consists of either energy produced domestically or imported fossil fuels, which are produced by Japanese companies.

$$EI = ESF + SFFS(100\% - ESF) \tag{4.22}$$

where,

EI = Energy independence ratio;

ESF = Energy self-sufficiency;

$SFFS$ = Self-developed fossil fuel supply.

To achieve this target, Japan would bring about a substantial change in its energy mix. The shares attributable to renewable energy sources and nuclear power would more than double. Renewables would increase from 6% in 2007 to 13% while nuclear power would increase from 10 to 24%. Meanwhile, the shares of most fossil fuels would decrease. Natural gas would decline from 18 to 16%, coal from 22 to 17%, and petroleum from 41 to 28%. Only LPG's small share of three percent would remain constant.

Target 2: The second and related target is to raise the "zero-emission power supply ratio" from the current 34 to 70%. The zero-emission power supply ratio concerns the percentage of electric power, which is generated by sources that produce little or no CO2. To achieve this goal, Japan will have to increase substantially the amount of electricity provided by nuclear power and renewable sources, especially "new" sources such as wind, solar, and biomass, because the country's hydroelectric potential has already been largely exploited. According to the BEP, the shares attributable to renewable and nuclear power will more than double. For renewables, this will mean going from 8 to 19% of electricity generated; for nuclear, from 26% to > 50%.

Targets 3–5: The remaining three targets can be stated much more briefly. One is to halve the CO_2 emissions of the residential sector. Another is to maintain and enhance the energy efficiency of the industrial sector. The final target is to maintain or obtain "top-class" shares of global markets for energy-related products and systems.

4.7.2.3 Energy Planning and Statistics for China [16]

China had become the leader in investments for low-carbon technologies, spending US\$34bil in 2010 against US\$17bil by the United States, US\$12bil by Britain, US\$11bil by Spain, US\$8bil by Brazil, US\$4bil by Germany, and US\$3bil each by Canada and India. Recently, the Chinese Goverment organized a conference on "Green Low-Carbon Development" in Beijing from June 22–24, 2011, bringing together international and local energy experts with national and international energy policy makers. China is one of the two largest Greenhouse Gas emitting countries in the world. As enumerated by Sun Cui Hua, deputy director-general of the Climate Change Department of the National Development and Reform Commission, these included 10 policy areas.

The first was implementing climate change macro policies. The targets in the recently unveiled 12th Five-Year Plan include:

- Non-fossil fuel to account for 11.4% of primary energy consumption

- A 30% cut of water consumption per unit of value-added industrial output

- A 16% reduction in energy consumption per unit of GDP

- A 17% cut in CO_2 emission per unit of GDP (en route to the pledged goal of 40%–45% reduction by 2020 compared with 2005)

- The forest coverage rate to rise to 21.7% and forest stock to increase by 600 million cubic meters

Not mentioned by Sun, but which will have equally important implications, is the new 7% average annual GDP growth target for the 2011–2015 period. This is a reduction from the annual growth of 10% plus per year that China has been used to.

A cut by three to four percentage points in GDP growth will in itself mean a large reduction in emissions growth, on top of the cuts in emissions intensity of GDP.

Other policies or actions Sun announced included:

- Establishing a fund in China to finance its climate actions

- Launching low-carbon pilot projects in selected cities and provinces

- Using market mechanisms, including new conditions for enterprises, and a pilot program on emissions trading

- A low carbon certification system to identify industries and products and encourage upgrading of enterprises

- Compiling an inventory of greenhouse gases, including building the capacity of local governments and having a guidebook for enterprises

- Strengthening legislation to accompany the policy measures

- Education and campaigns for low-carbon lifestyles

- Strengthening international cooperation through exchanges and South–South cooperation*

- Enacting policy measures in various sectors and improving forecasting and early warning for extreme weather events

Data on recent performance in China's energy and emissions were given by Wang Zhongmin of the China Institute of Standardization, who said that energy consumption per unit of GDP fell by a total of 19% in the 11th Five-Year Plan period (2006–2011).

Energy use per unit of copper smelting dropped 36%, and per metric ton of cement by 29%, while backward enterprises and technologies had been closed down.

During the period, the energy conserved was more than 600 million metric ton of standard coal, which meant there was an accumulated reduction of CO_2 by over 1.5 billion metric tons.

4.7.2.4 Energy Planning and Statistics for Europe

From the same recent conference on "Green Low-Carbon Development" in Beijing, Europe's climate policy and associated planning and statistics were presented by Jurgen Lefevere of the European Commission, who said that the EU countries had decoupled emissions from GDP growth, as domestic emissions had fallen 16% while GDP grew 40% between 1990 and 2009.

He reiterated the EU target of 20%–30% emissions reduction by 2020 (compared with 1990) with a reduction of 80%–95% by 2050 through a road map that includes emission reduction plans for various sectors, the use of key technologies, and investments.

The EC had identified additional investments needed for climate change actions of 270 billion euro a year in 2010–2050.

This would be more than offset by benefits including fuel savings of €175bil to €320bil a year; halving of imports by 2050, reducing the bill in that year by €400bil; health benefits of €88bil a year in 2050, and 1.5 million net jobs created in 2020.

Despite the positive domestic plans by China and the message from Europe that decoupling growth and emissions is possible, experts also highlighted the huge challenges facing developing countries in reducing their emissions growth while maintaining their ambition of high economic growth.

---

*South-South cooperation is a term used by the United Nations and many policy makers to describe the exchange of resources, technology, and knowledge between developing countries.

4.8   Applied Energy Modeling and Simulation

Applied energy modeling makes use of numerous computer-based tools to simulate the energy use of various systems.

According to Jebaraj and Iniyan (2006) [34], energy models can be broadly classified into a few main types, i.e., energy planning models, energy supply–demand models, forecasting models, renewable energy models, emission reduction models, and optimization models.

For the convenience of readers, the applied energy modeling and simulations are briefly explained in the following sub-sections:

4.8.1   Modeling Energy Supply and Demand

In modeling energy supply and demand, 2 broad classes of modeling approaches [29] are:

1. The economic or top down models and the technical/engineering or bottom-up models. The first approach, adopting a general perspective, described the economic linkages between energy demand and supply and the rest of the economic system, with the main goal of analyzing energy or wider economic policies.

2. The second approach, adopting a focused view of the energy sectors, explored the various technological options, with the main goal of highlighting low-cost energy production opportunities.

4.8.2   Energy Modeling and Simulation of Power Plants and Their Applications

There are variety of energy models for power plant applications, used for both traditional and renewable energy applications These include:.

1. Nuclear energy modeling and simulation applications

2. Geothermal power plants

3. Wind power

4. Hybrid energy applications, etc.

4.8.3   Energy Modeling and Simulation of Building Environment

Many models are availables for instance:

1. Energy consumption, e.g., in the residential sector application [30]

2. Thermal performance and ventilation, e.g., [31], etc.

4.8.4   Energy Modeling and Simulation of Transport and Vehicle Applications

Likewise, many models are available for transport applications, and some of them are as follows:

1. Energy consumption and emission of vehicles, e.g. Silva et al. [32]

2. Energy transport demand, e.g. Ceylan and Ceylan [33], etc.

Application Project

Linking knowledge with applied energy design and practice

Characteristic	Value
pH	4–4.5
Oil and Grease	~ 4,000 mg^{-1}
Biochemical Oxygen Demand (BOD)	~ 25,000 mg^{-1}
Chemical Oxygen Demand (COD)	≥ 51,000 mg^{-1}
Total Solid	40,000 mg^{-1}
Suspended Solid	18,000 mg^{-1}
Total Volatile Solid	34,000 mg^{-1}

Figure 4.18 (SEE COLOR INSERT)
(*Left*): Typical POME retention pond. (Courtesy of author.) (*Right*): Typical effluent characteristics.

1. Suppose you have read in the newspaper that the river and catchment in the town are heavily polluted from the nearby industrial effluents. Find out the possible impact to the environment and the people. Discuss the feasibility of a wastewater treatment plant to mitigate the problems. Design a prototype of the water treatment plant including a schematic diagram of the overall proposed plant structure.

2. Around the globe, about $50,000 \, \text{m}^3$ of palm oil mill effluent (POME) was produced from the processing of the fresh palm oil fruit bunches every year. We have visited a palm oil mill and a few of the POME retention ponds associated with the effluents. Your group is to construct a simple laboratory Microbial Fuel Cell (MFC) system, the laboratory prototype of which can be used to conduct a feasibility study on the use of POME for MFC application *cum* bioelectricity production. A photo of one of the ponds and typical initial POME effluent characteristics are as shown at right and left of Figure 4.18, respectively. Note: an MFC prototype built together with its schematic diagram is given in Figure 4.19.

   (a) Measure the waste water qualities before and after MFC treatment. Discuss the water treatment ability of your device.

   (b) The bioelectric and power current profile obtained from your setup.

   (c) Calculate the overall treatment efficiency of your system.

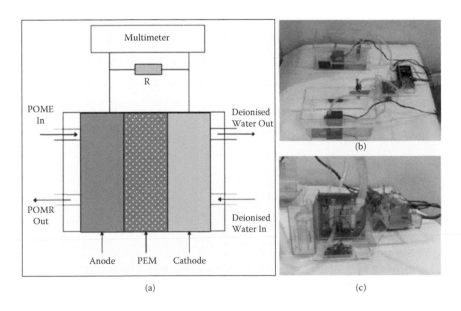

Figure 4.19
An MFC prototype built: (a) schematic diagram; (b) overall MFC system; and (c) side view of the MFC system. (Adopted from Abdul Wahab. With permission.)

Assignments

1. What are the main pollutants from vehicle industries? List at least five ways or techniques in which those pollutants can be minimized, and discuss in detail each of the techniques you have given.

2. Conduct a *detailed* LCA analysis on the environmental impacts due to the use of absorption cycle cooling powered by various energy resources (electricity from grid, LPG, and PV), the three different energy schemes of which were briefly discussed by Abdullah and Tang, 2010 [25]), page 1544 and are copied as follows:

   Scheme-I: "Conventional electricity from grid: Electricity supply obtained from grid is the daily facility that cannot be ignored in modern life. Electricity is the foremost class of energy — it can be used to power a fan motor, but unfortunately, it is non-renewable energy. Electricity itself does not have known impact on the environment, but the way to generate it, i.e., the energy resources and associated energy production processes, is the issue to be considered. Nowadays, there are many types of electricity generation, such as fossil fuels, hydro, or wind. In Malaysia, 42% is gas electric generation and 25% is coal electric generation, as in 2008."

   Scheme-II: "Liquid petroleum gas (LPG): LPG is a somewhat clean burning fuel compared with conventional fuels; it burns cleanly without soot, smoke, and smell. It is easy to ignite provides instant heat, and easy to control with a thermostat. It is non-poisonous and safe to use in daily life. It also is sulfur-free and very consistent in quality. By using LPG heating, the need for peak electric is avoided, and there is lessen need for more electric power from generation plants. The disadvantage of LPG, however, is that it is highly flammable. Storage and safety should be taken care of in its initial states, and it is also considered a non-renewable resource."

   Scheme-III: "Photovoltaic (PV): Photovoltaic systems have very little impact on the environment. It is perhaps one of the cleanest power-generating technologies available. It produces no known air pollution, global warming, hazardous waste, or noise, besides no transportable fuels. However, some toxic chemicals such as cadmium, are used to produce PV cells in the manufacturing system, and so it will generate a small amount of hazardous waste. The manufacturing process of PV systems, also emits some chemical gas and CO_2 which can contribute to the greenhouse effect."

Questions and Problems

Multiple choice questions

1. Which one of these is not the main concern of power plant pollutants from energy-intensive industries?

   (**A**) Thermal pollution

   (**B**) Radioactive waste

   (**C**) Carbon monoxide

   (**D**) Particulate matter

   (**E**) Nitrogen oxide

2. The waste effluent water from a typical industry was biochemically treated. What is the treatment efficiency and the reaction rate constant of the water treatment system? Given: Influent concentration = 0.55 mg/L; effluent concentration = 0.43 mg/L? Treatment period = 12 hours.

   (**A**) Efficiency = 20 %, k = - 0.238 day^{-1}

   (**B**) Efficiency = 55 %, k = - 0.251 day^{-1}

   (**C**) Efficiency = 22 %, k = - 0.238 day^{-1}

(D) Efficiency = 52 %, k = - 0.238 day^{-1}

(E) Efficiency = 62 %, k = - 0.238 day^{-1}

3. Which of the following pairs does *not* correctly relate the pollutant to health problems that are due to the pollutants from the transport and vehicle industries?

(A) TSP — allergic

(B) CO — brain damage

(C) CO_2 — global warming

(D) NOx — breath problems

(E) Sulfur dioxide (SO_2) — airway diseases

4. Which of the following are the general industrial practices to reduce thermal pollution?

I Using efficient condensers

II Using efficient cooling towers

III Building cooling ponds or lakes

IV Heat rejection at sea

(A) I and II only

(B) I and IV only

(C) I, II, and III only

(D) I, II, III, and IV

(E) All except II

5. Which of the following are the storage or disposal of radioactive waste materials?

I Nuclear waste storage tank

II Nuclear dry cask storage

III Cooling pond temporary storage

IV Suitable deep underground or ocean floor sites

(A) I and II only

(B) I and IV only

(C) III and IV only

(D) I and III only

(E) I, II, III, and IV

6. Which of the following are the energy planning and statistics for China, i.e., 12th Five-Year Plan (2011–2015) as revealed in the recent international conference on "Green Low-Carbon Development" in Beijing from June 22 to 24, 2011?

I Non-fossil fuel to account for 11.4% of primary energy consumption.

II A 16% reduction in energy consumption per unit of GDP.

III A 17% cut in CO_2 emission per unit of GDP (en route to the pledged goal of 40%–45% reduction by 2020 compared with 2005).

IV A 30% cut of water consumption per unit of value-added industrial output;

(A) I and II only

(B) I and III only

(C) II and III only

(D) I and IV only

(E) I, II, III, and IV

7. Which of the following are the energy planning and statistics for Europe (EU) as presented by Jurgen Lefevere of the European Commission, in the recent international conference on "Green Low-Carbon Development" in the Beijing conference mentioned above? (optional)

I EU target of 20%–30% emissions reduction by 2020.

II EU target of 80%–95% emissions reduction by 2050.

III Fuel saving of €175bil (RM756bil) to €320bil a year.

IV Clean Energy 2030 target of a total of 23 kWh/d/p of new renewables and use of natural gas contributing 4 kWh/d/p.

(A) I and II only

(B) I and III only

(C) I, II and III only

(D) I and IV only

(E) III and IV only

Theoretical questions

1. What are the benefits and applications of energy audits?

2. Basically, Life Cycle Assessment (LCA) is carried out in four phases or main components. List the four phases of the LCA, and briefly discuss each phase of the LCA.

3. List and explain the various benefits and applications of energy life-cycle analysis.

4. Discuss the impact of energy use on our society and environment. Aslo list the possible ways to mitigate the negative impact of energy applications.

5. What is *energy poverty*?

6. Briefly discuss *acid rain*.

7. What are the sources of offshore *oil spills* and what are the effects of *oil spills*?

8. What is *thermal pollution*, and how can it be reduced?

9. List three main factors leading to serious pollution from automobiles in developing countries, and suggest ways to minimize the environment a pollution from road transport applications.

10. Briefly describe the following policies, and explain their importance for energy management.

    (a) Fiscal policy

    (b) Energy standards

    (c) Energy regulations

11. Briefly describe various pollutants from power plants and explain ways to control them.

12. Briefly describe various pollutants from the automotive and transport industries, and suggest ways for mitigation and control.

13. Give and explain a typical energy plan and statistics (for example, China).

Homework problems

1. *Energy audit.* A typical energy audit for the building industry is given in Table 4.1.2, which shows various energy parameters such as output, fuel consumption, and CO_2 emissions as a function of various building materials in 2006. Determine the following:

   (a) Percentage contribution of cement in terms of total fuel consumption in 2006.

   (b) Percentage contribution of CO_2 from cement in terms of total CO_2 emissions of building materials in 2006.

   (c) The percentage energy consumption reduction required and the direct CO_2 emissions reduction necessary to achieve the energy savings target (in 2010) and the CO_2 emission target (in 2010), respectively.
   (Solution: 46 %; 71 %; 89.4 %)

2. *Thermal discharge index.* Compare the thermal discharge index (TDI) of the following power plant. *Plant A:* 300 MW conventional natural gas turbine, thermal output =225 MW. *Plant B:* 200,000 MW hydroelectric-steam combined power turbine, total thermal output =2150 MW, waste thermal output = 981 MW. *Plant C:* 50 MW CHP turbine, waste thermal output = 20MW. Determine the following:

   (a) Thermal discharge index for each of the plants.

   (b) Briefly compare the three types of thermal discharge potential from the plants.
   (Solution: 3; 0.005; 0.6)

3. *Coal, flue gas, and CO_2 emission.* A typical coal with calorific value of 5 kW/kg has ash content of 7.2 % by weight. For complete combustion of 1 kg of a typical coal, 8 kg of air is required for the combustion process.

   (a) What is the quantity of flue gas generated by burning 500 kg coal?

   (b) Estimate total CO_2 emissions if the above flue gas contain 3.5% of CO_2 by weight.
   (Answer: 4,464kg; 156.2 kg)

4. *Energy use and fuel consumption study.* An operator of a commercial building has decided to change to use of a higher efficiency fan (Model A), replacing the old fan (Model D). The data available from the fan manufacturer is given in Table 4.3.

   (a) Determine how much savings there will be per year if the fan operates 16 hours per day, and the electric rate is $0.1 per kW-hour. The drive loss is estimated at 0.45 hp with minimum efficiency of 91.7%.

   (b) What will be the savings if a unit of variable frequency drive (VFD) is employed with added drive loss of 5%? Assume VFD runs at reduced speeds with overall kW savings potential of 12% compared with a system without VFD.
   (Solution: $555 per year, $776 per year)

Answers to multiple choice questions:

1. (B) Radioactive waste normally emit from nuclear power plants.

2. (C)

   (C) $\qquad Efficiency = \left(\frac{(C_{in} - C_{out})}{C_{in}}\right) \times 100\,\% = \left(\frac{0.55mg/L - 0.43mg/L}{0.55mg/L}\right) = 22\%\#$

   $$C_t/C_0 = e^{-kt} \Rightarrow \frac{0.43}{0.55} = e^{-k(\frac{1}{2}day)}$$

   $$Therefore\ k = -0.238\,day^{-1}\#$$

3. (C) Note: Global warming is not a disease by itself, but it could cause other diseases.

4. (D) All — I, II, III, and IV

5. (E) I, II, III, and IV

6. (E) I, II, III, and IV

7. (C) I, II, and III only; (IV) Clean Energy 2030 is the energy planning and statistics for the United States of America.

Bibliography

[1] Cui, Y.-S., Wang, X., (2006), The State of the Chinese Cement Industry in 2005 and Its Future Prospects, *Cement International* 4/2006, pp. 36–43.

[2] Greening, L.A., BOYD, G., and Joseph, M.R., Modeling of industrial energy consumption. An introduction and context. *Energy Economics*, 29 (2007), pp. 599–608.

[3] Shafiee, S., Topal, E. An econometrics view of worldwide fossil fuel consumption and the role of US, Energy Policy Volume 36, Issue 2, February 2008, pp. 775–786.

[4] ISO 14040 (2006): Environmental management – Life cycle assessment – Principles and framework, International Organisation for Standardisation (ISO), Geneva, Switzerland.

[5] ISO 14044 (2006): Environmental management – Life cycle assessment – Requirements and guidelines, International Organisation for Standardisation (ISO), Geneva.

[6] Pradhan, A., Shrestha, D.S., McAloon, A., Yee, W., Haas, M., Duffield, J.A., Shapouri, H., Energy Life-Cycle Assessment of Soybean Biodiesel, United States Department of Agriculture (USDA), September 2009.

[7] Cardis, E., Gilbert, E.S., Carpenter, L., Effect of low doses and low dose rates of external ionizing radiation: cancer mortality among nuclear industry workers in three countries, *Radiation Research*, 142,1995:117–132.

[8] Department of Energy and Climate Change, UK, "Energy - its impact on the environment and Society, 2006," http://www.decc.gov.uk/en/content/cms/statistics/publications/energy_impact/energy_impact.aspx, accessed Sept 1, 2011.

[9] NB., Jacobs, Energy policy: economic effects, security aspects and environmental issues. Nova Science Publishers Inc.; 2009.

[10] Abdelaziz, E.A., et al. Renewable and Sustainable Energy Reviews 15 (2011) 150–168.

[11] Price, L., Worrell, E., International industrial sector energy efficiency policies. Lawrence Berkeley National Laboratory 2000.

[12] Xing-Ping Zhang, Xiao-MeiCheng, Jia-HaiYuan, Xiao-JunGao, Total-factor energy efficiency in developing countries, *Energy Policy* 39 (2011) 644–650.

[13] American Public Health Association, American Water Works Association, Water Environment Federation, (1999) Standard Methods for the Examination of Water and Wastewater, 1999.

[14] Kramer, U., Behrendt, H., Dolgner, R., Ranfit, U., Ring, J., Willer, H., Chlipkoter, H.-W., Airway diseases and allergies in East and West German children during the first 5 years after reunification: time trends and the impact of sulphur dioxide and total suspended particles, *International journal of Epidemiology* 1999:28, 865–873.

[15] International Energy Agency (IEA), Energy Poverty: how to make modern energy access universal? September 2010.

[16] The Star, Towards green low-carbon growth?, June 27, 2011, also available online: http://thestar.com.my/columnists/story.asp?col=globaltrends&file=/2011/6/27/columnists/globaltrends/8978879&sec=Global%20Trends, accessed June 27, 2011.

[17] Ackerman, K. V., Eric, T., Comparison of Two U.S. Power-Plant Carbon Dioxide Emissions Data Sets, *Environ. Sci. Technol.*, 2008, 42 (15), pp. 5688–5693.

[18] Cardis, E., Vrijheid, E., Blettner, M., Gilbert, E., Hakama, M., Hill, C., Howe, G., Kaldor, J., Muirhead, C. R., Schubauer-Berigan, M., Yoshimura, T., Bermann, F., Cowper, G., Fix, J., Hacker, C., Heinmiller, B., Marshall, M., Thierry-Chef, I., Utterback, D., Ahn, Y-O., Amoros, E., Ashmore, P., Auvinen, A., Bae, J-M., Bernar, J., Biau, A., Combalot, E., Deboodt, P., Diez Sacristan, A., Eklöf, M., Engels, H., Engholm, G., Gulis, G., Habib, R. R., Holan, K., Hyvonen, H., Kerekes, A., Kurtinaitis, J., Malker, H., Martuzzi, M., Mastauskas, A., Monnet, A., Moser, M., Pearce, M. S., Richardson, D. B., Rodriguez-Artalejo, F., Rogel, A., Tardy, H.,

Telle-Lamberton, M., Turai, I., Usel M., Veress, K., (2007) The 15-Country Collaborative Study of Cancer Risk among Radiation Workers in the Nuclear Industry: Estimates of Radiation-Related Cancer Risks. *Radiation Research*: April 2007, Vol. 167, No. 4, pp. 396-416.

[19] Garshick, E., Laden, F., Hart, J. E., Rosner, B., Davis, M. E., Eisen, E. A., Smith, T. J., Lung Cancer and Vehicle Exhaust in Trucking Industry Workers, *Environ Health Perspect.* 2008 October; 116(10): 1327–1332.

[20] H.W. Jie, Y. Zhang and L.Q. Yin, Present situation of research of acid rain. *Environmental Science & Technology*, 27 (2004), pp. 179–181.

[21] Wang, Y.-F., Chao, H.-R., Wang, L.-C., Chang-Chien, G.-P., Tsou, T.-C., Characteristics of Heavy Metals Emitted from a Heavy Oil-Fueled Power Plant in Northern Taiwan, *Aerosol and Air Quality Research*, 10: 111–118, 2010.

[22] Keown, A.J., Martin, J.D., Petty, J.W., Scott, D.F., *Financial management: principals and applications. 10th ed.* USA: Prentice Hall; 2005.

[23] Thumann, A., Mehta, D.P., *Handbook of energy engineering. 2nd ed.* Prentice Hall, 1991.

[24] Elettronica Veneta. Absorption refrigeration trainer Mod. TAR/EV teacher/student handbook. 1st ed. Elettronica Veneta, Anon; 2005.

[25] Abdullah, M. O., Hieng, T. C., Comparative analysis of performance and techno-economics for a $H_2O - NH_3 - H_2$ absorption refrigerator driven by different energy sources, *Applied Energy* 87 (2010) 1535–1545.

[26] Koornneef, J., van Keulen, T., Faaij, A., Turkenburg, W., Life cycle assessment of a pulverized coal power plant with post-combustion capture, transport and storage of CO_2, *International Journal of Greenhouse Gas Control* 2 (2008) 448-467.

[27] MacKay, D. J. C., Sustainable Energy without the hot air, available at www.withouthotair.com, accessed October 13, 2011.

[28] Duffield, J. S., Woodall, B., Japan's new basic energy plan, *Energy Policy*, Volume 39, Issue 6, June 2011, pp. 3741–3749.

[29] Lanza, A., Bosello, F., Modeling Energy Supply and Demand: A Comparison of Approaches, *Encyclopedia of Energy*, 2004, pp. 55-64.

[30] Swan, L. G., Ugursal, V. I., Modeling of end-use energy consumption in the residential sector: A review of modeling techniques, *Renewable and Sustainable Energy Reviews*, Volume 13, Issue 8, October 2009, pp. 1819-1835.

[31] Chen, Y., Athienitis, A.K., Galal, K., Modeling, design and thermal performance of a BIPV/T system thermally coupled with a ventilated concrete slab in a low energy solar house: Part 1, BIPV/T system and house energy concept, *Solar Energy*, Volume 84, Issue 11, November 2010, pp., 1892–1907.

[32] Silva, C., Ross M., Farias, T., Evaluation of energy consumption, emissions and cost of plug-in hybrid vehicles, *Energy Conversion and Management*, Volume 50, Issue 7, July 2009, pp. 1635–1643.

[33] Ceylan H., Ceylan, H., Harmony Search Algorithm for Transport Energy Demand Modeling, *Studies in Computational Intelligence*, Volume 191/2009, 163–172.

[34] S. Jebaraj, S. Iniyan, A review of energy models, *Renewable and Sustainable Energy*, Reviews 10 (2006) 281–311.

[35] Greenblatt, J., Clean Energy 2030, Google's Proposal for reducing U.S. dependence on fossil fuels, http://knol.google.com/k/clean-energy-2030# accessed October 13, 2011.

[36] Lackner, K. S., Comparative impacts of fossil fuels and alternative energy sources, *Issues in Environmental Science and Technology*, 29, 2009 page 29-39.

[37] Noraziah, A. W., Evaluation of a single chamber microbial fuel cell using graphite felt anode for waste water treatment from POME, MSc thesis, University of Malaysia, Sarawak, 2009.

5

Energy Saving, Recovery, and Storage

5.1 Introduction to Energy Saving Technologies and Energy Storage

Energy saving, energy recovery, and energy storage are all inter-related, and are the fundamental aspects of applied energy.

5.2 Energy Recovery

Energy recovery is the term widely used in various energy applications to refer to energy retrieval or energy revival of an energy system, which was initially thought not techno-economically viable.

In this section, we will consider various means of energy recovery in a comprehensive way, from enhanced oil recovery (EOR) in the energy power industries, to various energy intensive industries as well as industries pertaining to SME industry.

5.2.1 Energy Recovery in Oil and Gas (O&G) Industry

Improved Oil Recovery (IOR) methods in the O&G industries encompass oil and gas energy recovery techniques, such as:

1. Enhanced Oil Recovery (EOR) techniques

2. Advanced drilling and well technologies

3. Intelligent reservoir management and control

4. Advanced reservoir monitoring techniques

In this subsection, we shall emphasize and devote to describing the first two, i.e., EOR and advanced drilling technology (especially horizontal well drilling) for energy recovery; the other two techniques are basically more concerned with environments and control, the subjects of which are beyond the scope of the present chapter.

5.2.1.1 Enhanced Oil Recovery (EOR)

Enhanced Oil Recovery (EOR) or "tertiary recovery" is a term for techniques to increase the amount of crude oil extracted from an oil reservoir, which can add around 10–20% of oil recovery, after the primary and secondary recovery (see Section 6.1).

With the gradual decline in oil discoveries during the last decades, it is believed that EOR technologies will play a key role to meet the oil demand in the years to come [1].

EOR production worldwide is about 2.5×10^6 B/D, and almost all of it comes from the United States, Mexico, Venezuela, Canada, Indonesia, and China, as seen in Figure 5.1 [2].

Depending on well and reservoir conditions, and economic as well as technological aspects, various techniques are used by the O&G industries; for example, gas injection, chemical injection, thermal recovery, and microbial injection for EOR [3].

Alvarado and Manrique (2010) [1] classify EOR by lithology based on their study of a total of 1,507 projects (see Figure 5.2). Figure 5.2 shows that most EOR applications have been applied successfully in sandstone reservoirs, as

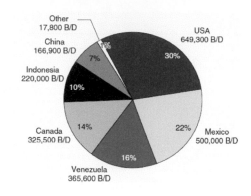

Figure 5.1
Current world EOR production. (After Thomas, 2008 [2].)

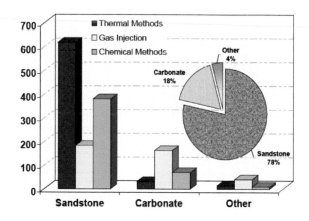

Figure 5.2
Classification of EOR by lithology. (After Alvarado and Manrique, 2010 [1].)

derived from a collection of 1,507 international EOR projects in a database consolidated by the authors during the last decade. From Figure 5.2, it is clear that EOR thermal and chemical projects are the most frequently used in sandstone reservoirs compared to other lithologies (e.g., carbonates and turbiditic formations).

Figure 5.3 depicts a simple classification of EOR methods* The explanations of the various EOR methods follow:

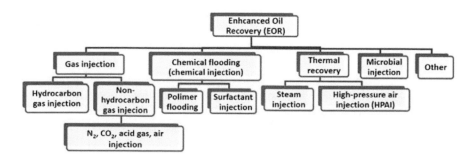

Figure 5.3
Classification of EOR methods based on injection type.

---

*For very detailed classification, readers can refer to Thomas, 2008 [2].

5.2.1.1.1 Gas Injection Apart from the effect of pressure increase, EOR gas injection aids recovery by reducing the viscosity of the crude oil as the gas mixes with it. EOR gas injection or "gas flooding" has been the most widely used recovery method for light, condensate, and volatile oil carbonate reservoirs. Basically, gas injection can be divided into either hydrocarbon gas injection or non-hydrocarbon gas injection. N_2, CO_2, acid gas, and air can be used for gas injection.

Interestingly, recently the CO_2 capture from coal industries has been widely used for EOR applications (see, e.g., Liner, 2010 [5] and *Oil and Gas Journal (OGJ)* editors, 2011 [4]).

We will cover the discussion of integrated gasification combined cycle (IGCC) technology in power stations for clean coal production *cum* CO_2 capture ability in Chapter 10, Section 10.3.

5.2.1.1.2 Chemical Flooding (Chemical Injection) Various chemicals, in particular, dilute solutions, can be injected into the oil reservoirs to improve oil recovery. Injection of alkaline or caustic solutions with oil that has organic acids components will produce soap to lower the interfacial tension; hence an increase in oil production. For instance, the injection of alkali, surfactant, alkali-polymer (AP), surfactant–polymer (SP), and Alkali–Surfactant–Polymer (ASP) have been tested in a number of fields, e.g., Pandey, 2010 [6] and Hongfu et al. (2003) [7]. Injecting a water soluble polymer increases the viscosity of the injected water, which causes an increase in the amount of oil recovered in suitable formations. Dilute solutions of surfactants such as petroleum sulfonates or biosurfactants such as rhamnolipids may be injected to lower the interfacial tension or capillary pressure that impedes oil droplets from moving through a reservoir.

Application of chemical methods is usually limited by the cost of the chemicals and their adsorption and loss into the rock of the oil-containing formation. Polymer flooding is widely reported for use, in particular, at the early stages of water flooding in carbonate reservoirs (Figure 5.2).

5.2.1.1.3 Thermal Recovery Various methods and techniques make use of thermal energy to heat the crude oil in the formation, which results in lower oil viscosity and causes vaporization of part of the volatile components of the oil. These processes improve the sweep efficiency and the displacement efficiency of the oil. Various approaches have been used that include cyclic steam injection, steam drive, and *in-situ* combustion.

For example, the Mene Grande and Tia Juana oil fields in Venezuela [8] and Yorba Linda and Kern River fields in California [9] are good examples of successful steam injection projects.

5.2.1.1.4 Microbial Injection Microbes react with a carbon source, such as oil, and produce surfactant, slimes (polymers), biomass, and gases such as CH_4, CO_2, N_2, and H_2 as well as solvents and certain organic acids. Oil recovery mechanisms in microbial EOR include IFT reduction, emulsification, wettability alteration, improved mobility ratio, selective plugging, viscosity reduction, oil swelling, and increased reservoir pressure due to the formation of gases. Increase in permeability can result from the acids formed. Microbes can be indigenous or exogenous. Exogenous microbes must adapt to reservoir temperature, salinity, and hardness. Nutrients, such as molasses or ammonium nitrate, are supplied to stimulate microbial growth in the reservoir. The process has advanced since its inception, and is receiving renewed interest in recent years [2].

Next, we shall consider the drilling technology in improving oil recovery from the reservoir.

5.2.1.2 Horizontal Wells and Related Drilling Technology for Maximum Oil Production/Recovery

The production and recovery of oil from the oil reservoirs depends on many factors, one of which is the coverage area of the drilling. To maximize the area of coverage, and hence oil recovery, horizontal well drilling is such a technology. Figure 5.4 is a schematic diagram showing a conventional well and a horizontal well. A horizontal well is drilled in parallel to the bedding plane or reservoir formation, with perforated holes along the horizontal sections, thus resulting in an overall larger volume of oil extraction.

The application of horizontal wells in improving oil production has been reported by many, for instance, Jacot et al. (2010) [10], Karlsson and Bitto (2008) [11], Joshi (1991) [12], and Louis (1989) [13].

The advantages of horizontal wells compared to conventional vertical wells:

1. By the use of horizontal wells, the possible added oil recovery is 5–10% or more.

2. Horizontal wells have been used for reservoirs with water and gas coning problems, and are thus able to minimize coning problems and enhance oil production.

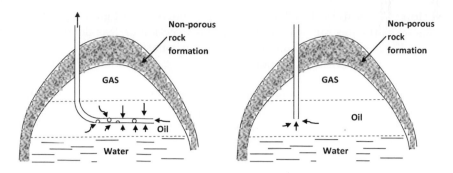

Figure 5.4
A simple schematic diagram showing (*Left*): a horizontal well, and (*Right*): a conventional vertical well. A horizontal well is drilled in parallel to the bedding plane of the reservoir formation.

3. Horizontal wells have been used for EOR production. A long horizontal well provides a large reservoir contact area, thus enhancing injectivity of an EOR injection well.

4. In environmentally sensitive areas and reservoirs under cities, horizontal wells can be employed to drain a reservoir with minimum surface disturbance.

5. Horizontal wells can save on cost, especially when used in remote areas where multiple vertical wells can incur high costs.

5.2.1.3   Shale Oil or Shale Gas Production/Recovery

Apart from oil and gas recovery, we have other hydrocarbon-based fossil fuels, such as shale oil, shale gas, and bituminous sands (tar sands) that still have much reserved around the world; however, they generally need expensive extraction and processing technologies. Such resources will be vital, and will become more economical in the future when the oil and gas resources become scarce. We have already discussed shale oil production in Chapter 3, Section 3.2.2.

In the next section, we will consider energy recovery from the energy-intensive industry aspects.

5.2.2   Energy Recovery (ER) in Energy-Intensive Industries

As discussed earlier, in Chapter 2, energy recovery potential in energy-intensive industries is enormous. We shall briefly review the individual industries.

5.2.2.1   Energy Recovery in the Aluminum Industry

Currently, high electrical costs and energy instability have led to the reduction of aluminum primary production in most parts of our world. The aluminum smelters have to explore ways to reduce energy consumption and overall cost, such as:

1. Seek energy power supplements to grid-supplied electricity, such as distributed generation and renewable energy, in particular, hydro power.

2. Improve energy efficiency and reduce costs.

3. Control and manage quality of carbon, coke, pitch, and the other raw materials for primary production.

4. Better instrumentation and control systems to produce optimum electromagnetic effects and metal stirring for alumina feeding requirements.

5. In the future, the aluminum industries may perhaps require replacement of the Bayer and Hall-Héroult processes.

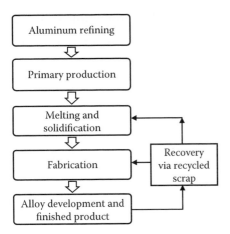

Figure 5.5
Aluminum process and recovery steps.

Figure 5.5 shows a general aluminum process diagram together with the recovery steps. Utilizing various ways to recycle process wastes, especially those generated during primary production, can help aluminum producers to minimize waste, eliminate waste streams, and improve waste heat recovery, especially from exit gases and from the cathode of the cell.

For melting, solidification, and recycling as well as energy recovery applications, alternative clean energy sources can help the industry meet its energy needs while further minimizing its impact on the environment, possibly using the following technologies:

- Combined heat and power (CHP)

- Distributed generation (DG)

- Using hydrogen fuel and fuel cells

- Induction melting using renewable electricity sources

5.2.2.2 Energy Recovery in the Chemical and Petrochemical Industries

5.2.2.2.1 Energy Efficiency Index (EEI) The concept of the *energy efficiency index* (EEI) based on BPT is used to derive the improvement potentials. Forty-nine products have been included in this aggregate indicator of chemicals and petrochemicals. These products represent more than 95% of all energy used in the chemical and petrochemical industries, (Tam and Gielen, 2006) [14]. Table 5.1 shows various country efficiency potentials derived from the EEI. They suggest an improvement potential that ranges from 6.2 (for Italy) to 29.8% (for the United States).

5.2.2.2.2 Enegy Recovery Methods Here are some of the energy recovery methods used in the chemical and petrochemical industries:

1. Optimum thermal energy recovery from process waste streams in all chemical and related processes.

2. Improvement of thermal performance together with minimization of greenhouse gas emissions from chemical units and process operations.

3. Installation of waste heat recovery systems allowing industrial facilities to achieve improved energy efficiency. Examples of waste heat recovery units are:

Table 5.1 Global chemical and petrochemical industry — energy use, EEI, and improvement potential

	Reported energy use PJ	BPT calculated energy use PJ	Energy efficiency index	Improvement potential
United States	6862	4887	0.70	29.8%
China	3740	2975	0.80	20.5%
Chinese Taipei	741	599	0.81	19.2%
Saudi Arabia	1115	917	0.82	17.8%
Netherlands	618	508	0.82	17.8%
Brazil	577	478	0.83	17.2%
India	1090	910	0.84	15.8%
France	654	582	0.88	11.0%
Japan	2130	1917	0.90	10.0%
Germany	1157	1044	0.90	9.8%
United Kingdom	490	460	0.94	6.2%
Italy	389	365	0.94	6.2%
World	**28 819**	**23 682**	**0.82**	**17.8%**

(From IEA [15].)

- Heat recovery boiler technology

- Heat exchanger technology

- Heat recovery steam generator (Ganapathy, 1996 [16])

- Nano-fluid coolant technology for cost-saving and efficient cooling where heat can be further captured and recycled, leading to reduced emissions and improving overall fuel usage efficiency.

5.2.2.3 Energy Recovery in Pulp and Paper (Forest Products) Industry

The pulp and paper industry produces pulp from natural fibers such as woods, agricultural residues, and waste paper. Pulp is then converted into different types of paper by a few methods, in particular, mechanical pulping and chemical pulping. Global best available technology (BAT) and energy usage in the paper and pulping industry is shown in Table 5.2. Various assumptions of BAT are applied to heat and electricity consumption in mechanical pulping, chemical pulping, waste paper pulping and seven different grades of paper making. In the IEA approach, figures for heat (steam) demand in each country are estimated based on reported fuel consumption in the industry, and assume 80% efficiency.

The world pulp, paper, and paperboard industries have been progressively faced with various challenges over the years. For example, some of the aspects, such as increasing overall energy and raw material costs as well as new environmental regulations, demonstrate severe threats for world pulp and paper production. The energy intensity for the pulp and paper industries is about 15–30% of total costs.

Some of the important factors contributing to the importance of energy recovery for the paper and pulp industry are as follows:

1. *Recycling.* The world paper and pulp industry has to fully commit to increasing *recycling* as a means to advance its energy recovery *cum* environmental performance.

2. Feedstock. Feedstock accounted for almost 60% of total energy products.

3. Saving of fuel and power. Fuel and power accounts for more than 40% of all sources of energy use.

5.2.2.4 Energy Recovery in Glass Industry

In order to allow energy recovery and less energy losses, glass manufacturers are implementing various methods as follows:

Table 5.2 Global best available technology (BAT) and energy usage in paper and pulping industry

	Reported energy use PJ	BPT calculated energy use PJ	Energy efficiency index	Improvement potential
United States	6862	4887	0.70	29.8%
China	3740	2975	0.80	20.5%
Chinese Taipei	741	599	0.81	19.2%
Saudi Arabia	1115	917	0.82	17.8%
Netherlands	618	508	0.82	17.8%
Brazil	577	478	0.83	17.2%
India	1090	910	0.84	15.8%
France	654	582	0.88	11.0%
Japan	2130	1917	0.90	10.0%
Germany	1157	1044	0.90	9.8%
United Kingdom	490	460	0.94	6.2%
Italy	389	365	0.94	6.2%
World	**28 819**	**23 682**	**0.82**	**17.8%**

(From IEA.)

1. *New innovative melting and forming technologies.* Continued research and innovative techniques are critical to improvement of overall glass production and energy recovery. More energy-efficient furnaces eliminate contamination, increase recovery while producing high-quality glass. Forming processes such as water quenching and plate forming are equally important to improving and providing recovery as much as possible.

2. *Optimization of operating conditions.* According to Beenkens (2009) [18], several measures, such as changes in batch composition (less batch humidity), optimization of operating conditions, and limiting the combustion air excess, can lead to typically 2–8% of energy savings for industrial glass furnaces.

3. *Better furnace design.* Larger energy savings could be accomplished by new furnace designs, better insulation, and improved heat recovery from the flue gases. The flue gas heat contents, downstream regenerators, or recuperators of air-fired furnaces or downstream the exhaust of oxygen-fired glass melting furnaces, can be exploited to preheat batch and cullet up to 275–350° C for regenerative furnaces and up to more than 500° C for recuperative and oxygen-fired furnaces. Above 550–600° C, the batches may start to show sticking behavior. Energy savings of 12–20% have been reported for regenerative air-fired glass furnaces after connecting a batch and pellet-preheat system. The highest savings in the consumption of specific energy (energy consumption per metric ton of molten glass) can be achieved by combining the application of batch and cullet preheating with an increased pull rate. Increased pull rates of more than 10% have been achieved and this potential of increasing production in the same furnace will improve the economics of batch/cullet preheaters [18].

4. *Use of preheating systems.* There are five or six different types of batch and/or cullet preheating systems applied in the glass industry or still in a testing phase [18].

5.2.2.5 Energy Recovery in Metal Processing Industry

Compared with energy-intensive industries such as basic metal (iron, steel, and aluminum) production, metal casting, the chemical industry, food, beverage, and tobacco industries, as well as the paper industry, the metal processing sector in Europe generally has a lower share in energy consumption and in CO_2-emissions. Within the metal processing industry, energy is mainly used for [52]:

- Drives (machine tools, logistics, ventilation, pumps)

- Compressed air (machine tools, hand workshop tools, logistics)

- Process heat (cleaning of parts, varnishing, and drying)

Studies in Germany (e.g., Seefeldt et al., 2007 [53]) assume that there are potential energy savings of up to 30% of the energy consumption in the metal processing industries. The studies consider only technologies that are currently available on the market, with 50% of the energy savings achieved in normal commercial and economical savings (regarding payback periods and return on investment).

In general, to allow energy recovery and less energy losses, the metal processing industries are implementing various methods and policies, which include:

1. Use of combined heat and power (CHP) units

2. Use of alternative/renewable energy

3. Energy policies and energy management

4. Address all metal processing steps, such as: planning, purchasing, manufacturing, delivering, maintenance, etc.

5.2.2.6 Energy Recovery in Steel Industries

Steel production takes place in two different processes, with various stages:

1. The blast furnace route involves integrated steel plants reducing iron ore into metal, including the following processing phases: raw materials storage and treatment; coke ovens; blast furnaces to produce pig iron; oxygen convertors to convert pig iron into steel; hot rolling mills; cold rolling mills; coating and/or other treatments.

2. Electric arc route, involving the melting of recovered scrap to produce steel, including the following processing plants: scrap treatment; electric arc furnace; secondary treatment; rolling mills.

- Energy input corresponds to approximately 15% of production value, and electricity to around 4%.

- The relatively big share of energy in the cost structure has always been a strong incentive for the industry to reduce energy consumption in the production processes. The EU steel industry cut its energy consumption by 47% per metric ton of finished steel between 1975 and 2000.

Energy type use:

- Electricity consumption of 0.25–0.65 MWh/t steel depending on the type of furnace (2004).

- Coal = main source (of which 95% goes to coke ovens and 5% to other uses such as on-site electricity power generation)

- Natural gas (consumption has increased substantially over the years)

- The third important energy source is electricity used primarily in electric arc furnaces.

- Fuel oil is also used

GHG Emission:

- Direct emissions in the EU-15 are approximately 0.7 t CO2 per metric ton of steel as of 2006.

- 95% of emissions are process emissions.

5.2.2.7 Energy Recovery in Cement Industry

Energy use in the cement industry is around 60–130 kg of fuel oil, and about 105 KWh of electricity per ton of cement production. GHG emission are on average 0.6–0.9 tCO_2 per ton. The two main steps within the cement value chain are clinker production and cement production [19].

Some of the recovery methods used in cement industries are:

- Improvement of cement process production technology, such as cement kiln forced-circulated waste and heat recovery boiler technology

- Material recycling and energy recovery in cement kilns

- Reduction of CO_2 emissions. The bulk of the CO_2 emissions in particular come from clinker production.

5.2.2.8 Summary of Energy Recovery Requirements for Energy-Intensive Industry

For best utilization of energy for providing maximum energy recovery in the energy-intensive industries, generally the following points must be observed:

1. Efficiency improvement and recovery priorities options must be implemented on a life cycle basis.

2. Improve and enhance energy materials recycling to reduce industrial energy use. Energy recovery from waste materials can provide a waste-to-wealth concept in energy-intensive industry.

3. The potential of energy recovery can be increased by optimizing energy management and the energy product process. This includes the use of more energy efficient facilities.

4. Implementation of energy policy to achieve the energy saving target. The energy recovery requirements also include the CO_2 emission reductions.

5. Techno-economical aspects of the recovery methods must be thoroughly measured and benchmarked with standard practices.

5.2.3 Energy Recovery in SME Energy-Related Industry

The methods of energy recovery in SMEs include the following:

1. Good insulation, lighting, and air-conditioning system for space heating/cooling conservation.

2. Good control system to control processes in SME industries to ensure suitable operating temperature and pressure.

3. Preventing any leakage, e.g., steam leakage, pipe leakage.

4. Use of efficient motors, compressors, heat exchangers, ovens, as well as other equipment and utilities such as combustion equipment.

5. Use suitable or optimum mode of process operation.

6. Use of suitable energy resources.

7. Use of other energy recovery methods.

5.3 Process Integration: Pinch Technology and Energy Optimization

Industrial processes usually deal with heat transfer from one process stream to another process stream or from a utility stream to a process stream. In those heat exchanging processes, many associated heat losses occur such that an energy recovery technique is required to maximize energy recovery or minimize energy requirements. That very useful technique is called pinch technology, the main emphasis of which is on heat exchanger network (HEN) design. Later on, pinch technology was further developed far more widely in the fields of overall process improvement and utility system design.

Pinch technology analyzes process utilities, particularly energy and water, to find optimum ways to use them, resulting in reduced operating cost and financial savings. It does this by making an inventory of all producers and consumers of these utilities and then systematically designing an optimal scheme of utility exchange between them.

5.3.1   What Is Pinch Technology and Process Integration?

The term *pinch technology* was introduced by Linhoff and Vredeveld, who employed thermodynamic and heat transfer methods that warrant/guarantee minimum energy application required in the design of heat exchanger networks. Its primary aim is to maximize heat recovery and to minimize the capital costs.

Process integration can be defined as "systematic and general methods for designing integrated production systems, ranging from individual processes to total sites, with special emphasis on the efficient use of energy and reducing environmental effects. Its scope is much wider than just heat recovery" (IEA, 2007 [15]). The target is to reduce energy and raw materials used, as well as minimizing overall environmental effects and providing sustainability.

5.3.2   Use of Process Integration

Process integration can be used for (Friedler, 2010 [54]):

1. Heat integration — to identify the economically optimal level of heat recovery and to design a corresponding heat exchanger network with minimum equipment cost.

2. Heat and power — to identify economically optimal loads and levels for steam consumption and/or production, and to identify opportunities for combined heat and power systems. Thermodynamically "correct" and economically optimal use of heat pumps can also be easily identified by using the systematic methods of process integration.

3. Plant productivity — remove bottlenecks for production throughput, e.g., where the energy system is limiting the mass flow through the process. This is certainly the case in many oil refineries where furnaces operate at maximum capacity.

4. Environment and sustainable development — minimize the investments required to comply with regulations and societal expectations, e.g., reduce emissions and water use.

5.3.3   General Application Advantages of Process Integration

In general, the application advantages of process integration can be enumerated as follows:

1. Heat recovery optimization

2. Minimize energy loss and maximize energy savings

3. Financial savings

4. Better environment and sustainability

5. Better plant productivity

Based on the survey of process integration conducted by IEA, 2002 [17]), we can summarize the process integration practices together with the application advantages in various industries, as in Table 5.3. As can be seen from the table, depending on the applications, with the use of process integration approach, energy savings of 10–50% can possibly be achieved!

5.3.4   Basis of Pinch Technology and Analysis

Pinch technology employs a simple methodology for systematically analyzing industrial processes, within and surrounding utility systems, by considering the First and the Second Law of Thermodynamics:

1st Law: Provides the energy balance for calculating the enthalpy changes ΔH in the streams passing through a heat exchanger.

2nd Law: Determines the direction of heat flow. That is, the heat energy may only flow in the direction of hot to cold. This prohibits "temperature crossovers" of the hot and cold stream profiles through the exchanger unit.

In the above context, therefore, the hot stream can only be cooled to a temperature defined by the "temperature approach" or "temperature limit" of the heat exchanger concept. It is the minimum allowable temperature difference

Table 5.3 Summary of some process integration practices and application advantages

Industry	Process integration practice	Application advantages
Petrochemical Industry	Heat recovery from product streams, e.g., distillation columns. Boiler water and feed preheating using waste heat.	Typical savings are a 20% reduction in energy consumption with some reports of savings as high as 50%.
Chemical Industry	Water pinch analysis, e.g., reusing water within processes and effluent streams.	Typical savings on the order of 20–50%. And, a reduction in fresh water and wastewater volumes of 30% with a 76% reduction in the chemical oxygen demand (COD) of the effluent.
Pulp and Paper Industry	Main pinch analysis, e.g., integration of the evaporation plant and the secondary heat system.	In chemical pulp mills efficient heat exchanger networks lead to energy savings on the order of 10–40%.
Food Industry	Process integrated heat pumps or CHP plants.	Fossil fuel savings of 25% have been achieved.

$\triangle T_{min}$ in the stream temperature profiles, for the heat exchanger unit. The temperature level at which $\triangle T_{min}$ is observed in the process is referred to as the "pinch point" or "pinch condition." To achieve the minimum energy targets for a process, the "pinch" defines the minimum driving force allowed in the exchanger unit, where:

- Heat must not be transferred across the pinch, i.e., no temperature crossovers.

- There must be no external cooling above the pinch.

- There must be no external heating below the pinch.

We discussed the use of process integration for energy saving and recovery in the previous subsection. In the next section, we will consider combined power plant systems and the associated energy saving methods or energy saving potential.

5.4   Energy Saving in Combined Cycles Power Plants

From industrial literature and thermodynamic books, we shall note that there is considerable energy waste in conventional power plants, where heat is rejected to the environment, and as a result of which the average power plant efficiency has a low efficiency range between 35% and 40%.

In combined cycle power plant schemes, the unwanted heat rejected is reused to supply heat to a lower cycle power system; hence enhanced energy savings. Depending on the energy fuel resources available, space location and techno-economic (fuel cost, operating cost, plant characteristics, etc.) considerations, the following are some of the combined cycles power plants use:

- Binary vapor cycle

- Combined gas and steam turbine cycle

- Others

The advantages of combined cycles power plants are:

1. Higher overall plant cycle efficiency.

2. Greater flexibility and reliability of resources used.

3. Avoid complete shut-down as in the conventional power plant, where when one unit/station fails, the consumers can be continuously fed from the other unit/station.

4. Greater output to meet demand, thus the amount of generating capacity required would be less as compared to stand-alone power plant systems.

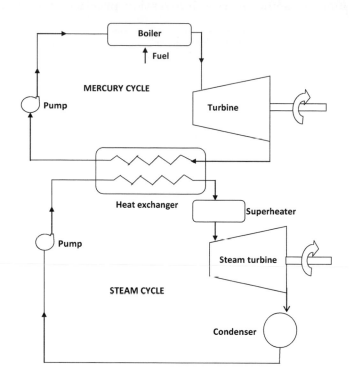

Figure 5.6
Mercury–water binary vapor cycle.

5.4.1 Binary Vapor Cycle

In a *binary vapor cycle* approach, two vapor cycles are arranged in which heat rejected from the condenser of the high-temperature cycle (topping cycle) provides the heat supply to the evaporator of the low-temperature cycle (bottoming cycle). A typical example of a binary vapor cycle is a mercury–water binary cycle. Other feasible combined cycles are, e.g., a sodium–water binary cycle and a potassium–water binary cycle. The typical schematic diagram and the associated *t-s* diagram is presented in Figure 5.6 and Figure 5.7, respectively.

Advantages:

1. *Improved efficiency.* The mercury cycle of the binary vapor approaches the Carnot cycle efficiency, i.e., $\eta_{carnot} = \frac{T_{max} - T_{min}}{T_{min}}$ where all or the bigger part of the vapor region is designed to operate at the wet region, rather than superheated. The efficiency therefore depends on the supplied temperature T_{max} since T_{min} is fixed by heat sink, the temperature level of which is also the supply temperature of the bottoming steam cycle; an overall plant combined efficiency $\geq 50\%$ can be achieved.

2. *Corrosion resistance.* Mercury does not cause corrosion to metals.

Disadvantages:

1. *High capital cost.* Binary vapor cycles implementation needs careful planning; too high capital cost compared with overall improvement in efficiency is not desired.

2. *Mercury has high toxicity.* The system must not have leakage.

5.4.2 Combined Gas and Steam Turbine Cycles

In a combined gas and steam turbine cycles plant, the gas turbine is the higher-temperature unit (topping cycle) while the steam turbine the lower-temperature component. Exhaust gas leaving the gas turbine at a high temperature

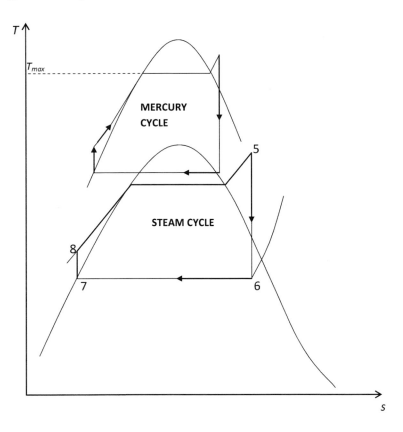

Figure 5.7
T-s diagram of mercury–water binary vapor cycle.

is used as a heat source for steam production in the steam turbine at suitable pressure and temperature range (Figure 5.8).

The combined cycle efficiency can be *estimated*[*] using Eq. 5.1:

$$\eta_{combined} = \frac{\left\{ \frac{\eta_{top} + \eta_{bottom}}{2} \right\} + \left\{ \eta_{top} + \eta_{bottom}(1 - \eta_{top}) \right\}}{2} \qquad (5.1)$$

where

$\eta_{combined}$ is the combined cycles efficiency;

η_{top}= the top higher-temperature cycle efficiency; and

η_{bottom} = the bottom lower-cycle efficiency.

Application advantage of higher efficiency is shown in the following simple example.

Example 5.1 In a combined gas–steam turbine, the top higher-temperature cycle efficiency = 26 % while the bottom lower-cycle efficiency = 35%. What is the estimated combined power efficiency?

Solution: From Eq. 5.1,

$$\eta_{top} = 0.26; \; \eta_{bottom} = 0.35$$

$$\eta_{combined} = \frac{\left\{ \frac{\eta_{top} + \eta_{bottom}}{2} \right\} + \left\{ \eta_{top} + \eta_{bottom}(1 - \eta_{top}) \right\}}{2}$$

---

[*]It is to be noted that in the present introductory text, we are only introducing the rough estimation method for combined cycle efficiency. Detailed calculations involving heat and mass transfer as well as advanced thermodynamics treatments of the various combined cycles would hopefully be covered in an advanced treatise in the future.

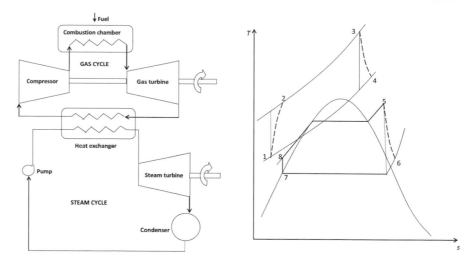

Figure 5.8
Combined gas–steam turbine power plant: (*left*) schematic diagram; (*right*) T-s diagram.

$$\eta_{combined} = \frac{\left\{\frac{0.26+0.35}{2}\right\} + \{0.26 + 0.35(1 - 0.26)\}}{2}$$

$$= \frac{0.305 + 0.519}{2} = 0.42 \text{ or } 42\%\#$$

5.5   Energy Saving in Combined Heat and Power (CHP) Plants

In the conventional power plant and engine systems, a great portion of energy is rejected to the environment as waste heat: a power plant such as that of a coal-fired power plant, or a nuclear power plant, will normally reject a vast amount of heat to the sea, river, or lake. Likewise, an internal combustion (IC) engine also rejects heat through the water jacket, cooling system, and the exhaust.

5.5.1   What Is Combined Heat and Power (Cogeneration)?

Combined heat and power (CHP) or *cogeneration* is the use of a heat engine or a power plant to simultaneously generate both electricity and useful energy from the waste heat, which is otherwise rejected to the environment. By making use of the waste heat, CHP is a way of recycling energy to improve the efficiency of the overall power plant or engine system.

The waste heat is normally rejected from the plant or engine systems in the form of *high grade* steam or hot water, i.e., steam or hot water at high temperature.

High grade hot water or steam can be used in many energy-intensive and SME industries, such as the chemical, food processing, pulp and paper, and textile industries. The thermal heat available from this high grade hot water when used in industries is normally called *process heat*. The facility using the high grade heat or process heat is normally known as a *process heater*, *water heater*, or *heat recovery steam generator*.

5.5.2   What Are the Similarities and Differences between Combined Cycle and CHP?

The similarity between combined cycle and CHP (cogeneration) are that both of the methods are energy recovery methods in which waste energy is recycled back to the systems.

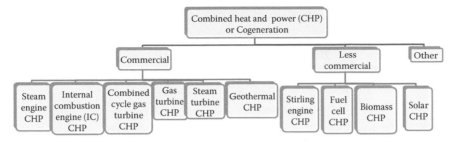

Figure 5.9
Classification of CHP systems.

In the CHP applications, steam produced from the process heater is mainly used for process applications, whereas in the combined cycle, normally 0.35 power is generated via a steam turbine generator. CHP systems usually result in more than one useful form of energy, that is, power and process heat.

5.5.3   Basis and Types of CHP

Unlike conventional power plants or engines, which reject heat generated as a by-product of electricity into the environment through cooling towers, flue gas, rivers, and so forth, a CHP captures the waste heat for heating purposes — either for heating the plant or as hot water for district heating or cooling. This results in energy saving and higher overall efficiency.

Generally, we can group CHP systems into two broad categories, i.e., either commercial or less commercial. The CHP systems can be, generally, further classified into a few types based on the heat-contributing plant or engines (Figure 5.9) such as steam engine CHP, internal combustion engine CHP, etc. The less commercial but advanced CHPs include turbine-fuel cell CHP, biomass CHP, solar concentrating power CHP, etc.

Some of the CHP variations are enumerated below:

1. Combined gas turbine–steam turbine for space heating or space cooling. Figure 5.10 shows a typical combined heat and power or co-generation plant for production of heat and electrical power. The high temperature exhaust gases (typically at 1000°C) from the gas turbine are used to raise steam in a steam generator; the gases leaving the steam generator then pass through a heat exchanger in the process heater to heat water for space heating via a heat pump; or space cooling via an absorption chiller. In the steam turbine, the steam could be bled off at an intermediate temperature and passed through an open-feed heater; the water leaving the heater is pumped to the steam generator.

2. Combined Brayton–Rankine cycle CHP (Figure 5.11)

3. Combined internal combustion engine–exhaust heat (Micro-CHP)

4. Rankine cycle for combined heat and power production

5. Renewable energy CHPs such as biomass-CHP, fuel cell-CHP, solar-CHP, and so on.

5.5.4   Application Advantages of CHP

As reported by IEA very recently (IEA, 2011 [20]), co-generation is attractive to policy makers, private users, and investors because it delivers a range of energy, environmental, and economic benefits, including:

1. Dramatically increased energy efficiency. According to IEA, CHP allows 75% to 80% of fuel inputs, and up to 90% in the most efficient plants, to be converted to useful energy.

2. Reduced CO_2 emissions and other pollutants

3. Increased energy security through reduced dependence on imported fuel

4. Cost savings for the energy consumer

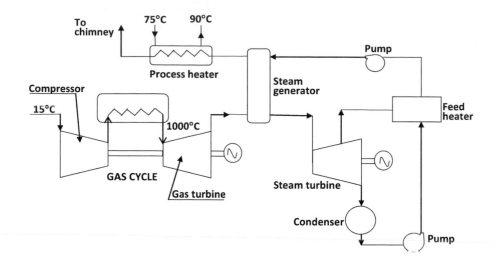

Figure 5.10
Combined gas–steam turbine for space heating or space cooling.

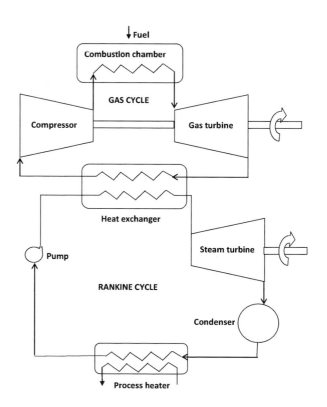

Figure 5.11
A typical schematic combined Brayton–Rankine CHP plant.

5. Reduced need for transmission and distribution networks

6. Beneficial use of local energy resources (particularly through the use of waste, biomass, and geothermal resources in district heating and cooling systems), providing a transition to a low carbon future.

5.5.5 *Case Study 1*: Concentrated Solar Power–Desalination CHP [21],[40]

Trieb et al. (2008, 2009) describes the schemes of concentric solar power–desalination CHP. The projects are being realized for electricity and also for combined heat and power for desalination, electricity, and cooling applications. The Jordanian/German consortium performed a technical and economical study for the production of 10 MW of power, 10,000 metric tons (t) per day of desalinated water, and 40 MW cooling capacity for the Ayla Oasis Hotel Resort in Aqaba, Jordan (Trieb et al., 2009). The team investigated both a conventional solution using gas as the only energy source and an integrated solution using gas and solar energy sources (CSP was the chosen solar technology). The study showed that the integrated process, using CSP and CHP (co-generation), required 34% less fossil fuel (gas) than the conventional solution (Table 5.4).

Table 5.4 Comparison options for CHP of power, cooling, and desalination for arid area applications

	Conventional	Conventional + CHP	Integrated (CHP + Solar)	Remarks
Energy input (MW):				
Gas	85 MW	70 MW	56 MW	
Solar	0	0	16 MW	
Desalination technology:				
Electricity driven reversed osmosis	✓			
Heat driven distillation		✓	✓	
Cooling technology use:				
Electricity driven compression chilling	✓	✓	✓	
Utilities output:		✓	✓	
Gas saving (compared with conventional heat-driven absorption cooling).	0%	18%	34%	

5.5.6 Future of CHP (Cogeneration) and Applications

The current CHP status already has proven effectiveness and success.

Due to increasing energy demand, future CHP technologies would be promising. This is especially needed by the energy intensive industries and the SME energy related sectors.

The compact size of CHP means that it may be used widely in both SMEs and private residential applications, mainly for energy and cost saving opportunities.

Renewable energy CHPs such as solar CHPs, biomass CHPs, and so forth may also gradually replace the fossil-based CHPs.

5.6 Energy Saving in Energy-Intensive and SMEs Energy-Driven Industries

Each day, both energy-intensive and SMEs energy-driven industries consume enormous amounts of energy. The energy saving methods and strategies in the energy-intensive industry are outlined in Sections 5.6.1 through Section 5.6.3. Section 5.6.4 outlines the saving strategies for the SMEs.

5.6.1 Energy Saving in Steel and Related Industries

Around the world, the overall energy specific consumptions were reduced over the years. This is primarily due to better and efficient energy conservation and energy recovery. Some of the approaches are generalized as follows:

1. Optimization of energy supply and consumption

2. Introduction of new energy-efficiency technologies

3. CO_2 emission reduction

For the steel industries, the following three steps of energy conservation are:

1. *Improving plant operational efficiency and savings*
   - Improve cost-effectiveness of energy recovery technologies
   - Optimization of regenerator flue size and mass-area ratio
   - Prevention of losses and air leakage provide better steam line insulation and promote steam conservation
   - Increase efficiency of heat transfer processes including the temperature of preheated combustion air
   - Application of advanced ceramic fiber in reheating and heat treatment sections of the plant and other operational methods such as adoption of efficient hydro jet cleaning

2. *Improving processes, instrumentations and control*
   - Minimize energy losses eliminating wastage, leakage, and better thermal insulation materials
   - Implement energy saving such as waste heat recovery, improving combustion, and refractory systems
   - Better fuel source utilization, instrumentation, and control

3. *Improving productivity with better overall energy management*
   - Proper energy audit for each processes
   - Better fuel use, e.g., blast furnace gas and coke oven gas utilization, for steel melting furnaces
   - Better combustion through progressive increase in hot blast temperatures to reduce blast furnace coke required. Other measures include better humidity control of blast, and smart rotating chutes and deliveries.

5.6.1.1 Example 5.2.

Waste heat recovery from furnace.

Waste exhaust gases are leaving a typical heat treatment furnace at 850° C and at a rate of 1735 m^3/hour. Given the density and specific heat of the flue gas substance is 1.2 kg/m^3 and 100.5 kJ/kg°C, respectively, estimate the heat recoverable at 190° C final exhaust. Maximum waste heat content can be calculated by,

$$Q = V \rho C_p \triangle t \tag{5.2}$$

where,

Q = Waste heat content, kJ;
V = flowrate of the substance, $m^3/hour$;
ρ = density of the flue gas, kg/m^3;
C_p = specific heat of the flue gas substance, kJ/kg°C;
$\triangle t$ = temperature difference, °C.

Heat recoverable,

$$Q = (1735 \frac{m^3}{hour})(1.2)(100.5 \frac{kJ}{kg°C})(850 - 190)°C$$

$$= 138,099,060 \, kJ/hour\#.$$

By installing a recuperator, this heat can be recovered to preheat the combustion air.

Accroding to [42] estimation, the fuel savings would be 33% (@ 1% fuel reduction for every 22ºC reduction in temperature of flue gas. Therefore fuel saving reduction would be

$$Fuel\ saving = \frac{(850 - 190)}{22} = 30\,reductions\ steps$$

which is around

$$Fuel\ saving = 30\ \times 1\% = 30\%\#$$

5.6.2 Energy Saving in Cement and Related Industries

Cement plants mainly comprise raw mill, pyroprocessing, coal mill, cement mill, and cement packing process units. The kiln process center is one of the largest energy consumers, one that consumes almost 100% thermal energy. Cement, concrete grinding, raw meal grinding, and kiln process centers are the major areas for electricity usage.

Energy saving steps in cement industries:

- Optimize kiln sizing and improve kiln combustion efficiency, e.g., use of multistage preheater [43].

- *Use of waste heat recovery systems.* Preheating raw materials at the raw mills with exit gases has been the effective waste heat conservation concept practiced in the cement plants [43].

- Practice of effective instrumentation, operation, and maintenance control of the plant.

- Optimization technology and improved energy management systems.

- *Utilizing alternative fuels.* Alternative fuels such as natural gas, and harvesting energy from biomass and waste fuel, can reduce pollution and minimize energy consumption.

- Use of CHP techniques [44].

5.6.3 Other Energy-Intensive and Related Industries

Other energy-intensive industries include the chemical industry, and pulp and paper industry, which deal with high energy consumption and thus have great energy-saving potential. Energy saving steps for chemical industries include:

- The reduction of power providing fuel consumption, i.e., reduction of reaction pressure [47].

- Water minimization and reuse of water [45].

- *Minimization of total energy losses,* e.g., for losses from a pulp and paper plant, total energy losses included boiler and electricity generation losses, distribution losses, and losses due to equipment inefficiency [46].

- Use of recovery methods and CHP.

5.6.4 Energy Savings in SME Energy-Driven Industries

Generally, SMEs deal with numerous energy consuming appliances such as electrical-driven machines, motor compressors, and other processes machines. The energy saving strategies in the SMEs include:

1. Use of high efficiency motors, compressors, and pumps.

2. Optimization use of heat exchanger, preheater, and heat pipe appliances.

3. Improvement of space heating, cooling, and ventilation.

4. Low carbon and sustainability for long-term energy savings.

5. Efficient energy management and control.

Next, we shall turn to the subject of the energy saving and energy recovery opportunities, specifically, in built environment applications, i.e., energy for building and space comfort (i.e., HVAC) environments.

5.7　Energy Saving in Building, Heating, Refrigerating and Air-Conditioning Systems

Many methods of energy savings could be accomplished by the building industry. The techniques are enumerated as follows:

1. ***Energy saving by thermal insulation.*** Thermal conductivity, hence the thermal insulation capability of an insulating material varies with its density, porosity, pore size in case of fibrous materials, fiber diameter, and non-fibrous (shot) content, i.e., non-fibcrized particles of glass/rock/mineral/wool. Cork, felt, and straw are natural and are relatively cheap construction materials but they are flammable at high temperature. Reflective insulation consists of numerous layers or random packing of low emissivity foils like aluminum. The choice for a thermal insulating material depends on various factors:

   - Thermal conductivity of the insulation
   - Building operating temperature
   - Space conservation requirements and providing a highly insulated building envelope
   - Resistance requirement for heat, fire, or chemical protection
   - Ability to withstand vibration, noise, and mechanical damage
   - Techno-economic consideration.

2. ***Energy saving by using renewable energy resources***

   - Consideration of renewable energy, e.g., geothermal energy, heat pump, and energy recovery, which offer many energy saving advantages.
   - Provide overall building space system optimization and control.

3. ***Energy saving by temperature control***

   - High efficiency task lighting (occupancy control*)
   - Low density ambient lighting (elctronic dimmable)
   - High performance packaged systems, including varible refrigerant flow (VRF) systems

4. ***Energy saving by passive cooling/heating***

   - Optimize use of daylighting
   - Improve ventilation and air-distribution systems
   - Use of new insulation and construction materials
   - Building orientation to suit climate zone
   - Coordinated siting, landscaping and building location

5. ***Energy saving in built environment by new technologies***

   - BIPV applications
   - Use of high efficiency motor and ventilation systems
   - Optimize use of daylighting
   - Super efficient HVAC systems
   - Energy saving LED lighthing technologies
   - Use of new insulation and construction materials

After discussing energy recovery and saving in the previous sections, we now turn the subject matter to energy storage.

---

*Occupancy control systems provide cost and energy savings by employing a sensor to detect when people are in the building spaces, and reduce unnecessary lighting when areas are empty.

5.8 Energy Storage

Energy storage is one of the most important topics in applied energy. In this section, we shall discuss the various energy storage methods and practices, from energy storage in the energy (power) industry to residential applications, and our daily energy usage such as battery storage of our mobile phones.

Energy storage can basically be classified into two categories:

(A) Energy grid storage in the electricity supply network

(B) Stand alone (small-scale, distributed generation of electricity)

5.8.1 Fossil Fuel (Petroleum and Gas) Storage

Liquid fossil fuel storage is normally accomplished with conventional fuel tanks, either underground or above ground. For a smaller volume of fuels or for fuel transportation, smaller tanks or cylinders are usually employed.

5.8.2 Ammonia Storage

Ammonia is a hazardous chemical widely used, for instance, in the manufacture of fertilizers. Worldwide, about 1000 ammonia tanks are in operation. Figure 5.12 is a photo of a typical double-walled, refrigerated anhydrous ammonia tank together with the tank's specifications.

Material of construction According to British guidelines, maximum yield strength (YP) should not exceed 350 MPa for ammonia storage tanks. American standards for storage and handling of anhydrous ammonia state that steels used should have a tensile strength less than 480MPa. Materials with yield strength between 290 and 360 MPa are often used (EFMA, 2002 [22]). Except for the sensitivity to SCC, carbon steel is fully acceptable in ammonia service (NPL, 1982 [23]).

Some of the possible parameters influencing ammonia storage tank failure are:

1. Stress corrosion cracking (SCC). Water, oxygen content and temperature are three main factors contributing to SCC. (See, e.g., Lunde and Nyborg, 1989 [24], McGowan, 2000 [25], Abdullah et al., 2011).

2. Stress. There is a threshold stress below which cracking does not occur [23].

3. Fatigue (negligible) [23].

4. General corrosion (negligible) [22]

Next, we shall discuss the storage and safety of nuclear fuel.

Ammonia tank specification	
Internal diameter	22.4 m
Height	24.6 m
Annular space	0.8 m
Storage temp.	−33°C
Material of construction	Carbon steel

Figure 5.12 (SEE COLOR INSERT)
A typical ammonia storage tank. This typical tank can store about 5,000 metric tons of ammonia. (After M.O. Abdullah et al. [51])

Figure 5.13 (SEE COLOR INSERT)
Typical dry cask storage containers. (Courtesy of the U.S. Nuclear Regulatory Commission.)

5.8.3 Nuclear Storage and Safety

The storage of nuclear fuel, in particular, spent nuclear fuel (SNF), is considered a critical issue. And, over the years, the complicated storage issues have been debated by many, including policy makers, scientists, and the general public.

Chang et al. (2000) [26] discussed the nuclear fuel storage facilities and inspection for a 200-MW nuclear heating reactor.

Saegusa et al. (2007) [27] discussed various challenges for realization of concrete cask storage of spent nuclear fuel in Japan, and made a comparison between metal cask storage and concrete cask storage. Nagano (2007) [28] carried out an assessment to draw quantitative prospects of SNF management with emphasis on uncertainty of storage needs for SNF up to the year 2050.

Generally, the storage of nuclear fuel can be conveniently classified into either *short-term storage* or *long term storage*.

5.8.3.1 Short-Term or Temporary Storage

Generally, the hot spent fuel rod assemblies from the nuclear reactor core are highly radioactive and are temporarily stored in special cooling ponds. The pool is filled with special water containing boric acid.

The water in the cooling ponds is used for three main purposes:

1. To provide cooling by dispersing the heat from the spent fuel.

2. To help absorb some of the radiation given off by the radioactive nuclei inside the spent rods.

3. To act as a barrier medium against radiation.

Apart from "wet" storage, we shall note that the spent fuel can also be moved and dry stored in specially engineered air-cooled containers called "dry storage," usually after 5 years cooling in the pool.

The dry cask storage is basically reinforced casks or concrete bunkers (e.g., see Figure 5.13). Typically, the casks are steel cylinders that are either welded or bolted closed. The steel cylinder provides a leak-tight containment of the spent fuel. Each cylinder is surrounded by additional steel, concrete, or other material to provide radiation shielding to workers and members of the public (U.S. Nuclear Regulatory Commission) [29].

5.8.3.2 Long-Term Storage

For long term nuclear waste storage, one has to find suitable storage sites, deep underground or on the ocean floor. The United States Department of Energy, for instance, has long term plans for nuclear spent fuel to be stored deep in the earth in a geological storage site, at Yucca Mountain, Nevada.

Nuclear safety We all witnessed the recent tsunami and the 9.0 earthquake that struck the coast of Northern Japan in March 2011, disabling the Fukushima Nuclear Power Plant. The devastating 14 m high tsunami knocked out the vital cooling systems, causing the nuclear reactor to overheat. This event has prompted the International Atomic Energy Agency (IAEA) and all nations for even higher standards of storage, safety, and security.

IAEA has published many safety standards, and here are some of the safety standards areas applicable to nuclear facilities and application activities:

1. Fundamental safety principles

2. Safety standards for protection against radiation

3. Emergency guides

4. Storage of radioactive waste safety guide

5.8.4 Battery Storage

A battery is an electrochemical system that stores electrical energy chemically. The amount of energy stored in batteries can be determined by Eq. 5.8.4 and Eq. 5.8.4:

$$Energy\ stored,\ E_{store}(joules) = P \times t \times 3600\ sec/hour \tag{5.3}$$

$$E_{store}(joules) = V \times I \times t \times 3600\ sec/hour \tag{5.4}$$

where
E_{store} = energy stored (joules);
P = battery power (watt);
t = time (hour);
V = voltage across the device (Volts); and
I = current flowing into the device (Ampere).

5.8.4.1 Battery for Everyday Applications

1. Mobile phone. Rechargeable batteries for mobile phone applications have been widely improved with high energy storage capacity of > 3.8 *Wh*, desired light weight, and compact size suitable for wide applications (see Figure 5.14).

Figure 5.14
A typical mobile phone battery. Depending on applications, when fully charged, this rechargeable Li-ion battery could normally power a mobile phone for two to three days of average applications. (Courtesy of author.)

2. Automotive batteries, especially lead acid batteries, are widely used in cars, trucks, boats, etc., for engine starting and other duties [2].

3. Leisure batteries*, as used in caravans, boats, etc. to supply continuous auxiliary power, are an upmarket form of flat-plate battery that may experience regular discharges of moderate depth.

4. Industrial (stationary) batteries are employed in uninterrupted power supplies in many different situations where a loss of mains power would be undesirable. Tubular-plate traction batteries are used to power electric vehicles, e.g., tugs boats, tractors, fork-lift trucks, and some road vehicles. And, "valve-regulated" (or "sealed") lead–acid batteries are assuming increasing importance as they do not require water additions and may be used in any orientation [2].

5. *Batteries for battery-operated-vehicles (BOVs)*. Batteries for BOVs are designed for lightweight vehicles like motorcycles and scooters. Batteries used in BOVs are normally of two types:

   - Lithium batteries with discharge capacity > 80 mAh/g;
   - High-energy lithium polymer batteries of 1 Ah capacity with life cycle of 350, and cell efficiency $> 60\%$

   Advantages of BOVs include:

   - They consume no oil and are thus environmentally friendly
   - Noise-free. Similar to fuel cell-powered vehicles, BOVs do not produce a high level of noise at all.

6. Mini computer and lap top

7. Camera and battery-operated toys

8. Medical mobile apparatus, and so on.

More discussions on batteries are given in Chapter 4, Section 7.5. Various issues associated with battery applications and different types of batteries are outlined in Chapter 12, Section 12.8.

5.8.5 Flywheel Storage

A *flywheel* is an electromechanical storage system in which energy is stored in the kinetic energy of a rotating mass. This is usually called *rotational energy*, from rotation at a very high speed in a vacuum. Flywheels have high energy storage density, with sizes ranging from 40kW to 1.6MW or more, for times of 5–120 seconds. Flywheel systems under development include those with steel flywheel rotors and resin/glass or resin/carbon-fiber composite rotors. The mechanics of energy storage in a flywheel system are common to both steel- and composite-rotor flywheels. In both systems, the momentum of the rotating rotor stores energy. The rotor contains a motor/generator that converts energy between electrical and mechanical forms. In both types of systems, the rotor operates in a vacuum and spins on bearings to reduce friction and increase efficiency. Steel-rotor systems rely mostly on the mass of the rotor to store energy while composite flywheels rely mostly on speed (Boyes and Clark 2000 [31]) .

The amount of stored energy in a flywheel (see Figure 5.15) is,

$$\text{Stored energy}, E_{stored} = \tfrac{1}{2} I \omega^2 \sim \frac{1}{4} M r^2 \omega^2 \tag{5.5}$$

where,
$M =$ mass,
$\omega =$ tip velocity,
$r =$ radius.

One example of flywheel application is the car engine that stores energy in a flywheel between cylinder strokes. When the potential energy is used and extracted from the system by a load, the flywheel's rotational speed is reduced accordingly due to the principle of conservation of energy.

---

*Unlike automotive batteries, which are "starting batteries" designed to supply a burst of immediate power, leisure batteries are "auxiliary power batteries" designed to supply a steady rate of continuous power.

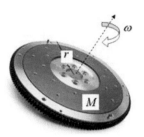

Figure 5.15
A flywheel.

Flywheels are generally mounted in vacuum enclosures, to eliminate air drag, and on low-friction bearings or magnetic suspension systems. Flywheels offer a long list of application advantages:

1. Flywheels have high energy storage density [31];

2. They can provide efficient energy recovery.

3. Flywheels have exceptionally long service lives and low life-cycle costs as a result of minimal operations and maintenance (O&M) requirements, i.e., frequency of cycling (charge–discharge) or by the rates of uptake and release of energy (Boyes and Clark, 2000 [31]).

4. Flywheels do not have the electrical inefficiencies associated with electrochemical devices [2].

5. Flywheels have flexibility in design and unit size.

6. Flywheels require no maintenance (unlike many batteries) [2].

7. Flywheels can be constructed from readily available materials [2].

8. They can store energy with an efficiency of 85%, and can spin up and down for numerous cycles during their working life, making them far more durable than batteries.

9. Flywheels, in principle, can be mass-produced at reasonable cost (especially when expressed on a per kW, rather than a per kWh, basis) [2].

10. They create no environmental impact in use or in recycling [2].

5.8.6   Solar Energy Storage

Depending on applications and other requirements, solar energy storage generally has a few techniques.

For industrial process applications, a solar water heater (SWH) system normally employs a hot water storage tank (see Figure 5.16a), or an integration of solar collectors (see, e.g., Figure 5.16b).

For applications such as that of Small Solar Power System (SSPS) projects sponsored by the International Energy Agency (IEA), a 5-MWht big thermocline storage tank was used. This was a vertical cylindrical shell about 15 m high and 4.2 m in diameter with a working volume of 115 m^3. The second storage system used was a *dual-medium* storage tank where thermal energy was specially stored by means of *thermal oil* and *cast-iron slabs*, consisting mainly of a vertical-steel vessel that contained a stack of 115 cast-iron slabs [49].

For high energy capacity applications, Concentrated Solar Power (CSP) plants are normally designed to have, for example, a 6-h *molten-salt thermal storage system* as reported by Taggart (2008) [50]. In Spain, the CSP plants were designed as *hybrid storage systems*: these basically consist of a 5-MWe molten salt medium with two storage tanks operating at different temperatures (290° C and 550° C) together with a superheated steam generator.

Apart from that, solar energy could be stored by means of pumped hydroelectric storage (see Subsection 5.8.8) or compressed air storage methods. Superconducting magnetic energy storage (SMES) technologies also can be used in parallel to solar energy storage.

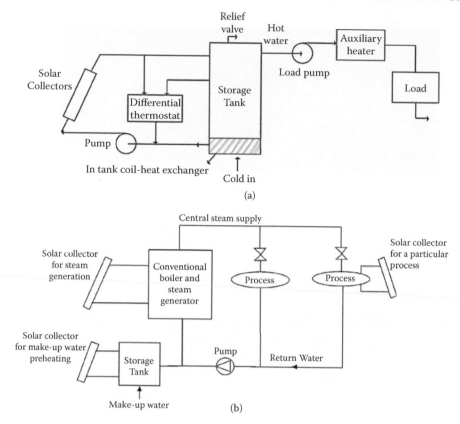

Figure 5.16

(a) Schematic of SWH system with storage. (After Govind et al., 2008 [32].) (b) Schematic of an integration of solar collectors for a typical industrial thermal powered system application. (After Kalogirou, 2003 [33].)

5.8.7 Wind Energy Storage

Wind energy storage methods are usually quite similar to those of solar photovoltaic (PV) storage, such as pumped hydroelectric, compressed air storage, batteries, and so on. Like solar energy, advanced technology such as SMES also could be employed for wind energy storage applications.

5.8.8 Pump Hydroelectric Storage

Pump hydroelectric storage is one of the most economical means of energy recovery available to electric utilities. In a pump hydroelectric storage system, there may be an affordable water supply at a lower level and a reservoir is constructed at a very high level usually on a mountain to create pressure "head." Unlike the conventional hydro energy generation (see Chapter 8), pump hydroelectric storage gets the benefit of high head without the need of building a dam. Pump hydroelectric storage does not require a bigger area as in the case of conventional hydroelectric generation. Also, pressure head is caused primarily by the height of the mountain, not by the depth of water in the reservoir.

During the period when the power demand exceeds the power supply, the water is allowed to flow back down through a hydraulic turbine. This energy recovery method is suitable for both solar and wind energy storage applications where there is some advantage of avoiding the inherent intermittent* characteristics of solar or wind energy.

The disadvantages is that pump work is necessary to deliver the water from the lower level to the mountain top. Thus pump hydroelectric is techno-economical for storage application, in particular for wind and solar energy

---

*Wind and solar energy are often considered as "intermittent" in nature as the energy is often inconsistent, unreliable, or unpredictable.

storage, and not for direct power generation. The efficiency of pump hydroelectric is in the range of 70 to 80 %.
Pump hydroelectric plant overall efficiency is

$$\eta_{overall} = \frac{effective\,head}{total\,head} \times \eta_{pump} \times \eta_{generation} \tag{5.6}$$

where,

$\eta_{overall}$ = plant overall efficiency;
η_{pump} = pumping efficiency;
$\eta_{generation}$ = generation efficiency.

To determine the loss of head due to friction, the following formula can be used:

$$Head\,loss\,due\,to\,friction\,(h_f) = \frac{f\,L\,v^2}{2\,g\,D} \tag{5.7}$$

The head loss due to friction can also be estimated from a Moody diagram.

Example 5.1 It is suggested to build a pump hydroelectric storage facility at a typical mountain of 1000 m height to provide storage for a 500 MW solar energy system. The reservoir will be located on the mountain with reservoir depth of only 50 m. What area and volume of water are required in the reservoir if it is to supply the desired power for at least three days' backup application? Assume overall efficiency = 80 %.

Solution:

From $Power_{net} = g \times Q \times h \times \eta_{overall}$

$$= 9.81 m/s^2 \times Q\frac{m^3}{s} \times 1000\,m \times 0.8 = 500,000\,kW\#$$

Therefore, the volume flow rate is

$$Q = 63.71\,\frac{m^3}{s}\#$$

And, for three days, we need an area of

$$63.71\,\frac{m^3}{s} \times 3days\left(\frac{24 \times 60 \times 60s}{day}\right) = 16,513,632\,m^3\#$$

And, since

$$Volume = area \times height \Rightarrow area\,needed = \frac{16,513,632\,m^3}{50m} = 330,272.6\,m^2\#$$

The pump hydroelectric storage system can provide means of storage as well as energy recovery during the cloudy period for solar power, or for the wind power backup during low wind periods.

5.8.9   Fuel Cell Storage

A fuel cell is an electrochemical system that can store energy chemically, like a primary battery, but unlike a battery, as long as there is a flow of chemicals into the cell, the fuel cell can dynamically store and produce electricity continuously. Therefore, it is to be noted that fuel cell storage is considered only as a temporary storage method. *

Like batteries, fuel cells are electrochemical devices that can provide direct production of low-voltage DC electricity. Generally, hydrogen gas (used as a fuel) is supplied to the anode (negative electrode) of the cell and oxygen to the cathode (positive electrode), resulting in an electrochemical reaction that produces electricity and pure water as by-products. In other words, a fuel cell is more similar to a primary battery than to a secondary battery in that it does not store electricity by recharging, and the hydrogen H_2 fuel has to be:

(a) ideally in the form of pure H_2 gas;

---

*A fuel cell is primarily designed for clean and efficient energy conversion.

(b) produced externally or via internal reforming; and

(C) input into the fuel cell continuously.

More detailed discussion of hydrogen fuel (H_2 gas) storage is given in Subsection 5.8.10; while fuel cell technology and associated applications are given in Chapter 7.

5.8.10   Hydrogen Storage

Hydrogen (H_2) has become popular in recent years due to its inherent high energy and clean properties; for instance, to use for fuel cell applications. The use of hydrogen as an energy fuel has a few advantages and disadvantages, which include the following.

5.8.10.1   Application Advantages

1. Hydrogen has high specific energy.

2. Hydrogen is a clean energy.

5.8.10.2   Application Disadvantages

1. *Hydrogen has very low density.* This implies that hydrogen gas is very difficult to store: to get a large mass of hydrogen gas into a small space, extremely high pressure has to be used.

2. *Hydrogen is difficult to liquify.* Unlike LPG, hydrogen gas cannot be easily compressed and has to be cooled to about 22 K.

3. *Safety.* Hydrogen is highly volatile and highly flammable; it explodes easily when subjected to a spark.

4. *Hydrogen has very low power density.* Thus it is relatively slow in reaction for power generation.

Hydrogen as a fuel can be stored for subsequent use. The storage methods have many approaches, including high pressures, cryogenics, and chemical compounds that reversibly release H_2 upon heating.

The various hydrogen storage technologies can be broadly classify into compressed hydrogen, liquid hydrogen, reversible metal hydride hydrogen, and alkali metal hydride:

1. *Compressed hydrogen storage.* For a small amount of hydrogen gas, it can be conveniently stored in compressed vessels or cylinders available in a a wide range of sizes. The pressurized storage for hydrogen usually is at 200 bar.

2. *Liquid hydrogen storage* or cryogenic hydrogen storage. This is the widely use storage method for big quantities of hydrogen. Basically, the hydrogen is contained in a large, strongly reinforced tank at high pressure and low temperature conditions, i.e., 3–5 bar; 22 K.

3. *Reversible metal hydride hydrogen storage.* Metal hydride hydrogen containers, such as those made of titanium iron hydride, are suitable for hydrogen storage for portable power applications. When needed, the hydrogen can be released by a small amount of heat. This method has the advantage of safety, as it is operating at low pressure, typically ≤ 2 bar.

4. *Alkali metal hydride.* Alkali metal hydride reacts with water to release hydrogen and produce metal hydroxide, e.g., reaction in a calcium hydroxide storage tank (Bossel, 1999 [30]):

$$CaH_2 + 2H_2O \rightarrow Ca(OH)_2 + 2H_2 \qquad (5.8)$$

5.8.11   Thermal Energy Storage

Thermal energy storage can be performed by heating, melting, or vaporization of some materials. By raising the temperature of the material, sensible heat storage takes place. When phase change occurs, like the transition from solid to liquid or from liquid to vapor, the thermal energy thus stored is called latent heat storage. Water and rock can be used as the sensible heat storage material. For latent heat storage purposes, Glauber's salt ($Na_2SO_410H_2O$) is used. However, for high temperature storage on the order of 200–300° C, other salts such as MgO and CaO are usually used ([34], [35]).

5.8.12 Inductor, Capacitor, and Supercapacitor Storage

5.8.12.1 Inductor

The energy stored in an inductor of an electrical system is equal to the amount of work required to establish the current through the inductor and the magnetic field:

$$E_{stored} = \frac{1}{2} L I^2 \tag{5.9}$$

where

E_{stored} = energy stored (J);
L = inductance, henries (H)
I = current through the inductor (Ampere).
The inductance L can be determined from the following relationship:

$$L = \frac{N^2 \mu A}{l} \tag{5.10}$$

where

L = inductance, henries (H)
N = number of turns
μ = permeability, henries per meter (H/m)
A = cross-sectional area, m^2
l = core length, m.
Inductors are less versatile than capacitors and limited in their applications due to size and cost factors. They are used for electromagnetic applications.

5.8.12.2 Capacitor

While an inductor stores electrical energy at low voltage and high current, a capacitor stores electrical charge at high voltage and low current. A typical capacitor is as shown in Figure 5.17. The energy stored in a capacitor can be calculated by

$$E_{stored} = \frac{1}{2} C V^2 = \frac{1}{2} V Q \tag{5.11}$$

where,

E_{stored} = energy stored (J);
C = Capacitance (F);
V = voltage (V); and
Q = charge stored on each plate of the capacitor (Coulomb).

5.8.12.3 Supercapacitor or Ultracapacitor

A *supercapacitor* or *ultracapacitor*, also called an *electrochemical* capacitor, is a special electrochemical capacitor that has very high energy density compared to a conventional capacitor.

Supercapacitors can store energy using either ion adsorption (electrochemical double layer capacitors) or fast surface redox reactions (pseudo-capacitors). They can complement or replace batteries in electrical energy storage and harvesting applications, when high power delivery or uptake is needed (Simon and Gogotsi, 2008 [36]).

Specifically, they are used for supplementary storage for battery driven electrical vehicles as well as backup storage for wind and solar energy.

According to Simon and Gogotsi [36], the discovery that ion desolvation occurs in pores smaller than the solvated ions has led to higher capacitance for electrochemical double layer capacitors using carbon electrodes with sub-nanometer pores, and opened the door to designing high-energy density devices using a variety of electrolytes. The combination of pseudo-capacitive nanomaterials, including oxides, nitrides, and polymers, with the latest generation of nanostructured lithium electrodes, has brought the energy density of electrochemical capacitors closer to that of batteries.

Figure 5.17 (SEE COLOR INSERT)
A typical capacitor. This capacitor is designed with a capacity of $35\mu F$ at 450 VAC, suitable for air-conditioning unit application. (Courtesy of author.)

5.8.12.3.1   Advantages of Supercapacitor (Ultracapacitor)

1. A supercapacitor has high charge and discharge rate capability for energy storage applications (see Table 5.5). Typically, a supercapacitor merely requires a second to several minutes for charging.

2. A supercapacitor has high power density than a battery (see Table 5.5).

3. Unlike a battery, a supercapacitor is non-toxic.

Table 5.5 Comparison of supercapacitor (ultracapacitor) with battery

Attribute	Ultracapacitor	Batteries
Power Density	> 1000 W/kg	< 500 W/kg
Energy Density	< 5 Wh/kg	10–100 Wh/kg
Cold Temperature	< –40°C	< –20°C
Hot Temperature	+65°C	+40°C
Efficiency	98%	95%
Charging time	Fraction of a second to several minutes	Several hours
Charging/discharging efficiency	88%–98%	70%–85%
Self discharging	Hours to days	Weeks to several months
Cycle life	10^6–10^8	200–1000
Lifetime	8–14 years	1–5 years
Toxicity	Not-toxic	Lead, Strong Acid
Monitoring	Not required, simple voltage, current measurement	Sophisticated
Handling	Human handling	Requires equipment

(After Mallika and Kumar, 2011 [37].)

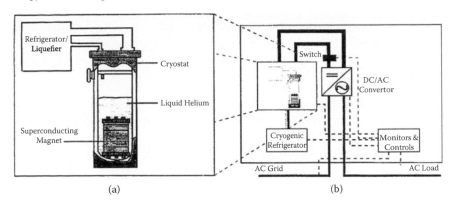

Figure 5.18
a) Cutaway of cryostat. b) SMES system. (After Boyes and Clark, 2000 [31].)

5.8.13 Superconducting Magnetic Energy Storage (SMES) and Superconducting Coil

A *Superconducting Magnetic Energy Store* (SMES) is an energy storage device (see Figure 5.18) that stores electrical energy in a magnetic field (also called inductive energy storage) without conversion to chemical or mechanical forms. In SMES, a coil of superconducting wire allows a direct electrical current to flow through it. It makes use of wire coil made from superconducting material* that gives very small resistance. This current creates the magnetic field that stores the energy. On discharge, switches tap the circulating current and release it to serve a load. The DC current is converted to 3-phase AC output using a solid-state power-conditioning system. To remain superconductive, the SMES coil must operate at cryogenic temperatures (this is one of its disadvantages, where proper operation and maintenance of a cryogenic plant is required, which in turn requires liquid helium production at a very low super conductivity temperature, maintained in the electromagnetic storage system). Therefore, SMES devices require cryogenic refrigerators and related subsystems in addition to the solid-state power conditioning devices, monitors, controls, climate controls, utility and user interface equipment, safety devices, and transportation features (Boyes and Clark, 2000 [31]).

Most SMES units can provide 1MW for 1 second and can be paralleled for more power. In terms of applications, SMES have been used in power quality applications to correct voltage sags and dips at industrial facilities. It is also a utility that can be used to stabilize a large transmission network.

5.8.13.1 Application Advantages of SMES Systems (Boyes and Clark, 2000 [31]; Shikimachi et al. 2011 [38]):

1. An SMES coil has the ability to release large quantities of power within a fraction of a cycle, and then fully recharge in just minutes. This quick, high-power response is very efficient and economical.

2. SMES manufacturers cite controllability and reliability and no degradation in performance over the life of the system as prime advantages of SMES systems.

3. SMES systems are compact, self-contained, and highly mobile; a single semitrailer or equivalent space can deliver megawatts of power. It can be kept at remote locations.

4. Also, SMES units contain no hazardous chemicals and produce no flammable gases. The estimated life of a typical system is at least 20 years.

5. SMES are suitable to use as a hybrid system with fuel cell devices, e.g. [39], large-scale photovoltaics and wind systems.

5.8.13.2 Disadvantages

1. The major obstacle for SMES technology is high initial cost (Ali et al., 2010 [48]).

---

*Superconductors like niobium titanium alloy or its compounds are usually used for SMES.

2. The main problem with the application of ultracapacitors is that maximum voltage of each cell in the SMES's stack should not be be exceeded [48].

5.8.14   Compressed Air Storage

Compressed air storage is an energy storage method where energy is generated by compressing air to a particular container during periods of low energy demand (off-peak periods). The compressed gas is released for use during higher demand, high peak load periods.

Examples of compressed air storage systems:

1. The Huntorf plant, a typical compressd air storage plant near Huntorf, Germany, has been in operation since 1978. It can supply almost 300 MW of reserve power for up to three hours, and comes into operation about 100 times a year: When demand for electricity in the local grid is low, the plant uses excess power to compress air and pump it into two underground salt caverns with a combined volume of more than 300,000 cubic metres. Then, at times of high demand, the compressed air is allowed to expand through turbines on the surface to regenerate the electricity.

2. A similar but smaller plant built in McIntosh, Alabama, operated since 1991.

3. The Paris compressed air energy system (built in 1896) supplies 2.2 MW compressed gas capacity, with distribution at 550 kPa in 50 km of air pipes for various light and heavy industry applications.

Advantages and applications: Compressed air storage is suitable for wind turbine and solar energy storage applications. In a windmill plant, the excess energy from the wind turbine can be used to provide storage by pumping air into a suitable pressurized storage tank. Likewise, solar energy can be stored in a volume of compressed air for energy backup purposes.

5.9   Energy Storage Comparison and Energy Density

A Ragone chart (see Figure 5.19) also known as a Ragone Plot, is a "bubble chart" widely used for performance comparison of various energy storing devices, in which the values of energy density (in Wh/kg) are plotted against

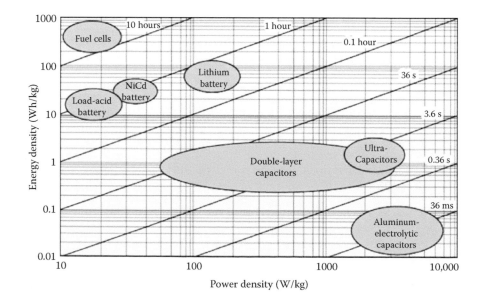

Figure 5.19
Ragone chart illustrating various energy storage devices in terms of their relative power and energy densities. (Courtesy of the U.S. Defense Logistics Agency.)

power density (in W/kg). Both axes are logarithmic, which allows comparison of a wide range of energy densities and power densities of storage facilities. The vertical axis essentially describes *how much* energy is available, while the horizontal axis shows *how quickly* that power can be delivered by the respective devices.

For example, a hydrogen-driven fuel cell may require relatively low amounts of power, but the power should be delivered slowly and sufficiently to operate a fuel cell-powered engine for long hours of usage. Conversely, an ultracapacitor may require very little energy to activate; yet it has the advantage of rapidly delivering enough power for electric power generation. Hence, these two devices would find themselves at different locations of the Ragone chart.

One advantage of such a chart can be shown in a typcial example, described by Sutanto and Ding in 2009 [39], where a new hybrid fuel cell and energy storage system was reported, the system of which is essentially a high temperature SMES combined with a fuel cell system. Making use of the individual advantage of SMES and that of fuel cells, the energy storage of such a combined system therefore can be used to supply the fast changing part of the load while the fuel cell system can continuously supply a constant full load demand.

5.10 *Application Projects, Assignments, Questions, and Problems*

Application Project

Linking knowledge with applied energy design and practice

1. *Gas turbine CHP/cogeneration* with space cooling. Given a combined heat and power (CHP) or co-generation plant, the exhaust gases of which are suggested to be used to produce steam via a steam generator; the gases leave the steam generator at a mass flow rate of 7 kg/s, then pass through a heat exchanger to heat water for space cooling, employing an absorption air-conditioning unit. For energy recovery, steam is bled off at an intermediate temperature and passed through a feed heater. Your student team is required to design the CHP plant system. Ignore feed pump work and any losses, and provide realistic assumptions in your design. Given: for gas turbine cycle, pressure ratio for air compressor = 12; air inlet conditions = 1 bar and 25° C; maximum cycle temperature = 1100° C; specific heat and γ for gases = 1.15 kJ/kg K and 1.33, respectively. For steam turbine cycle: steam conditions at entry to turbine $\leq$ 40 bar; bleed pressure to feed heater = 2 bar. Space cooling for absorption chiller application: flow temperature = 86° C; return temperature = 65° C, mass flow rate of water = 180 kg/s. Provide a group report with the following design information:

   - Draw a schematic diagram and construct a T-s diagram showing your proposed combined CHP plant.
   - Estimate or calculate the power output from the plant.
   - Total cooling effect and COP of your absorption chiller.
   - Estimate the percentage saving in fuel using your CHP system.

Questions and Problems

Multiple choice questions

1. CHP or Co-generation is attractive to policy makers, private users, and investors because it:

   I Dramatically increased energy efficiency.

   II Reduced CO_2 emissions and other pollutants.

   III Cost savings for the energy consumer.

IV Reduced need for transmission and distribution networks.

V Increased energy security through reduced dependence on imported fuel.

(A) I only

(B) I, II, and IV

(C) I and III

(D) All except (V)

(E) All are correct

2. All energy systems must have a provision for energy storage application. Which one of the following is *not* an energy storage method?

(A) Power transformer

(B) Fuel cell

(C) Flywheel

(D) Ultracapacitor

(E) Superconducting coil

3. Which one of the following statements does *not* give a correct definition of the energy storage given?

(A) A supercapacitor is a special electromagnetic capacitor that has very high energy density compared to a conventional capacitor.

(B) A superconducting magnetic energy storage (SMES) is an energy storage device that stores electrical energy in a magnetic field (also called inductive energy storage) without conversion to chemical or mechanical forms.

(C) A flywheel is an electromechanical storage system in which energy is stored in the kinetic energy of a rotating mass, usually called rotational energy, rotating at a very high speed in a vacuum.

(D) A battery is an electrochemical system that stores electrical energy chemically.

(E) A fuel cell is an electrochemical system that can store electrical energy chemically similar to the electrochemical storage in a battery, but unlike a battery, as long as there is a flow of chemicals into a cell, fuel cell can store and produce electricity continuously.

4. Which of the following statements is/are *true* in relation to chemical flooding for EOR application?

I Various chemicals in the form of dilute solutions can be injected into the oil reservoirs to improve oil recovery.

II Injection of alkaline or caustic solutions with oil that has organic acid components will produce soap to lower the interfacial tension, hence increase oil production.

III The injection of alkali, surfactant, alkali-polymer (AP), surfactant-polymer (SP), and Alkali-Surfactant-Polymer (ASP) has been tested in a number of fields

IV Injecting a water soluble polymer would increase the viscosity of the injected water, which causes an increase in the amount of oil recovery in suitable formations.

V Dilute solutions of surfactants such as petroleum sulfonates or biosurfactants such as rhamnolipids may be injected to lower the interfacial tension or capillary pressure that impedes oil droplets from moving through a reservoir.

(A) I only

(B) I, II, and IV

(C) I and III

(D) All except (V)

(E) All are correct

5. Which of the following are the possible influencing parameters for ammonia industrial storage tanks failure?

I Stress corrosion cracking

II Water, oxygen content, and temperature

III Stress

IV Fatigue

V Storage tank size

(A) I only

(B) I, II, and IV

(C) I and III

(D) All except (V)

6. Which of the following are the main differences between conventional hydro generation and pump hydroelectric storage?

I Pump hydroelectric storage gets the benefit of high head without the need for constructing a dam.

II Pump hydroelectric storage does not requires bigger area as in the case of conventional hydro-electric generation.

III Pump hydroelectric requires less input energy to operate.

IV Pump hydroelectric storage has higher efficiency than the conventional hydro-electric systems.

V Pressure head in pump hydroelectric storage is caused primarily by the height of the mountain, not by the depth of water in the reservoir.

(A) I only

(B) I, II, and IV only

(C) I, II, and V only

(D) I and II only

(E) II, III, and IV only

Theoretical questions

1. Define enhanced oil recovery, and explain briefly how EOR can improve oil production.

2. Discuss *any two* of the following EOR methods:

   (a) Gas injection

   (b) Chemical flooding

   (c) Thermal recovery

   (d) Microbial injection

3. Discuss the use of CO_2 capture from coal industries for EOR application.

4. With the aid of a simple schematic diagram, show the difference in between a horizontal well and a vertical well. What are the advantages of horizontal wells compared to vertical wells in terms of overall oil recovery?

5. Discuss various methods that steel industries can adopt for energy conservation and recovery.

6. Briefly discuss the possible techniques that can be used for energy recovery in SME energy industries.

7. Briefly discuss on nuclear storage and safety.

8. What are the problems with hydrogen storage? Name and discuss *any three* of the following hydrogen storage methods:

(a) Compressed hydrogen

(b) Liquid hydrogen

(c) Reversible metal hydride hydrogen

(d) Alkali metal hydride

9. What are the main functions of the water of the spent fuel cooling pool located near a nuclear power plant?

10. Explain briefly "wet storage" and "dry storage" of the spent fuel used in the nuclear industry.

Homework problems

1. *Energy saving and energy recycling from furnace.* Waste exhaust gases are leaving a typical heat treatment furnace at 850 $^{\circ}$C and at a rate of 1735 $m^3/hour$. Given that the density and specific heat of the flue gas substance is 1.2 kg/m^3 and 100.5 kJ/kg $^{\circ}$C, respectively:

    (a) Determine the heat recoverable at 190 $^{\circ}$C at the final exhaust

    (b) Estimate fuel saving in percentage
    (*Answer:* 138, 099, 060 kJ/hour; *30%*)

2. *Micro-CHP application* : A typical SME smoked-fish factory has an electrical demand of 150 kW and a heat requirement of 200 kW, which is currently supplied using power from grid and a gas boiler with an efficiency of 80%. Taking advantage of lower fuel price, it is proposed to install a micro-CHP system employing a biomass-driven engine capable of producing 70 kW electrical power and a heat output of 95 kW; the overall efficiency of the engine is 75%. Assume gas cost is 25% of the electricity cost, and under the micro-CHP system the excess power requirements will be met from the grid.

    (a) What is the total equivalent energy cost in kW before introduction of the CHP unit?

    (b) Determine total equivalent energy cost in kW after introduction of the CHP unit.

    (c) Calculate the percentage saving in fuel cost in changing to the new (CHP) system.

    (*Solution:* 850 kW; 540kW; 36.5 %)

3. *Micro-CHP application* : A typical factory has an electrical demand of 150 kW and a heat requirement of 200 kW, which is currently supplied using power from grid and a fossil fuel boiler with an efficiency of 80%. Taking advantage of environmentally green fuel and incentives from the government, it is proposed to install a micro-CHP system employing a fuel cell engine system capable of producing 70 kW electrical power, and the overall efficiency of the fuel cell system is 75%. Assume fossil-fuel gas cost is 25% of the electricity cost. The H_2 gas cost for the fuel cell system is twice the electricity cost. Under the micro-CHP system, the excess power requirements will be met from the grid. Given that the total equivalent energy cost after introduction of the CHP unit is 740 kW:

    (a) What is the total equivalent energy cost in kW before introduction of the CHP unit?

    (b) What is the percentage energy saving after introduction of the CHP unit?

    (c) Calculate the heat output from the fuel cell system.
    (*Solution:* 850 kW; 12.9 %; 365 kW)

4. *Pumped hydroelectric storage and recovery.* It is suggested to build a pump hydroelectric storage system at a typical mountain of 320 m height to provide storage for a 25 MW wind energy system.

    Data: Diameter of headrace tunnel = 4.2 m; length of headrace tunnel = 625 m; friction factor f =0.02; flow velocity = 6.5 m/s; pump efficiency = 87%; generation efficiency = 92%.

    Also given: Head loss due to friction $(h_f) = \frac{fLv^2}{2gD}$

    (a) What is the plant's overall efficiency?

    (b) How much maximum energy recovery (in MW) can be obtained from the pumped hydroelectric system?

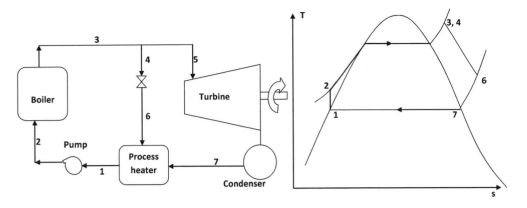

Figure 5.20
Schematic diagram of the CHP system and its T-s diagram.

(c) How much energy recovery is there over two days during the non-solar sunshine period (in MWh)?
(Answer: 78.5 %; 19.6 *MW*; 3, 386, 880 MWh*)*

5. *Combined cycle power plant estimation.* In a combined gas-steam turbine, the top higher-temperature cycle efficiency = 27%, while the bottom lower-cycle efficiency = 41%. What is the estimated combined power efficiency?
(Answer: 45%*)*

6. *CHP or co-generation.* A steam turbine CHP (cogeneration) plant (Figure 5.20) will generate power and meet the steam temperature requirement of 116 °C for a typical chemical industrial application. You are required to design suitable operating conditions for the chemical plant applications.

(a) What is the net power produced from the plant?

(b) What is the rate of the process heat supply?

(c) What is the utilization factor of this CHP plant?

Design data: t_1= 116 °C ; P_3= 41 bar.

Answers to multiple choice questions:

1. *(E) All are correct*

2. *(A) Power transformer is not an energy storage system).*

3. *(A) A supercapacitor is an electrochemical capacitor.*

4. *(E) All are correct.*

5. *(D) All except (V).*

6. *(C) I, II, and V only.*

Bibliography

[1] Alvarado, V., Manrique, E., Enhanced Oil Recovery: An Update Review, *Energies* 2010, 3, 1529–1575; doi:10.3390/en3091529.

[2] Thomas, S., Enhanced Oil Recovery – An Overview, *Oil and Gas Science and Technology* – Rev. IFP, Vol. 63 (2008), No. 1, pp. 9–19.

[3] Moritis, G., 2008 Worldwide EOR Survey. *Oil Gas J.*, 2008, 106, 41–42, 44–59.

[4] OGJ Editors, Whiting to buy power plant carbon dioxide for EOR, Houston, July 22, 2011 or http://www.ogj.com/content/ogj/en/articles/2011/07/whiting-to-buy-power.html.

[5] Liner, C., Carbon Capture and Sequestration: Overview and Offshore Aspects (OTC-21071). In Proceedings of Offshore Technology Conference, Houston, TX, USA, May 3–6, 2010.

[6] Pandey, A., Refinement of Chemical Selection for the Planned ASP Pilot in Mangala Field — Additional Phase Behaviour and Coreflood Studies (SPE-129046). In Proceedings of SPE Oil and Gas India Conference and Exhibition, Mumbai, India, January 20–22, 2010.

[7] Hongfu, L., Guangzhi, L., Peihui, H., Zhenyu, Y., Xiaolin, W., Chen Guangyu, C., Dianping, X., Peiqiang, J., Alkaline/Surfactant/Polymer (ASP) Commercial Flooding Test in Central Xing2 Area of Daqing Oilfield (SPE-84896). In Proceedings of SPE Asia Pacific International Improved Oil Recovery Conference, Kuala Lumpur, Malaysia, October 20–21, 2003.

[8] Ernandez, J., EOR Projects in Venezuela: Past and Future. Presented at the ACI Optimising EOR Strategy 2009, London, UK, March 11–12, 2009.

[9] Hanzlik, E.J., Forty Years of Steam Injection in California—The Evolution of Heat Management (SPE-84848). In Proceedings of SPE International Improved Oil Recovery Conference in Asia Pacific, Kuala Lumpur, Malaysia, October 20–21, 2003.

[10] Jacot, R. H., Bazan, L. W., Meyer, B. R., Technology Integration - A Methodology to Enhance Production and Maximize Economics in Horizontal Marcellus Shale Wells, SPE Annual Technical Conference and Exhibition, September 19–22, 2010, Florence, Italy.

[11] Karlsson, H., Bitto, R., Worldwide experience shows horizontal well success, *World Oil*, 2008.3, pp. 51–60.

[12] Joshi, S. D., 1991, *Horizontal well technology*, PennWell Publishing Company: Tulsa, OK, USA.

[13] Bruckert, L., Horizontal well improves oil recovery from polymer flood, *Oil and Gas Journal*, pp. 35–39, Dec. 18, 1989.

[14] Tam, C., Gielen, D.J., Petrochemical Indicators, IEA/CEFIC Workshop. In: Proceedings: Feedstock substitutes, energy efficient technology and CO2 reduction for petrochemical products, December 12–13, Paris, France; 2006.

[15] IEA. Tracking industrial energy use and CO2 emissions. Paris: IEA/OECD; 2007.

[16] Ganapathy, V., Heat-Recovery Steam Generators: Understand the Basics, Chemical Engineering, August 1996, pp. 32–45.

[17] IEA Process Integration Implementing Agreement, (2002), A Briefing Package on Process Integration, http://www.tev.ntnu.no/iea/pi/

[18] Beerkens, R., Energy saving options for glass furnaces & recovery of heat from their flue gases and experiences with batch & cullet pre-heaters applied in the glass industry, *Ceramic Engineering and Science Proceedings* Volume 30, Issue 1, 2009, pp. 143–162.

[19] Worrell, E., Martin, N., Price, L., Potentials for energy efficiency improvement in the US cement industry, *Energy* 25 (2000) 1189–1214.

[20] International Energy Agency, Co-generation and renewables. Solution for a low-carbon energy future, OECD/IEA, 2011.

[21] Trieb, F., Muller-Steinhagen, H. (2008), Concentrating solar power for seawater desalination in the Middle East and North Africa, *Desalination*, pp.165–183, Elsevier, The Netherlands.

[22] EFMA (2002) European Fertilizer Manufacturers' Association (EFMA) Working Group document, Common Recommendations for Safe and Reliable Inspection of Atmospheric, Refrigerated Ammonia Storage Tanks, Brussels, Belgium.

[23] NPL (1982) National Physical Laboratory, Stress Corrosion Cracking: *Guides to good practice in Corrosion Control*, United Kingdom.

[24] Lunde, L., Nyborg. R., (1989) SCC of Carbon Steel in Ammonia – Crack growth studies and means to prevent cracking, American Chemical Institute of Engineers 1989 Ammonia Symposium: Safety in ammonia plants and related facilities, San Francisco, CA, USA. Paper No. 238 c.

[25] McGowan, P.A., (2000) Advanced Safety features of an ammonia tank, *International Journal of Pressure Vessels and Piping*, Vol. 77, No. 13, November 2000, pp. 783–789.

[26] Chang, H., Liu, J., Wang, J., 2000. Fuel storage facilities for 200-MW nuclear heating reactor. *Nuclear Engineering and Design* 197, 375–379.

[27] Saegusa, T., Yagawa, G., Aritomi, M., 2008. Topics of research and development on concrete cask storage of spent nuclear fuel. *Nuclear Engineering and Design* 238, 5, 1168–1174.

[28] Nagano, K., 2008. An assessment of spent nuclear fuel storage demands under uncertainty. *Nuclear Engineering and Design* 238, 5, 1175–1180.

[29] United States Nuclear Regulatory Commission, http://www.nrc.gov/waste/spent-fuel-storage.html, accessed Aug. 3, 2011.

[30] Bossel U.G. (1999) Portable fuel cell battery charger with integrated hydrogen generator, Proceedings of the European Fuel Cell Forum Portable Fuel Cells Conference, Lucerne, Switzerland, pp. 79–84.

[31] Boyes, J.D., Clark, N.H., Technologies for energy storage. Flywheels and super conducting magnetic energy storage, 2000, Power Engineering Society Summer Meeting, 2000. IEEE, pp. 1548–1550 vol. 3.

[32] Kulkarni, G. N., Kedare, S. B., Bandyopadhyay, S., Design of solar thermal systems utilizing pressurized hot water storage for industrial applications. *Solar Energy* 2008; 82 (August (8)): 686–99.

[33] Kalogirou, S., The potential of solar industrial process heat applications. *Applied Energy* 2003; 76 (December (4)): 337–61.

[34] Hasnain, S.M., Review on sustainable thermal energy storage technologies, Part I: heat storage materials and techniques. *Energy Conversion and Management*, Volume 39, Issue 11, 1 August 1998, pp. 1127–1138.

[35] Dinçer, I., Rosen, M. A., Thermal energy storage: systems and applications, 2nd Edition, John Wiley & Sons., Ltd, Chichester, U.K., 2010.

[36] Simon, P., Gogotsi, Y., Materials for electrochemical capacitors, *Nature Materials* 7, 845–854 (2008).

[37] Mallika, S., Kumar, R. S., Review on Ultracapacitor- Battery Interface for Energy Management System, *International Journal of Engineering and Technology* Vol. 3 (1), 2011, 37-43.

[38] Shikimachi, K., Tamada, T., Naruse, M., Hirano, N., Nagaya, S., Awaji, S., Nishijima, G., Watanabe, K., Hanai, S., Kawashima, S., Ishii, Y., Unit Coil Development for Y-SMES, IEEE Transactions on Applied Superconductivity, Vol. 21, No. 3, June 2011.

[39] Sutanto, D., Ding, K., Hybrid Fuel Cell and Energy Storage Systems Using Superconducting Coil or Batteries for Clean Electricity Generation, Proceedings of 2009 IEEE International Conference on Applied Superconductivity and Electromagnetic Devices Chengdu, China, September 25–27, 2009.

[40] Trieb, F., Müller-Steinhagen, H., Kern, J., Scharfe, J., Kabariti, M., Al Taher, A., Technologies for large scale seawater desalination using concentrated solar radiation, *Desalination*, Volume 235, Issues 1–3, January 15 2009, pp. 33–43.

[41] Lönnberg, M., Variable speed drives for energy savings in hospitals. *World Pumps* 2007 (494): 20–4.

[42] Bureau of Energy Efficiency. Waste heat recovery, 2009, Available online at: http://www.em-ea.org/Guide%20Books/book-2/2.8%20Waste%20Heat%20 Recovery.pdf, accessed Oct. 10, 2011.

[43] Kabir, G., Abubakar, A.I., El-Nafaty, U.A., Energy audit and conservation opportunities for pyroprocessing unit of a typical dry process cement plant, *Energy*, Volume 35, Issue 3, March 2010, pp. 1237–1243.

[44] Shaleen, K., Rangam, B., Uday, G., 2002. Energy balance and cogeneration for a cement plant. *Applied Thermal Engineering*, 22, 485–494.

[45] Abd El-Salam, M. M., El-Naggar, H. M., In-plant control for water minimization and wastewater reuse: a case study in pasta plants of Alexandria Flour Mills and Bakeries Company, Egypt, *Journal of Cleaner Production*, Volume 18, Issue 14, September 2010, pp. 1403–1412.

[46] Hong, G.-B., Ma, C.-M., Chen, H.-W., Chuang, K.-J., Chang, C.-T., Su, T.-L., Energy flow analysis in pulp and paper industry, *Energy*, Volume 36, Issue 5, May 2011, pp. 3063–3068.

[47] LIU Huazhang, Analysis of energy saving in ammonia synthesis industry, *Chemical Industry and Engineering Progress*, 2011.

[48] Ali, M. H., Wu, B., Dougal, R. A., Hybrid Fuel Cell and Energy Storage Systems Using Superconducting Coil or Batteries for Clean Electricity Generation, IEEE Transactions on Sustainable Energy, Vol. 1, No. 1, April 2010.

[49] Kalt, A., Loosme, M., Dehne, H., Distributed Collector System Plant Construction Rep. Tech. Rep. No. IEA-SSPS-SR-1. Cologne: IEA; 1981.(Kesselring, P., Selvage, C.S. The IEA/SSPS solar thermal power plants. vol. 2: Distributed collector system (DCS). Berlin: Springer-Verlag; 1986.

[50] Taggart, S., Parabolic troughs: CSP's quiet achiever. Renewable Energy Focus; March/April 2008, pgs. 51–54.

[51] Abdullah, M. O., Zen, J., Yusof, M., 2011, On the Stress Corrosion Cracking, Crack Growth Prediction and Risk-based Inspection of Industrial Refrigerated Ammonia Tanks, *Corrosion*, 67 (4) National Association of Corrosion Engineers (NACE) International, Houston, Texas.

[52] IECSME, Sectorial report: Metal processing, also available at www.iecsme.eu, accessed October 23, 2011.

[53] Seefeldt, F., Wünsch, M., Michelsen, C., et al., (2007) Potenziale für Energieeinsparung und Energieeffizienz im Lichte aktueller Preisentwicklungen. Prognos, Berlin.

[54] Friedler, F., Process integration, modelling and optimisation for energy saving and pollution reduction, *Applied Thermal Engineering*, Volume 30, Issue 16, November 2010, pp. 2270–2280.

6

Energy from Fossil Fuels versus Alternative Energy

In this chapter, we will consider fossil fuels, in particular, petroleum, gas, and coal. We will then study the environmental effects of fossil fuels. We shall end the chapter with the introduction of alternative (renewable) energy as the alternative fuels and favored fuels of the future.

6.1   Introduction to Fossil Fuels (Petroleum, Natural Gas, and Coal)

Fossil fuels are non-renewable energy sources that formed more than 300 million years ago during the Carboniferous Period. Today, most of the energy we consume for our daily applications is produced from fossil fuels, mainly derived from petroleum, natural gas, and coal. Oil was formed from the remains of animals and plants in a marine environment. When plants and animals died, their bodies decomposed and were buried under layers of earth, sand, and silt. High temperature (heat) and high pressure from these layers helped the remains turn into what we today call crude oil.

6.1.1   Petroleum

The word "petroleum" means "rock oil" or "oil from the earth." Petroleum, sometimes better known just as "oil," is a complex mixture of hydrocarbons, which may be either gas, liquid, or solid, depending on its own unique composition and the pressure and temperature at which it is confined [1].

Petroleum is found way down in the ground, usually between layers of rock in spaces called oil and gas reservoirs. The oil and gas are trapped in the porous and permeable bed of rocks (reservoir rocks). To get oil out, a well is dug by drilling through the rock formations. The oil is then pumped out from the well using an oil rig. There are three main steps practiced in oil and gas industries in recovering petroleum, normally called oil production (see Figure 6.1). The first step is the *primary recovery*, where the oil flows by natural pressure, (i.e., pressure difference due to high pressure in the borehole in the well and the atmospheric pressure) or simple pumping. The maximum recovery is about 30% of the oil available in the well. The next step is called *secondary recovery*, where water or gas is injected into the well at the respective water or gas formation in order to provide additional pressure forcing oil out from the oil formation. This adds 10–20% to oil recovery. The third step is called the *tertiary recovery* or *enhanced recovery*, where chemicals, CO_2, and hot gases are pumped into the well to make the oil less viscous for

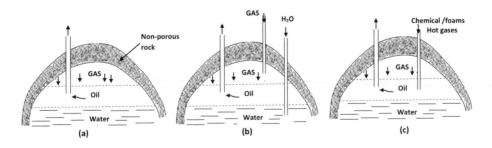

Figure 6.1
Oil recovery steps: (a) primary or normal recovery; (b) secondary recovery; and (c) tertiary recovery or enhanced recovery.

easier pumping. The reserves to production ratio (R/P ratio) is a term used to describe the rate of production of a reserve measured in years.

6.1.1.1 Volumetric Equation of Oil in Place (OIP)

In the oil and gas industry worldwide, two different systems of units are frequently used: SI units and the so-called "oil field" units. The SI units are based on international definitions, whereas the use of "oil field units" may vary from one country to another.[*]

For an oil reservoir zone of A acres and h feet thickness, the volume of oil in place (OIP) is

$$V_{OIP} = 7758\,A\,h\,\phi S_o = 7758\,A\,h\,\phi(1 - S_w) \tag{6.1}$$

Where,

V_{OIP} = tank oil in place, bbl

A = acres

h = thickness, feet

S_o = the oil saturation (fraction of pore space occupied by oil)

S_w = the water saturation (fraction of pore space occupied by water)

The volume of oil and gas changes as the temperature and pressure conditions change from that in the reservoir to that in the stock tank during production (the volume change depends upon the changes in pressure and temperature and the composition of the oil or gas). The oil formation volume factor B_o is the ratio of the volume of a standard mass of oil at reservoir conditions to that at stock tank conditions, usually expressed in reservoir barrel per stock tank barrel. [†] Hence, we can now calculate the total hydrocarbon originally in place in the reservoir when measured at the pressure and temperature conditions prevailing in the stock tank (i.e., during production) by the well-known volumetric equation of oil in place,

$$V_{STOOIP} = \frac{7758\,V_b\phi S_o}{B_o} = \frac{7758 V_b\phi(1 - S_w)}{B_o} \tag{6.2}$$

Where,

V_{STOOIP} = stock tank oil originally in place, bbl/acre-ft

V_b = bulk (rock) volume, acre-ft = area A×thickness, h

ϕ = fluid-filled porosity of the rock (fraction)

S_o = the oil saturation (fraction of pore space occupied by oil)

B_o = the formation volume factor for the oil at the reservoir pressure (reservoir bbl per stock tank bbl)

S_w = the water saturation (fraction of pore space occupied by water)

Note that B_o is a dimensionless factor for the change in volume between reservoir and standard conditions at the surface.

6.1.1.2 Example 6.1

Determine the amount of tank oil existing in the following oil field in Mbbl and cc:

- Area of reservoir field = 535 acres

- Average sand thickness = 18 ft

- Average porosity = 16 %

---

[*]In this subsection of this book, for calculating the oil in place, we are adopting the oil field units that are widely used and that are recommended by the Society of Petroleum Engineers (SPE). The final calculated numbers, however, are also given in SI.

[†]We shall note that volume of oil at the surface (called *stock tank barrel*) is less than the oil volume in the reservoir (called *reservoir barrels*) such that $Bo > 1$. When hot pressurized oil in the reservoir is brought to the surface at cooler and lower pressure, it contracts (because the temperature reduction causes oil to contract more than to expand due to the pressure reduction). In other words, the oil compressibility effect with pressure (for the range of pressures usually found in reservoirs) is greater then the effect of thermal expansion of the oil (caused by reservoir temperature ranges).

- Average connate water saturation = 30%

- Formation volume factor = 1.2

Note: 1 barrel =159,000 cc

6.1.1.3 Solution

$$V_{STOOIP} = \frac{7758(535)(18)(0.16)(1 - 0.30)}{1.20}$$

$$= 6.97 \times 10^6 \, bbl = 6.97 \, Mbbl\# \, or \, 1.11 \times 10^{12} cc\#$$

The porosity of the reservoir rock is usually determined from laboratory core sample analysis. The general formula for porosity is

$$\phi = \frac{V_p}{V_b} = \frac{V_b - V_s}{V_b} \tag{6.3}$$

where,
$\phi =$ porosity (fraction);
$V_p =$ pore volume of the rock;
$V_b =$ bulk volume of the rock;
$V_s =$ net volume occupied by solids or grains (fraction).

6.1.1.4 Example 6.2

The data available from a typical cylindrical core sample from a particular reservoir are as follows:

- Weight of dry clean sample = 314 g

- Weight of sample with pores completely filled, i.e., 100% saturated with brine = 335 g

- Sample diameter = 4.0 cm

- Sample length = 10.0 cm

Compute the reservoir rock's porosity if the specific gravity of brine = 1.05.

6.1.1.5 Solution

$$V_p = \frac{335 - 314}{1.05} = 20 \, cm^3$$

$$V_b = \pi r^2 L = \pi (\frac{4}{2})^2 (10) = 40\pi \, cm^3$$

From porosity, $\phi = \frac{V_p}{V_b}$

$$\phi = \frac{20}{40\pi} = 0.159 \, or \, 15.9\%\#$$

6.1.1.6 Chemical Composition of Petroleum

The principal hydrocarbon series found in petroleum are of three types: paraffins (saturated hydrocarbons/alkanes), cycloparaffins (naphthenes), and aromatics (benzene series). The series are briefly discussed as follows [1]:

Paraffins. Paraffins have the general formula $C_n H_{2n+2}$. These compounds are chemically stable and can have either straight or branch chains of structure. All crude oils contain some paraffins; the first few members of the series are given in Figure 6.2.

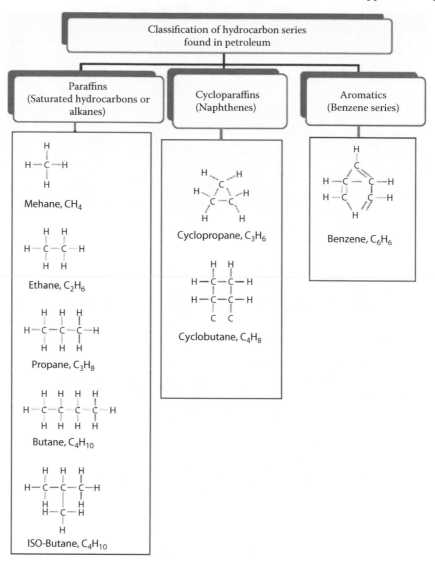

Figure 6.2
Classification of hydrocarbon series found in petroleum. (Modified from Carl Gatlin, *Petroleum Engineering: Drilling and Well Completion*, USA: Prentice-Hall Inc., Englewood Cliffs, NJ, 1960.)

Cycloparaffins Cycloparaffins or naphthenes have the general formula C_nH_{2n}. These compounds have a ring structure, typical members of the group are cyclopropane and cyclobutane (see Figure 6.2).

Aromatics Aromatics or benzene series have the general formula C_nH_{2n-6}. These compounds are chemically active and contain the benzene ring. The simplest member is benzene C_6H_6 (see Figure 6.2).

Apart from hydrocarbons, petroleum also may contain impurities such as carbon dioxide (CO_2), hydrogen sulfide (H_2S), and other complex compounds of nitrogen, sulfur, and oxygen.

6.1.1.7 General Classifications of Petroleum

Petroleum is often classified by a base designation, either as a paraffin base, asphalt base, or mixed based crude.

Paraffin based crude oil A paraffin based crude is an oil that contains primarily paraffins, and which, upon being completely distilled, leaves a solid residue in the form of wax.

Asphalt based crude oil An asphalt based crude oil is an oil whose main components are cyclic compounds, and which, when distilled, leaves a solid residue in the form of asphalt.

We shall also note that petroleum can also be classified according to its viscosity and sulfur content. For example, Middle Eastern and Malaysian crude oil are usually low in sulfur; Venezuela's crude oil is generally high in sulfur and viscosity. Low sulfur petroleum has a higher price in the crude oil markets.

Another widely used indicator of crude oil's worth is American Petroleum Institute (API) Gravity, which is essentially a measure of the crude oil's density based on the following simple formula,

$$\text{API gravity (degrees)} = \frac{141.5}{\text{Specific gravity}} - 131.5 \tag{6.4}$$

6.1.1.8 Example 6.3

What is the specific gravity of an $30°$ API gravity of oil?

6.1.1.9 Solution

From Eq.6.4, the specific gravity of the oil is,

$$\text{Specific gravity} = \frac{141.5}{(30° + 131.5)} = 0.88\#$$

Now, how about transportation and storage of oil? For long distances, oil is carried in pipelines across continents, and in large tanker ships across oceans. For short distances, we can transport the oil by oil truck and by rail, etc. As to the oil storage, it is normally stored in big storage tanks. We already discussed the more detailed aspects of oil storage in Section 5.8.

6.1.1.10 Petroleum and Refining Process

In the refinery, the crude oil is converted into products like gasoline (petrol), jet fuel, and diesel fuel. Petroleum refining has three major processes (see Figure 6.3). The first process is a physical process called *distillation*, which separates components according to their boiling points. The second step is the *cracking* process, a chemical process using a catalyst which breaks down long hydrocarbon chains to make more gasoline, diesel, jet fuels, etc. The third

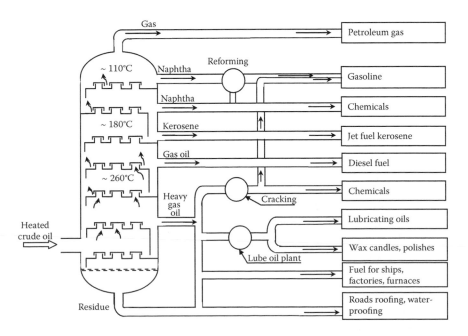

Figure 6.3
Crude oil fractional distillation.

Table 6.1 Some common fuels and their compositions (by weight %)

Fuel	Carbon, C	Hydrogen, H	Sulfur, S	Ash
Gasoline	85.5	14.4	0.1	-
Benzene	91.7	8	0.3	-
Kerosene	86.3	13.6	0.1	-
Diesel	86.3	12.8	0.9	-
Light fuel oil	86.2	12.4	1.4	-
Heavy fuel oil	88.3	9.5	1.2	1.0

process is a *reforming* process, where the process converts straight hydrocarbon chains into branch chains for better performance in gasoline engines, etc.

Now, from an application aspect, the fuels from petroleum products are burned in factories and power plants to make electricity. The fuel is burned, which produces gases that turn a turbine, for example a gas turbine, to create electricity. We can use fuel to boil water to produce steam that also turns turbines, for instance, in a steam turbine. In the internal combustion engines of our vehicles, the fuel, normally gasoline or diesel, is put in the combustion cylinder with the piston: a spark ignites the gasoline engines, while compression ignites the diesel engines. We use fuels for our residential applications too, such as for heating, cooling, cooking, etc.

6.1.1.11 How Is Petroleum Processed?

Petroleum originated from organic matter. It is believed that during the prehistoric period, e.g., the Carboniferous Period about 360 to 286 million years ago, enormous numbers of animals, especially primary producers (see Chapter 1) and plants in excess of consumption accumulated in swamps. This was buried and compressed to form coal, oil deposits, and gas, via bacteria actions and mega distillation under suitable high pressure and temperature (HPHT) conditions. Today, we drill down through layers of silt, sandstones, and rock formations to reach the reservoirs containing petroleum and gas.

Petroleum or crude oil is made up of almost entirely liquid hydrocarbons, together with some solid hydrocarbons and small amounts of gases, nitrogen, and sulfur compounds. Petroleum or crude oil becomes useful only when it is split up into different fractions in a fractionating tower by a process called crude oil fractional distillation (see Figure 6.10).

First the crude oil is heated in a furnace to a temperature of around 400° C where much of it turns into vapor. The hydrocarbons in oil vapor separate out at different levels according to their respective boiling points. We should note that the refining of crude oil in refineries supplies one of the most important products — our gasoline. Now, the heavy bottom fractions are often subjected a cracking process where the residue is changed into lighter, useful chemical products. Table 6.1 shows the common liquid fuels, together with their composition (by weight), used in our daily residential and industrial applications.

6.1.2 Petroleum Fuel or Products' Properties Required for Energy Applications

We shall note that the requisite fuel properties vary from one application to another, and from one energy system to another. Generally the fuels should be:

1. *Low ash content*, The ash value is related to the inorganic material in the fuel oil. Excessive ash in liquid fuels can cause fouling deposits in the combustion equipment. Ash has erosive effect on the burner tips, causes damage to the refractories at high temperatures, and gives rise to high temperature corrosion and fouling of equipment.

2. *High heating value or calorific value*, The calorific value is the measurement of heat or energy produced per kg, and is measured either as gross calorific value or net calorific value. Gross calorific value (GCV) assumes all vapor produced during the combustion process is fully condensed. Net calorific value (NCV) assumes the water leaves with the combustion products without being fully condensed. Fuels are compared based on the net calorific value.

3. *Suitable specific heat.* Specific heat is a term referring to the amount of energy (kcals) needed to raise the temperature of 1 kg of oil by 1° C. The unit of specific heat is kcal

4. /kg°C. The specific heat determines how much steam or electrical energy it takes to heat oil to a desired temperature. Light oils have a low specific heat, whereas heavier oils have a higher specific heat.

5. *Less corrosive tendency.* This property is vital for metal and energy systems' life.

6. *Low sulfur content.* Crude oil with lower sulfur content is desired. Generally, the amount of sulfur in the refining fuels depends mainly on the source of the crude oil and to a lesser extent on the refining process. The typical sulfur content range for kerosene, diesel oil, and furnace oil is 0.05–0.2, 0.05–0.25, and 2.0–4.0, respectively.

7. *Low pour point.* The pour point of a fuel is the lowest temperature at which it will pour or flow when cooled under prescribed low temperature conditions. It is an indication of the lowest temperature at which fuel oil is readily pumpable. It also means that a fuel with low pour point will not solidify easily with temperature decrease.

8. *High flash point.* The flash point of a fuel is the lowest temperature at which the fuel can be heated so that the vapor gives off flashes momentarily when an open flame is passed over it. The higher the flash point, the safer the fuel will be, especially in energy applications that deal with high surrounding temperature.

9. *Low or suitable viscosity.* The viscosity of a fuel is a measure of its internal resistance to flow. Viscosity depends on temperature and decreases as the temperature increases. It is the most important characteristic in the storage, transport, and use of fuel oil. It influences the degree of preheating required for handling, storage, and satisfactory atomization. If the oil is too viscous, it may become difficult to pump, hard to light in the burner, and result in poor fuel atomization. Poor atomization may in turn, cause the formation of carbon deposits on the burner tips or on the walls. Therefore, preheating of high-viscosity oil in industrial heating is necessary for proper atomization.

10. *Safety, ease of transportation and storage.* All fuel oil should be relatively safe and easy for transportation (such as for vehicle internal combustion engine applications). The fuel should be relatively easy to store and retrieve.

11. *Other suitable rheological, physical, and chemical properties.* For example, for power plant combustion applications, normally the fuel should possess higher octane number in order to achieve a higher compression ratio, and thus thermal efficiency. The cetane* of a diesel oil should be as high as possible.

6.1.2.1 Energy from Combustion of Petroleum

The amount of energy released by burning hydrocarbons depends very much on the energy required to break the chemical bond of the particular combustion hydrocarbon. Basically, the higher the number of chemical bonds, the harder the molecule is to break apart or released.

Since bond energies are small, we usually use units of energy per "mole," where one mole is equivalent to 6.022×10^{23} molecules (atom of bonds).

6.1.2.1.1 What Is Avogadro's Number?

Avogadro's number is the number of atoms in which the number of grams of a substance equals the atomic mass of the substance,

$$N_A = 6.022 \times 10^{23} mol^{-1} \tag{6.5}$$

Therefore, a mole of hydrogen (H) atoms is 1 g, a mole of hydrogen molecules is 2 g, a mole of carbon-12 atoms is 12 g, and 1 mole of oxygen is around 16 g. Table 6.2 displays the common hydrocarbon bonds and the associated energy required to break the bonding.

---

*Cetane number indicates the ignition quality of a diesel fuel. It is essentially a measure of a fuel ignition's delay. For instance in diesel engines, the use of high cetane fuels will have shorter ignition delays than low cetane fuels.

**Table 6.2 Common hydrocarbon bonds and the associated energy
needed to break the bonds**

Bond	Energy (kJ/mole)	Remarks
C-C	347	
C-O	360	
C-H	410	
O-H	460	
O=O	494	
C=C	519	More energy required than for a single bond
C=O	799	More energy required than for a single bond

6.1.2.1.2   What Is Enthalpy of Formation?   Enthalpy of Formation or "molecular binding energy" is the difference between the energy of a molecule and its elements in their most stable state.

6.1.3   Natural Gas (Gaseous Petroleum)

Gaseous petroleum, better known as *natural gas*, is a colorless and odorless mixture of hydrocarbon gases. Gas usually is formed in our Earth along with petroleum, and it usually occupies the upper portion of the reservoir as it is lighter then petroleum.

6.1.3.1   Gas Originally in Place (GOIP)

Similar to oil, the volume of gas changes as the temperature and pressure conditions change from that in the reservoir to that in the stock tank during production (the volume change depends upon the changes in pressure and temperature and the composition of the oil or gas). The gas formation volume factor B_g is a term to describe the ratio of the volume of a standard mass of gas at reservoir conditions to that at stock tank conditions, usually expressed in reservoir cu.ft per stock tank cu.ft.* Therefore, we can calculate the amount of gas originally in place in the reservoir when measured at the pressure and temperature conditions prevailing in the stock tank by the well-known volumetric equation of gas originally in place,

$$V_{STGOIP} = \frac{43560 \, Ah\phi \, (1 - S_w)}{B_g} \tag{6.6}$$

Where,

V_{STGOIP} = Stock tank gas originally in place, cu.ft. or (ft^3)
V_b = bulk (rock) volume, acre-ft = Area A × thickness, h
ϕ = fluid-filled porosity of the rock (fraction)
B_g = the gas formation volume factor at the reservoir pressure (reservoir cu.ft per stock tank cu.ft)
S_w = the water saturation (fraction of pore space occupied by water)

6.1.3.2   Gas Composition

Now, while natural gas is formed primarily of methane (70–90%), it can also include a mixture of ethane, propane, butane, and other gases. The composition of natural gas can vary widely, but Table 6.3 shows the typical composition of natural gas before it is refined. Natural gas is often described as a "clean burning" or "relatively green" fuel because it produces fewer undesirable by-products than petroleum. Like that of other fossil fuels, combustion of natural gas emits CO_2, but at about half the rate of coal.

---

*We shall note that unlike with oil, the volume of gas at the surface (called *stock tan cu.fu.*) is more than the gas volume in the reservoir (called *reservoir* cu. ft) such that $B_g \ll 1$. This is because when hot pressurized gas at reservoir conditions is brought to the surface at cooler and lower pressure, the effect of compressibility of gas with pressure (for the range of pressures usually found in reservoirs) is large compared to the effect of thermal expansion of the gas caused by reservoir temperatures.

Table 6.3 Typical composition of natural gas

Components	Chemical formula	Percentage (%)
Methane	CH_4	70–90 %
Ethane	C_2H_6	
Propone	C_3H_8	0–20 %
Buthane	C_4H_{10}	
Carbon dioxide	CO_2	0–8 %
Oxygen	O_2	0–0.2 %
Nitrogen	N_2	0–5 %
Hydrogen sulphide	H_2S	0–5 %
Rare gases	A, He, Ne, Xe	trace

(Courtesy of NaturalGas.org.)

6.1.3.3   The Ideal Gas Equation and the Gas Law

In around 1812, two Frenchmen, J. Charles and J. Gay-Lussac, experimentally determined that at low pressure, the volume of a gas is proportional to its temperature,

$$Pv \propto RT \tag{6.7}$$

where,

P is the absolute pressure;
v is the specific volume;
R is the gas constant; and
T the absolute temperature.

Equation $Pv = RT$ is normally termed the *Ideal gas equation*.
The gas law of a natural gas is commonly given as,

$$PV = ZnRT \tag{6.8}$$

where P is the absolute pressure; V is the volume; n is number of moles, R is the gas constant, and T the absolute temperature; Z is the gas compressibility factor or deviation factor.

The R constant is different for different gases, and it can be determined from the universal gas constant R_u as follows,

$$R = \frac{R_u}{M} \tag{6.9}$$

where, R is the gas constant, R_u is universal gas constant; and M is the molar mass.

The universal gas constant R_u is the same for all gases, the values (in different units) of which are as follows:

$$R_u = 8.314 \, \text{kJ/(kmol} \cdot \text{K)} \tag{6.10}$$

$$R_u = 8.314 \, \text{kPa} \cdot \text{m}^3/(\text{kmol} \cdot \text{K}) \tag{6.11}$$

$$R_u = 0.08314 \, \text{bar} \cdot \text{m}^3/(\text{kmol} \cdot \text{K}) \tag{6.12}$$

$$R_u = 1.986 \, \text{Btu/(Ibmol} \cdot \text{R)} \tag{6.13}$$

$$R_u = 10.73 \, \text{psia} \cdot \text{ft}^3/(\text{Ibmol} \cdot \text{R}) \tag{6.14}$$

$$R_u = 1545 \, \text{ft} \cdot \text{Ibf/(Ibmol} \cdot \text{R)} \tag{6.15}$$

6.1.3.4   Example 6.4.

Given the gas law equation as $PV = ZnRT$ as in Eq. 6.8, if v is the specific volume of gas, and M is the total weight of gas, show that the gas density $\rho = \frac{1}{v} = \frac{pM}{ZRT}$.

6.1.3.5   Solution

Let W denote the total weight of gas, and since $n = \frac{W}{M}$,

$$PV = Z\left(\frac{W}{M}\right)RT \tag{6.16}$$

Since $v = \frac{V}{W}$, Eq. 6.16 becomes,

$$\Rightarrow \rho = \frac{1}{v} = \frac{W}{V} = \frac{pM}{ZRT}\#$$

Now, if we consider the properties of gas at two different states of P-V-T, with constant number of moles of gas, Eq. 6.8 becomes,

$$\frac{P_1 V_1}{Z_1 T_1} = \frac{P_2 V_2}{Z_2 T_2} = nR \tag{6.17}$$

The *principle of corresponding states* says that the Z factors for all gases are approximately the same at the same reduced pressure P_r and temperature T_r. We can find the values of Z factors as a function of reduced pressure P_r and temperature T_r for natural gases at various gas gravities in standard references, in particular Brown et al., 1946 [8].

6.1.3.6   Natural Gas and Refining Process

The natural gas is refined in refineries to remove impurities such as water, other compounds, and gases. Some hydrocarbons, such as propane and butane, are removed and sold separately. Impurities such as hydrogen sulfide are also removed (like propane and butane, after refining, hydrogen sulfide can be processed to produce sulfur). After complete refining, the clean natural gas is transmitted through a network of gas distribution pipelines to end users.

Natural gas is considered "dry" when it is almost pure methane, having had most of the other commonly associated hydrocarbons removed. When other hydrocarbons are present, the natural gas is called "wet." The natural gas that is delivered for our daily applications is almost pure methane. Chemically, methane is a molecule made up of 1 C atom and 4 H atoms, and is chemically referred to as CH_4. We note that natural gas is lighter than air and highly flammable, it gives off a great deal of energy and few emissions, and is much cleaner than petroleum and other fossil fuels.

Most of our world production of natural gas is transported by pipelines. An alternative for gas transportation is liquified natural gas (LNG), where the natural gas is liquified by cooling to around 161 °C, and then carried by refrigerating at 161 °C in specially designed LNG refrigerated tanks, and subject to regasification at the receiving gas terminals (regas terminals).

6.1.3.7   How Is Natural Gas Processed?

Natural gas is found near oil in the ground. It is pumped, just like oil, from wells that tap into the source and send it to large pipelines.

Because you cannot smell or see natural gas, it is mixed with a chemical to give it a stinky smell — like rotten eggs. That way, it is easy to tell if there is a leak. After the stinky chemical is added, the natural gas is sent through underground pipes, which go to your home so you can cook food and heat your house.

It is also sent to factories and power plants to make electricity. Natural gas is burned to produce heat, which boils water, creating steam, which passes through a turbine to generate electricity.

6.1.3.8   Energy from Combustion of Natural Gas

The *combustion energy* (i.e., the energy source per kg of the gaseous fuels like methane) can be estimated based on the combustion reaction together with the binding energy and its molar mass. Table 6.4 lists some common molecules together with their binding energy.

Considering methane (CH_4) gas, the principal component of natural gas, the combustion reaction is

$$CH_4(g) + 2O_2(g) \rightarrow CO_2(g) + 2H_2O\,(l) + energy \tag{6.18}$$

The combustion energy can then be calculated and obtained as 50.2 MJ/kg as shown in Example 6.5.

Table 6.4 Common molecules and their binding energy

Molecule	Binding Energy (kJ/mole)	Molar mass (kg/mole)
CH_4	-75	0.0160
C_2H_6	-85	0.0301
C_3H_8	-104	0.0441
C_4H_{10}	-125	0.0581
CO	-111	0.0280
CO_2	-394	0.0450
NO	+91	0.0300
NO_2	799	0.0400
H_2O	-242	0.0180

6.1.3.8.1 Example 6.5 Combustion energy estimation: Using Table 6.4, compute the combustion energy (heat content) from methane and propane. What is the relation between the combustion energy with the molar mass and the associated binding energy of the gaseous fuels?

6.1.3.8.2 Solution: From Eq. 8.2, for methane CH_4:

$$CH_4(g) + 2O_2(g) \rightarrow CO_2(g) + 2H_2O\,(l) + energy$$

$$\Rightarrow -75 + 0 \rightarrow -394 + 2(-242) + \text{energy}$$

$$\therefore \text{Energy} = 803\,\text{kJ/mole}$$

$$\therefore \text{Energy sourceper kg} = \frac{803\,\text{kJ/mole}}{0.016\,\text{kg/mole}} = 50.2\,\text{MJ/kg}\#$$

For propane C_3H_8,

$$C_3H_8(g) + 5O_2(g) \rightarrow 3CO_2(g) + 4H_2O\,(l) + energy$$

$$\Rightarrow -104 + 0 \rightarrow 3(-394) + 4(-242) + \text{energy}$$

$$\therefore \text{Energy} = 2046\,\text{kJ/mole}$$

$$\therefore \text{Energy sourceper kg} = \frac{2046\,\text{kJ/mole}}{(3)\text{mole} \times 0.0441\,\text{kg/mole}} = 15.5\,\text{MJ/kg}\#$$

We note that the energy content, i.e., the combustion energy,* decreases with the higher molar mass and binding energy.

6.1.3.9 Uses and Application Advantages of Natural Gas

6.1.3.9.1 Advantages:

1. Natural gas is more environmentally friendly than coal or oil. It is composed of methane, which has just one carbon, producing very low carbon emissions. Natural gas emits an estimated 70% less carbon dioxide than other fossil fuels. Natural gas burns cleaner than heating oil, and does not leave products, like ash behind.

2. Like oil, natural gas is a popular fuel, and it is the major source of energy for many consumers. Normally, it is provided through local providers and utility companies. It is conveniently pumped to homes across the country through a network of underground pipelines, or transported via gas tanks. Also, for huge fuel volume requirements, it can be transported in the form of compressed natural gas (CNG) across countries and continents via big tankers.

3. Natural gas is more cost-effective because it is in abundant supply. Generally, natural gas is more economically viable than oil or coal (although coal reserves are greater than gas reserves).

---

*We shall also note that the energy is in the form of heat and light, with heat energy as the dominant form.

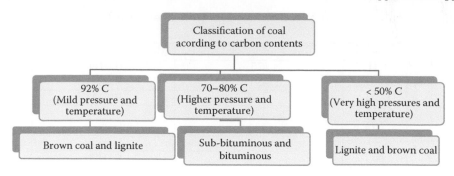

Figure 6.4
Classification of coal according to carbon contents.

6.1.3.9.2 Disadvantages

1. *Environmental impact*: Natural gas can still cause environmental impact, although it is much more environmentally friendly than oil and coal.

2. *Combustible*: Natural gas contains methane and is combustible and easily explosive if handled improperly. With a leak, the gas can build up within a room or structure. When that gas is ignited it causes an explosion. The severity of the explosion depends upon the amount of the leak. Natural gas is toxic if inhaled, which may lead to severe health risk or even death.

3. *Safety*: In the pipeline an odorant is added to the gas with a rotten-egg smell, so leaks may be detected. Homes that use natural gas should be equipped with a gas detector. Natural gas use is one of the most common causes of carbon monoxide deaths each year. Once the colorless and odorless gas is released, victims may suffocate without warning.

6.1.4 Coal

Coal is a sedimentary, organic rock normally occurring in layers called coal beds. Coal formation began during the Carboniferous Period, also known as the first coal age, around 360 million to 290 million years ago. Coal, normally hard black or brownish-black in color, is composed primarily of carbon along with variable quantities of other elements such as hydrogen, oxygen, sulfur, and nitrogen. Coal is a fossil fuel but it is far more plentiful than oil or gas, in terms of energy applications purposes. According to the World Coal Association, our Earth still has plenty of coal for our worlds consumption and other applications, with around 119 years of coal remaining worldwide (World Coal Organization) [2].

Coal is the most abundant and economical of fossil fuels; on the basis of proved reserves at end-2008, coal has a reserves-to-production ratio of about 128 years, compared with 54 for natural gas and 41 for oil [12].

6.1.4.1 Classification of Coal

We can rank coals according to their carbon content (see Figure 6.4). The lowest rank of coals are formed under mild pressure and temperature, for example, brown coal and lignite. Sub-bituminous and bituminous coal are formed at higher pressure, and under very high pressures, anthracites are formed. The anthracites contain more than 92% C, and 2–3% H together with O_2, volatile matter, and impurities such as sulfur.* Bituminous coal contained has a carbon content of 70–80% C, and about 5% H. The lowest ranks of lignite and brown coal may have less than 50% C content.

Accordingly, we must also note that we can also classify carbon based on heat content, coking properties of mechanical strength, ash content, and volatile matter content.†

---

*Sulfur (S) is an impurity in coal that can contribute to SO_2 emission during combustion and that can pollute the environment.

†Chemical composition of the coal can be determined via standard proximate and ultimate analysis. The parameters of proximate analysis are moisture, volatile matter, ash content, and fixed carbon in accordance with, e.g., the ASTM D3172 testing standard. Elemental analysis includes the quantitative determination of C, H, N, S, and oxygen within the coal.

6.1.4.2   How Is Coal Processed?

Coal is extracted from the ground by mining. Some forms of coal burn hotter and cleaner than others. Most coal is transported by trains to power plants where it is burned to make steam. The steam turns turbines, which produce electricity.

Coal is the largest source of energy for electricity generation worldwide. Unfortunately, it is also one of the largest worldwide sources of carbon dioxide emissions.

6.1.4.3   Energy and Combustion of Coal

The energy density of coal can be described by its heating value, which is around 24 MJ/kg. Combustion of coal provides electricity for many energy applications. It is an essential fuel for steel and cement production, as well as other industrial activities.

Typically, 1 kg of coal requires 7 to 8 kg of air depending upon the coal content (carbon, hydrogen, nitrogen, oxygen, sulfur, ash content, and caloric value) and combustion effectiveness for complete combustion. In practice, complete combustion will be achieved with an excess of air supplied, and it depends on the type of coal combustion equipment used.

Generally, there are three types of coal combustion equipment:

1. *Stoker fired boiler.* Stoker fired boilers (see Figure 6.5) place sized coal (coal graded according to mesh size) in a stationary or fixed grate furnace where primary air is supplied below the grate and secondary air is supplied over the grate to enhance complete coal combustion.

2. *Fluidized bed combustion* (FBC). Fluidized bed combustion is a combustion technique where fuel is fed into a solid fluidized bed, in which turbulence mixing is created and leads to better mixing of air and fuel, resulting in effective chemical reactions and heat transfer. FBC has high combustion efficiency operating at around 750–950 °C, the advantages of which include minimum thermal NOx generation and high efficiency removal of SO_2 from the products of combustion. The combustion may take place under atmospheric or high pressure in either a bubbling or a circulating fluidized bed boiler. For detailed treatment of the subject, the reader can refer to, e.g., Basu (1984) [3]. A schematic diagram of a turbo-charged fluidized bed combustion system is as given in Figure 6.6 (Elliott et al., 1998 [4]). Combustors with capacity of 70–80 MW are used widely in coal power plants.

3. *Pulverized fuel firing.* Pulverized fuel firing is a solid fuel burning technique in which powdered coal is fired and pulverized before being ignited in the flue gas stream. Thus this method can provide the maximum excess air (due to the high surface area of the coal) to ensure the most efficient and complete combustion.

6.1.4.4   Uses of Coal

Different types of coal have different uses.

Steam coal (also called *thermal coal*) is mainly used in power generation, while coking coal (*metallurgical coal*) is mainly used in steel production applications. Coal has a close relationship with electricity, steel, and cement in terms of worldwide energy applications:

- *Coal and Electricity generation.* Coal fuels over 40% of electricity worldwide.

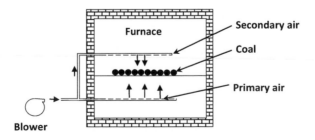

Figure 6.5
Schematic diagram of a simple stoker fired boiler.

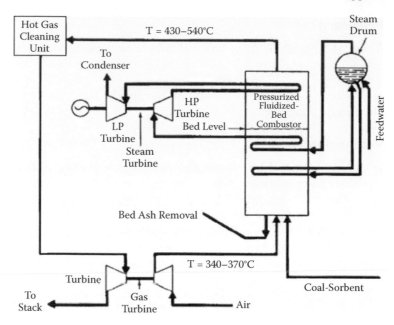

Figure 6.6
Fluidized bed combustion (FBC) system. This typical FBC has a turbo-charged pressurized fluidized bed boiler.
(After Elliott et al. [4].)

- *Coal and Steel.* Global steel production is dependent on coal. About 70–80 % of the steel produced today uses coal.

- *Coal and Cement.* Coal is an important source of the energy required in the manufacturing process of cement. It is used as a main fuel for rotary kilns, and coal ash is used for cement production.

Coal can also be used as a liquid fuel. Around 5.9 billion metric tons of hard coal were used worldwide last year and 909 million metric tons of brown coal. Since 2000, global coal consumption has grown faster than that of any other fuel. The five largest coal users — China, the United States, India, Japan, and South Africa — account for 82% of total global coal use [39].

Other important users of coal include alumina refineries, paper manufacturers, and the chemical and pharmaceutical industries. Several chemical products can be produced from the by-products of coal. Refined coal tar is used in the manufacture of chemicals, such as creosote oil, naphthalene, phenol, and benzene. Ammonia gas recovered from coke ovens is used to manufacture ammonia salts, nitric acid, and agricultural fertilizers. Thousands of different products have coal or coal by-products as components: soap, aspirins, solvents, dyes, plastics, and fibers, such as rayon and nylon.

Coal is also an essential ingredient for the following products:

- Activated carbon — used in filters for water and air purification and in kidney dialysis machines.

- Carbon fiber — an extremely strong but lightweight reinforcement material used in construction and other strong *cum* lightweight applications.

- Silicon metal — used to produce silicons and silanes, which are in turn used to make lubricants, water repellents, resins, cosmetics, hair shampoos, and toothpastes.

6.1.4.5   Emission from Combustion of Coal

During coal mining, the mining operation can disturb the land and modify the chemical properties of runoff rainwater, and affects ground water, stream, and river water quality.

During energy generation by burning coal in coal-fired power plants, the combustion of coal is a major source of CO_2 emissions from the flue gas generated. The power plant emissions also include other pollutants such as oxides of nitrogen, sulfur dioxide, particulate matter, and heavy metals (such as mercury) that affect air quality and human health (see Chapter 4, Section 4.4).

In response to increasingly stringent environmental policies and regulations, "clean coal" or "low carbon" technologies are being developed to reduce harmful emissions and improve the efficiency of these power plants. A brief discussion of the use of integrated gasification combined-cycle (IGCC) power generation for coal pollution treatment is given in Chapter 10.

6.1.5   Other Fossil Fuels

6.1.5.1   Oil Shales

"Oil shale" has potential for extraction of shale oil and combustible gas, e.g., for burning as a fuel.

Most of the oil shales are fine-grained sedimentary rocks containing relatively large amounts of organic matter usually known as "kerogen." The kerogen contains significant amounts of shale oil and combustible gas, which can be extracted by *destructive distillation*, a process involving the decomposition of the kerogen by high temperature heating to break up large molecules of the organic matter.

The organic matter in oil shale is composed mainly of carbon, hydrogen, oxygen, and small amounts of sulfur and nitrogen. It forms a complex macromolecular structure that is insoluble in common organic solvents (e.g. carbon disulphide). The organic matter (OM) is mixed with varied amounts of mineral matter (MM) consisting of fine-grained silicate and carbonate minerals. The ratio of OM:MM for commercial grades of oil shale is about 0.75:5 to 1.5:5. Small amounts of bitumen that are soluble in organic solvents are present in some oil shales. Because of its insolubility, the organic matter must be retorted at temperatures of about 500°C to decompose it into shale oil and gas. Some organic carbon remains with the shale residue after retorting but can be burned to obtain additional energy [39].

Oil shale differs from coal in that the organic matter in coal has a lower atomic H:C ratio, and the OM:MM ratio of coal is usually greater than 4.75:5.

6.1.5.1.1   Classification of Oil Shales   In 1987, Hutton developed a workable scheme for classifying oil shales on the basis of their depositional environments and by differentiating components of the organic matter with the aid of ultraviolet/blue fluorescent microscopy. His classification has proved useful in correlating components of the organic matter with the yields and chemistry of the oil obtained by retorting (Figure 6.7).

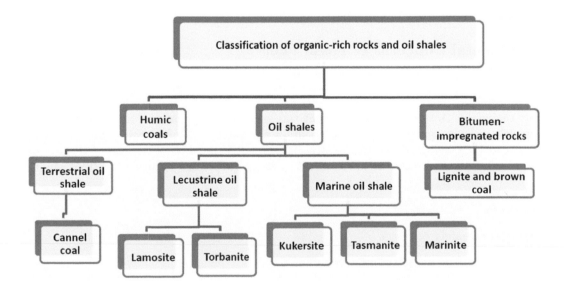

Figure 6.7
Classification of organic-rich soils and oil shales. (Modified from Hutton, 1987 [5].)

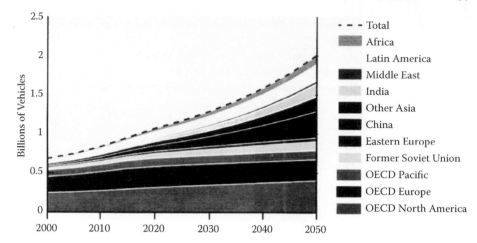

Figure 6.8
Worldwide increase in vehicles during the years 2000 to 2050. (From Kromer and Heywood 2007, [6].)

6.1.5.2 Natural Bitumen and Extra-Heavy Oil

Natural bitumen and extra-heavy oil are the remnants of very large volumes of conventional oils that have been generated and degraded, principally by bacterial action. Chemically and texturally, bitumen and extra-heavy oil resemble the residuum generated by refinery distillation of light oil. Natural bitumen and extra-heavy oil are characterized by high viscosity, high density (low API gravity), and high concentrations of nitrogen, oxygen, sulfur, and heavy metals.

According to the World Energy Council, the resource base of natural bitumen and extra-heavy oil is immense and not a constraint on the expansion of production. These resources can make an important contribution to future oil supplies if they can be extracted and transformed into usable refinery feedstock at sufficiently high rates and at costs that are competitive with alternative sources.

6.2 Environmental Concerns of Fossil Fuels

Concerns on the nature of the impact from fossil fuels are largely dependent on the specific application technology used, and include:

1. *Pollutant emissions.* According to the current rate of energy applications, the associated pollutant emission problems will worsen in the future, as a result of the rapid rate of motorization and industrialization, in particular, in China and India as well as other Asian countries. Figure 6.8 shows the worldwide increase in vehicles during the years 2000 to 2050. Table 6.5 presents the top six pollutant emitting countries in the world.

2. *Waste generation*

3. Public health and safety concerns

4. Concerns over land and water resource use

6.2.1 Technologies to Tackle Environmental Challenges from Fossil Fuels

The deployment of all energy generating technologies invariably leads to some degree of environmental impact.

Table 6.5 Top 6 emitting countries worldwide

Country	Share GHG (2005)	Share CO_2(2005)	Share CO_2(2009)
China	17 %	20 %	26 %
USA	15 %	20 %	17%
EU-27	11 %	14 %	12 %
Russia	5 %	6 %	5 %
India	5 %	4 %	5 %
Japan	3 %	5 %	4 %

(From http://www.oecd.org.)

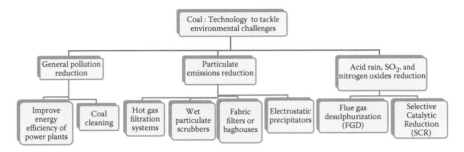

Figure 6.9
Coal techniques to tackle environmental challenges.

6.2.1.1 Coal Pollution

There are various technologies that are used by the coal industries to tackle environmental challenges associated with coal production and coal usage. The steps taken could be summarized in Figure 6.9. For general pollution reduction, two main techniques are (1) to improve the energy efficiency of coal-power plants; and (2) to provide efficient coal cleaning. Particulate emission reduction is carried out by employing a range of particulate reduction machines such as hot gas filtration systems, wet particulate scrubbers, fabric filters, and electrostatic precipitators. As to acid rain, SO2, and NOx reductions, these reductions are done by using specialized flue gas desulphurization (FGD), selective catalytic reduction (SCR) systems, etc.

General pollution reduction [2]   Technologies are now available to improve the environmental performance of coal-fired power stations for a range of pollutants. In many cases a number of technologies are available to mitigate any given environmental impact. Which technology option is selected for a power plant will vary depending on its specific characteristics such as location, age, and fuel source. The maturity of environmental technologies varies substantially, with some being widely deployed and available "off the shelf" to new innovative technologies that are still in the demonstration phase. A key strategy in the mitigation of coal's environmental impact is to improve the energy efficiency of power plants. Efficient plants burn less coal per unit of energy produced and consequently have lower associated environmental impact. Efficiency improvements, particularly those related to combustion technologies, are an active area of research and an important component of a climate change mitigation strategy.

Generally, the pollution reduction steps can be broadly divided into two main steps:

1. Coal cleaning

2. Particulate emissions reduction

Coal cleaning [2]   Mined coal is of variable quality and is frequently associated with mineral and chemical material including clay, sand, sulfur, and trace elements. Coal cleaning by washing and beneficiation* removes this

---

*Coal beneficiation refers to all types of coal cleaning. Coal beneficiation can be classified into three areas, namely: physical, chemical and biological (biochemical). Physical coal cleaning separates unwanted materials (such as dirt and sands) from the coal; chemical cleaning uses chemical to remove, in particular, organic sulfur compounds from the coal; and, biological (biochemical) cleaning involves the use of microbes, bacterias, and fungi to eat the organic (sulfur) compounds.

associated material, prepares the coal to customer specifications, and is an important step in reducing emissions from coal use. Coal cleaning reduces the ash content of coal by over 50%, resulting in less waste, lower sulfur dioxide (SO_2) emissions, and improved thermal efficiencies, which in turn leads to lower CO_2 emissions. While coal preparation is standard practice in many countries, greater uptake in developing countries is needed as a low-cost way to improve the environmental performance of coal.

Particulate emissions reduction   Particulate emissions are finely divided solid and liquid (other than water) substances that are emitted from power stations. Particulates can affect people's respiratory systems, impact local visibility, and cause dust problems.

A number of technologies have been developed to control particulate emissions and are widely deployed in both developed and developing countries [2]. These technologies include:

- Electrostatic precipitators

- Fabric filters or baghouses

- Wet particulate scrubbers

- Hot gas filtration systems

Electrostatic precipitators (ESP) are the most widely used particulate control technology and use an electrical field to create a charge on particles in the flue gas in order to attract them to collecting plates.

Fabric filters collect particulates from the flue gas as it passes through the tightly woven fabric of the bag. Both ESP and fabric filters are highly efficient, removing over 99.5% of particulate emissions.

Wet scrubbers are used to capture both particulates and SO2 by injecting water droplets into the flue gas to form a wet byproduct. The addition of lime to the water helps to increase SO2 removal.

Hot gas filtration systems operate at higher temperatures (260–900° C) and pressures (1–3 MPa) than conventional particulate removal technologies, eliminating the need for cooling of the gas, and making them suitable for modern combined-cycle power plants such as Integrated Gasification Combined Cycle (IGCC). A range of hot gas filtration technologies have been under development for a number of years but further research is needed to enable widespread commercial deployment.

Acid rain   The formation of SO_2 occurs during combustion of coals containing sulfur and can lead to acid rain and acidic aerosols, i.e., extremely fine air-borne particles. The combustion of coal in the presence of nitrogen, from either the fuel or air, leads to the formation of nitrogen oxides. The release of NOx to the atmosphere can contribute to smog, ground level ozone, acid rain, and GHG emissions (see Chapter 4.4).

During the late 20th century, rising global concerns over the effects of acid rain led to the development and utilization of technologies to reduce emissions of SO_2 and nitrogen oxides. A number of technologies, collectively known as flue gas desulfurization (FGD), have been developed to reduce SO_2 emissions. These typically use a chemical sorbent, usually lime or limestone, to remove SO_2 from the flue gas. FGD technologies have been installed in many countries and have led to enormous reductions in emissions (see, e.g., [7], [2]).

Technologies to reduce NOx emissions are referred to as either primary abatement and control methods or as flue gas treatment. Primary measures include the use of low NOx burners and burner optimization techniques to minimize the formation of NOx during combustion. These primary control measures are routinely included in newly built power stations and may also be retrofitted when reductions in NOx emissions are required. Alternatively, technologies such as Selective Catalytic Reduction (SCR) and Selective Non-Catalytic Reduction (SNCR) lower NOx emissions by treating the NOx post-combustion in the flue gas. SCR technology has been used commercially for almost 30 years and is now deployed throughout the world, removing between 80–90% of NOx emissions at a given plant [2].

Research is underway to develop combined SO_2/NOx removal technologies. Such technologies are technically challenging and expensive but new advances hold the promise of overcoming these issues.

The use of integrated gasification combined-cycle (IGCC) for coal pollution treatment is given in Chapter 10.

Other waste [2]   The combustion of coal generates waste consisting primarily of non-combustible mineral matter along with a small amount of unreacted carbon. The production of this waste can be minimized by coal cleaning prior to combustion. This represents a cost-effective method of providing high quality coal, while helping to reduce power station waste and increasing efficiency. Waste can be further minimized through the use of high efficiency coal combustion technologies. There is increasing awareness of the opportunities to reprocess power station waste into valuable materials for use primarily in the construction and civil engineering industries.

A wide variety of uses have been developed for coal waste, including boiler slag for road surfacing, fluidized bed combustion waste as an agricultural lime, and the addition of fly ash to cement.

6.3   Introduction to Alternative/Renewable Energy Resources

6.3.1   What Is Alternative or "Renewable" Energy?

Alternative energy or *renewable energy* is energy that could be derived from natural sources that are available to us, to create various energy applications. The term *natural sources* here refers to the sun, water, wind, geothermal sources, and so on.

Throughout history, humans have depended on the Sun or solar energy, both in the form of light and heat, for living; but also to provide electricity by means of solar conversion techniques in the modern world. Different methods have been used to harness energy from the Sun. For the light part of solar energy, photovoltaic cells can trap the sun's energy and convert it directly into electricity. As to the heat part of the solar energy, means can be used to convert the thermal heat to boil fluids at high temperatures, which produces steam for use in a steam generator for providing electricity (see Chapter 7). Or, the lower grade of thermal heat can be used for hot water applications. It could also be used to build up heat to generate steam in a solar-driven steam generator, for instance.

Water is also one of the important sources for clean renewable energy. The most common form of using water to get electricity is hydroelectric energy that is acquired from large river dams. Hydroelectric power stations are usually built in large river systems that have big quantifies of water with high heads and volume flow rates to provide rotation to the hydro turbine. The rotation energy from the water turbine drives the generator system to produce electricity (see Section 8.1). Smaller turbines are also used to harness smaller streams and streams from industrial waste waters (see, e.g., Section 8.1.6 for micro-hydro application). Beside hydroelectric energy there are also some other energy sources that harness water to get electricity from our ocean (Section 8.2) in the forms of ocean waves, tidal power, and ocean thermal energy.

Like most other renewable energy, wind power is also one of the rising renewable energy sectors, especially in some European countries like Denmark, the United Kingdom and Germany. Basically wind power uses the same principle found in hydroelectric turbines to convert the wind kinetic energy to electricity. There are lots of windy areas across the globe, and in many parts of the world people are trying to harness wind energy as much as possible (Section 8.3). Apart from on-land wind farms, many offshore windmills have been installed to take advantage of high wind density in our wide oceans which can produce high energy capacity in the megawatts range (see introduction on offshore windmills in Section 8.3.4).

Geothermal power is another form of renewable energy system. Geothermal energy uses the heat from deep inside the earth to produce electricity. Various methods have been developed to harness the steam from underground (see Section 8.4).

Now, nuclear energy — although often not classified as a renewable energy — is a far cleaner form of energy production than the fossil fuels. Moreover, nuclear energy can provide high MWs range capacity of energy. Currently, the use of nuclear power is on the increase around the globe, but its usage demands high energy management and safety requirements (see Section 9.1).

Like nuclear energy, biomass energy or bio-energy sometime does not classify as a pure renewable energy although biomass can be replenished quickly compared with fossil fuels. The application of biomass would be interesting, in particular, for transport fuel applications (see Section 9.2).

As the fossil fuel reserves are estimated to last 160 years or less, renewable energy systems are expected to positively and gradually eliminate our dependence on fossil fuels.

6.4   Fossil Fuels versus Alternative/Renewable Fuels for Energy Applications

Although renewables are essentially attractive, and can offer a big part of worldwide energy policy and associated programs development, the use of existing methods of energy supply, which are based on fossil (oil, gas, and coal) and nuclear fuels, are unavoidable due to the following reasons:

- The techno-economics of fossil and nuclear fuels are well-established.

- The energy density of the fossil and nuclear fuels is high, and flexible in various forms of energy applications.

- The resources are known to a reliable extent, at least for some years to come.

- The use of advanced technologies such as the horizontal well drilling technique and other enhanced recovery methods continue to improve fossil fuel production.

Some of the comparisons between fossil fuels and alternative/renewable fuels in terms of energy applications are enumerated as follows:

1. *Sustainability.* Fossil fuels are finite resources and non-renewable, and will not be sustainable according to scientific findings and estimations; see, e.g., [40]; [39]; [13].

2. *Energy density comparison.* The conventional fossil fuels have many application advantages. For instance, petroleum, in its various refined products like petroleum fuel (gasoline), kerosene, and diesel fuel, has a combination of many desirable and useful properties for our energy applications. These include wide availability with known reliable resources, high energy density (see Table 6.6), ease of transportation and storage, relative safety, as well as great versatility and flexibility suitable for almost any energy applications. Table 6.6 shows the energy density of the fossil fuels, which is high. For transportation applications, for example, biofuels such as bio-ethanol have far less energy per volume than does gasoline, so when used as a 10–20% mix with gasoline (called gasohol), more gasohol has to be purchased to make up the difference.

3. *Current energy supply, demand, and consumption scenarios*: At present, fossil fuels are far superior, with much bigger shares in terms of energy supply, demand, and consumptions. Alternative energy sources collectively supply only about 5–9% of the world's energy needs, while fossil fuels are supplying 91–95% of the world's energy resources. It is to be noted that the above figures may change in the future.

4. *Clean fuels*: Burning fossil fuels creates carbon dioxide and other greenhouse gases in the atmosphere, which contributes to global warming.

5. *Energy storage.* The challenge of alternative energy is to develop the capability to effectively and economically capture, store, and use the energy when needed. Fossil fuels have excellent means of storage.

6. *Fuel price comparison.* In general, the technology of alternative energy is still relatively expensive to replace the current energy infrastructure used for fossil fuel energy. There is, however, a gradually changing scenario on the use of renewables ([18]; [40].)

Table 6.6 Energy density of fossil fuels

Fossil fuel	Energy density (MJ/kg)
Crude oil	46.3
Diesel	46.2
Gasoline	46.4
Coal	24
Natural gas	53.6

6.5   Main Advantages of Using Alternative/Renewable Energy Sources

There are three main advantages of alternative/renewable energy, as far as the author is concerned.

1. *Sustainability.* At the moment, most of the energy we consume is produced from fossil fuels, which are not sustainable. Unlike fossil fuels, alternative/renewables promise sustainability and energy balance in terms of global energy supply and demand. Sustainable energy for the future could only be accomplished with the help of renewable energy sources, and this really is the main advantage that these energy sources have over traditional fossil fuels.

2. *Environmentally friendly.* Using renewable energy sources is not only highly beneficial from an energy applications point of view, but also from an ecological point of view, because by utilizing renewable energy we can save our environment for future generations. Unlike fossil fuels, alternative energy sources do not release large quantities of CO_2 and other harmful greenhouse gases into the atmosphere. From Chapter 4, we have already learned that excessive CO_2 emission in our environment could cause unwanted global warming and climate change.

3. *Impact on human health.* From various resources and scientific findings, it has been noted that various health problems were derived from uncontrolled emissions of fossil fuels (see Chapter 4). Unlike fossil-based fuels, alternative/renewables are clean energy with minimum emissions.

6.6   Present Limitations of Alternative/Renewable Energy

The challenge of alternative energy now is to develop the capability to effectively and economically capture, store, and use the energy when needed. The technology of alternative/renewable energy currently imposes a few main limitations, notably:

1. *Low overall capacity.* The installed alternative/renewable energy systems are still in developments around the globe; the total energy supply from renewables is far less than the world energy demand.

2. *Limitation in energy storage.* The energy density and storage capacity of renewable energy is still low, but with technology advancement is expected to grow.

3. *Expensive in energy cost.* Renewable energy applications are generally still techno-economically expensive to replace the current energy infrastructure used for conventional fossil fuels.

4. *Efficiency* and effectiveness. Generally, renewable energy is less efficient. For instance, solar power is significantly less effective in cloudy weather, and wind power highly ineffective to use during calm days.

6.7   How Can Alternative Energy Replace Fossil Fuels for Our Near Future Energy Applications?

1. In order to succeed in this world, all nations need to go hand-in-hand to stop relying solely on fossil fuels to satisfy their energy demand, and will have to focus on alternative energy sources, especially renewable energy sources, and make them more efficient and effective.

2. Energy science and research will have to continue to play its role making sure that all renewable energy sources become competitive to currently dominant fossil fuels. This is because many people still, despite the environmental disaster that threatens us, are not ready to pay more for energy than they are currently

paying. This is where science should contribute and offer solutions through different technologies that should transform renewable energy sources into highly competitive energy sources. Take for instance, that solar power is significantly less effective in cloudy weather and wind power is highly ineffective to use during calm days. But by combining a couple of renewable energy sources via hybrid technology, we can improve these drawbacks. For instance, home solar power has the ability to generate power on sunny days while home wind power can generate power only when strong wind blows, so these systems can compliment each other and make up for each other's weaknesses.

3. Energy politics, energy policy, and energy programs of the world should continue and strengthen new rules and regulations to enhance and encourage the renewable energy applications and reduce current emission levels.

4. Such change in dominance will of course not happen overnight. In some parts of the world renewable energy sources are gradually becoming dominant energy sources, but big countries, that are also big polluters, are still heavily relying on fossil fuels to meet energy demand. However, even in these big countries, the significance of renewable energy sources is constantly growing (though not as fast as many were expecting) as there are more and more funds available for new energy technologies.

5. Renewables' prices must be made competitive in relation to fossil fuel costs.

6.8 Application Projects, Questions, and Problems

Application Project

Linking knowledge with applied energy design and practice

1. *Group team project.* Suppose that your company ABC is an O&G company active in oil drilling and production activities. Your company has found two potential reservoirs and the data obtained from the geological studies are as follows:

   Reservoir A (Oil field):

   Area of reservoir field = 790 acres

   Average sand thickness = 19 ft

   Average porosity = 18%

   Average connate water saturation = 30%

   Oil formation volume factor $B_o = 1.6$

   Reservoir B (Gas field):

   Area of zone = 5000 acres

   Thickness = 200 ft

   Porosity = 15%

   Water saturation = 30%

   Gas formation volume factor $B_g = 0.0035$

   (a) Estimate how much annual earnings per year if the net selling oil price and net selling gas price are X \$ per barrel and Y \$ per stock tank cu. ft, respectively.

   (b) Due to economical constraints and prioritization of requirements, your company is going to choose *only one* project from the two potential projects. Which project you are going to choose? State and explain the reasons for your team's decision to pursue the project that you have chosen.

   Note: In attempting this team project, your company should analyze the current global market situations, and your team has to come up with latest market figures as well as realistic assumptions.

Questions and Problems

Multiple choice questions

1. What is the specific gravity of a 25°API gravity of oil?

   (A) 0.90
   (B) 0.88
   (C) 0.86
   (D) 0.85
   (E) 0.83

2. Which of the following are *not* the general petroleum fuel required properties for energy applications?

   (A) Low ash content
   (B) High heating value or Calorific Value
   (C) Low flash point
   (C) Less corrosive tendency
   (D) Low sulfur content

3. Which of the following statements does *not* correctly describe coal?

   (A) Coal is an igneous rock normally occurring in layers of non-porous rocks due to magma heat.
   (B) Coal formation began during the Carboniferous Period, also known as the first coal age, around 360 million to 290 million years ago.
   (C) Coal, normally hard black or brownish-black in color, is composed primarily of carbon along with variable quantities of other elements such as hydrogen, oxygen, sulfur, and nitrogen.
   (D) Coal is a fossil fuel but it is far more plentiful than oil or gas, in terms of energy applications purposes.
   (E) We can rank coals according to their carbon content, heat content, coking properties of mechanical strength, ash content, and volatile matter content.

4. Which of the following statements is *correct* in describing our petroleum or crude oil as fossil fuel?

   I The word "petroleum" means "rock oil" or "oil from the earth."
   II Oil, also called petroleum, is a thick, black, gooey liquid found way down in the ground, trapped in the porous and permeable bed of rock called reservoirs.
   III Petroleum or crude oil is made up almost entirely of liquid hydrocarbons, together with some solid hydrocarbons, and small amount of gases, nitrogen, and sulfur compounds.
   IV Petroleum or crude oil is split into different fractions via a process called crude oil fractional distillation in a fractionating tower.

   (A) I and II only
   (B) I and IV only
   (C) II and III only
   (D) None of the above
   (E) I, II, III, and IV

5. Which of the following statements *correctly* describes natural gas?

   (A) Natural gas is colorless with a stinky smell like rotten eggs due to mixture of hydrocarbon gases.

(B) Natural gas is formed purely of methane but it can also include a mixture of ethane, propane, butane, and other gases.

(C) Natural gas is considered "wet" when it is almost pure methane.

(D) Natural gas is heavier than air and highly flammable, which is efficient for combustion applications.

6. Fossil fuels are the conventional fuels that come mainly from oil, gas, and coal. Which of the following statements *correctly* compares the alternative fuels with fossil fuels?

I Fossil fuels are non-renewable, limited in supply, and will not be sustainable, according to scientific findings.

II Burning fossil fuels releases carbon dioxide and other greenhouse gases into the atmosphere, which contributes to global warming.

III Alternative energy sources collectively meet only about 5–9% of the world's energy needs, while fossil fuels supply most of the world's energy resources.

IV The challenge of alternative energy is to develop the capability to effectively and economically capture, store, and use the energy when needed.

(A) I and II only

(B) I and IV only

(C) II and III only

(D) None of the above

(E) All the above

7. Petroleum or crude oil is made up almost entirely of liquid hydrocarbons. But it becomes useful only when it is split into different fractions by a process called crude oil fractional distillation (see Figure 6.10). First the crude oil is heated in a furnace to a temperature of around 400° C where much of it turns into vapor. The hydrocarbons in oil vapor separate out at different levels according to their respective boiling points. Which of the following is the *correct* identity for products 1, 2, 3, 4, and 5, respectively?

(A) Petroleum gas, gasoline (petrol), jet fuel kerosene, diesel fuel heating oil, lubricating oil.

(B) Petroleum gas, gasoline (petrol), diesel fuel heating oil, jet fuel kerosene, lubricating oil.

(C) Gasoline (petrol), petroleum gas, fuel oil, diesel oil, kerosene.

(D) Gasoline (petrol), petroleum gas, fuel oil, wax candles, kerosene.

(E) Gasoline (petrol), petroleum gas, fuel oil, fuel for ships, wax candles.

Theoretical questions

1. Briefly explain the three main steps engaged in by the oil industry during oil production where oil is recovered from the oil well.

2. Name three types of principal hydrocarbon series found in petroleum, and sketch one example chemical structure for each showing the C-H structure arrangements.

3. What are the processes involved in a refinery for converting crude oil into other useful oil products? Briefly discuss the processes.

4. Name the Product 2 and Product 8 of the crude oil fractional distillator as given in Figure 6.10. Briefly explain their common applications.

5. Briefly explain the three main steps or the processes involved in a refinery for oil conversion into various oil products.

6. Explain how coals can be classified according to their carbon content.

7. Discuss any *two* of the following particulate emissions techniques.

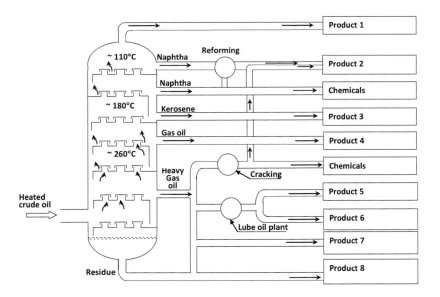

Figure 6.10
Crude oil fractional distillation. (Diagram for multiple choice question 7 and theoretical question 4.)

(**I**) Electrostatic precipitators

(**II**) Fabric filters or baghouses

(**III**) Wet particulate scrubbers

(**IV**) Hot gas filtration systems

Homework problems

1. A typical oil field has the following characteristics:

   - Area of zone = 1750 acres
   - Average thickness = 150 ft
   - Porosity = 18%
   - Water saturation = 30%
   - Oil formation volume factor $B_o = 1.65$ (reservoir bbl per stock tank bbl)

   How much volume of oil in place exists in the oil field; express your answers in oil field unit (Mbbl) and S.I. unit (cc), respectively?

   (***Answer:*** 155.51 *Mbbl*; $2.47 \times 10^{13} cc$)

2. A typical gas field has the following characteristics:

   - Area of zone = 2400 acres
   - Thickness = 180 ft
   - Porosity = 15%
   - Water saturation = 30%
   - Gas formation volume factor $B_g = 0.0035$ (reservoir cu.ft per stock tank cu. ft)

   Note: 1 ft = 30.48 cm

   Calculate the V_{STGOIP} in oil field unit ft^3 and SI unit cm^3.

(**Answer:** $5.65 \times 10^{11} ft^3$; $1.6 \times 10^{16} cm^3$)

3. A typical gas field has the following characteristics:

- Area of zone = 2500 acres
- Thickness = 150 ft
- Water saturation = 30%
- Gas formation volume factor B_g = 0.0035 (reservoir cu.ft per stock tank cu. ft)

The data available from the cylindrical core sample from the reservoir are as follows:

- Weight of dry clean sample = 311 g
- Weight of sample with pores completely filled, i.e., 100% saturated with brine = 331 g
- Sample diameter = 4.0 cm
- Sample length = 10.0 cm
- Specific gravity of brine = 1.05

(a) Compute the reservoir rock's porosity.

(b) Determine the V_{STGOIP}

(**Answer:** 15.2 %; $1.4 \times 10^{16} cm^3$)

4. The combustion energy, i.e., the energy source per kg of the gaseous fuels depends on combustion reaction together with the binding energy and its molar mass. Using Table 6.2, compute the combustion energy of the following gas fuels:

(a) methane CH_4

(b) ethane C_2H_6

(c) propane C_3H_8

(d) buthane C_4H_{10}

(e) What is the relation between the combustion energy (energy content) with increasing molar mass and the binding energy of the gaseous fuels?

(**Answer:** 50.2 MJ/kg; 23.7 MJ/kg; 15.5 MJ/kg; 11.5 MJ/kg)

Answers to multiple choice questions:

1. (A) Sp. gr.$= \frac{141.5}{(25+131.5)} = 0.90\#$

2. (C) High flash point of fuel is a required property.

3. Statement (A) Coal is a sedimentary, organic rock normally occurring in layers called coal beds.

4. (E) All the above I, II, III, and IV.

5. Statement (B) is correct. (A) Natural gas is a colorless and odorless mixture of hydrocarbon gases. (C) Natural gas is considered "dry" when it is almost pure methane. (D) Natural gas is lighter than air and highly flammable.

6. (E) All the above I, II, III, and IV.

7. (A) is correct

Bibliography

[1] Gatlin, C., (1960), *Petroleum Engineering: Drilling and Well Completion.* Englewood Cliffs, NJ: Prentice-Hall Inc.

[2] http://www.worldcoal.org/coal/, accessed on July 11, 2011.

[3] Basu, P., editor (1984), *Fluidised Bed Boilers: Design and Application.* New York: Pergamon Press.

[4] Elliott, T.C., Chen, K., Swanekamp, R.C., (1998), *Standard Handbook of Powerplant Engineering.* New York: McGraw-Hill.

[5] Hutton, A.C., 1987, Petrographic classification of oil shales. *International Journal of Coal Geology* 8, 203–231.

[6] Kromer, M. A., Heywood, J. B., Electric Power Trains Opportunities and Challenges in the U.S. Light-Duty Vehicle Fleet, Sloan Automotive Laboratory, MIT, May 2007, pp. 16, publication no. IEEE 2007-03RP.

[7] Galos, K.A., Smakowski, T.S., Szlugaj, J., Flue-gas desulphurisation products from Polish coal-fired power-plants, *Applied Energy*, Volume 75, Issues 3-4, July-August 2003, pp. 257–265.

[8] Brown, G.G., Katz, D.L., Oberfell, G.G., Alden, R.C., (1948), *Natural Gasoline and the Volatile Hydrocarbons, Section 1.* Tulsa: Mid-West Printing Co.

[9] Renewables 2010 Global Status Report. Renewable Energy Policy Network for the 21st Century, Paris.

[10] Renewables 2011 Global Status Report. Renewable Energy Policy Network for the 21st Century, Paris.

[11] Pimentel, D., 1998, Energy and Dollar Costs of Ethanol Production with Corn: *Hubbert Center Newsletter*, 98/2, M. King Hubbert Center for Petroleum Supply Studies, Golden, Colorado, p. 7.

[12] World Energy Council, 2010 Survey of Energy Resources, London.

[13] BP Energy Outlook 2030, London, January 2011.

7

Solar and Electrochemical Energy Conversion

In this chapter, we consider solar energy and electromechanical energy conversion applications. Solar energy can be divided into two parts: the light part — solar photovoltaic (PV), and the heat part — solar thermal energy. As to the electromechanical conversion applications, we shall first discuss the battery systems, and then the fuel cell.

- Solar light (Solar photovoltaic)

- Solar heat (Solar thermal)

- Battery

- Fuel cell

7.1 Introduction to Solar Energy

The solar radiance of our Sun is extremely high at its source, estimated to be about 63 MW/m^2. The solar constant, i.e., the energy from our Sun, was measured by satellite roughly at about 1353–1366 W/m^2 (Willson and Mordvinov 2003 [1]). The solar radiation arrives on the Earth's surface at a power density of only approximately 1 kW/m^2.

The useful energy can be collected, depending on location, hours of sunshine, efficiency of the solar collector, and other environmental conditions, such as cloud cover, etc. Although solar energy is free and available abundantly, the collection is not.

There are a few obstacles for solar energy harnessing:

1. Solar energy is not constantly available, therefore battery storage needs to be used for backup during low solar radiance and at night.

2. Solar energy is *diffused* such that the collection of solar energy requires a large area for effective collection and energy conversion applications (see Figure 7.1).

Solar energy can generally be classified into two types based on the sources, namely solar thermal and solar photovoltaic.

7.2 Solar Photovoltaic (PV)

7.2.1 What Is Solar Photovoltaic?

We can define solar photovoltaic as the generation of an electromotive force due to the absorption of ionizing radiation in converting solar energy to electrical energy, the process of which is usually called *photovoltaic effect*, usually denoted by PV.

7.2.1.1 What Is a Solar PV Cell?

A solar PV cell is an energy conversion device employed to produce the photovoltaic effect, i.e., via conversion of sunlight to electric energy.

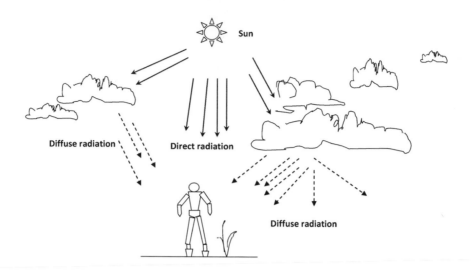

Figure 7.1
A sketch of solar radiation: direct vs. diffuse.

7.2.1.2   What Is a Solar PV Module or "Solar Array"?

Solar PV module or solar array (see Figure 7.2) refer to the combination of many individual solar PV cells into a modular form in order to increase the total electrical power output.

7.2.2   The Working Principle of a Basic Solar Photovoltaic (PV) Generator

We are all quite familiar with PV generators nowadays; the basic of the components of the systems are:

1. Solar PV array

2. Battery storage

3. Blocking diode

Figure 7.2
A typical solar module. (Courtesy of author.)

4. Inverter/convertor

5. Switches and circuit breakers

7.2.2.1 Brief Description of Components

1. The solar PV array converts the solar radiation to electric power.

2. A battery stores the electrical power generated, and is useful for power backup during periods of no solar radiation.

3. A blocking diode confines the electrical power generated to flow forward in one direction only, that is, to the battery, grid, and utility, but not vice versa.

4. An inverter/convertor device, such as a DC–AC convertor, converts the battery bus voltage to AC so that the frequency and phase are compatible and match with that of the utility grid.

5. Switches and circuit breakers allow system isolating control when necessary.

7.2.3 General Classifications and Applications of PV Systems

Based on the current worldwide market, PV technology tends to be classified into three main categories:

- *Small PV* systems, usually stand-alone, with capacity of 1–5 kW for private homes, rural electrification, communications, and other residential applications

- *Medium PV generators*, usually integrated in commercial, industrial, and office buildings, with a capacity range of between 10 and 250 kW

- *Centralized PV power plants* with capacity from 100 kW up to 5 MW or more

7.2.4 Application Advantages of a Solar Photovoltaic (PV) Power System

1. No moving parts

2. Free from carbon pollution

3. The energy source is unlimited and available globally

4. PV system can be fabricated easily

Some disadvantages of PV power systems are:

1. High initial (capital) cost

2. Low energy density, amount of energy generated per area is small

3. Output of PV is not consistent, i.e., it varies with the day and sun radiation availability.

7.2.5 Recent Advancements and Future Applications of PV Power Systems

Recent advancements and future applications of PV power can be described briefly in the following subparagraphs ([28]; [27]).

Application — Building integration **PV (or BIPV).** A grid connected PV system or BIPV decreases the price of the kWh generated with solar energy [28].

BIPV on a large scale is already technically feasible and does not present any significant hindrance in most places. Enhanced structural integration of PV systems in new buildings as well as during restoration and/or renewal of older buildings will significantly contribute to the reduction of the BOS energy and monetary costs in all decentralized installations. Innovative design and engineering solutions are being developed in order to facilitate the visual integration of PV.

***Cost reduction*:** PV costs have been declining steadily over the last two decades, with an average progress ratio of 80% (i.e., a cost reduction of 20% every doubling of production). It has been predicted that if a similar trend is maintained, PV electricity may become economically competitive in the very near future.

***Efficiency increase*:** Even if the energy return on investment of photovoltaic technologies is already quite good, a further efficiency increase is feasible and desirable for all PV technologies. In particular, a target efficiency of around 25% is reported to be foreseeable for the mid-term future of crystalline Si. As for non-silicon thin films technologies, recent estimates for the practical maximum efficiency of hetero-junction solar cells are somewhat lower than previously reported, and a practical maximum for CdTe may be achieved at approximately 16–17%; ultra-high efficiency third-generation devices are, on the other hand, expected to significantly exceed this figure by the middle of the twenty-first century.

***Storage network*:** Integration of PV with large energy storage systems will be mandatory in order to warrant the necessary stability of the network if PV is to provide a large percentage of the total electricity supply. One option that is currently being considered in this sense is represented by electrolytically produced hydrogen gas. The latter could be used as an energy buffer whereby the surplus energy generated by PV systems during peak irradiation hours is stored, only to be converted back to electricity by means of fuel-cell devices when the need arises. Another available energy storage option is pumped hydroelectric and compressed air energy storage (CAES); in fact, it has been argued that CAES may be the most readily viable large-scale energy storage solution in the medium term. Progress is also being made in the development of efficient high-speed flywheel systems whereby electric energy is converted into kinetic energy in a cylindrical or ringed mass, levitated by magnets and spinning at very high speeds ($\sim$10,000–20,000 rpm) in a vacuum chamber.

7.3   Solar Thermal Energy

7.3.1   Classification of Solar Thermal Energy Based on Motion Type of the Solar Collector

Table 7.1 shows a classification based on the three main solar collector features, i.e., motion type, concentration ratio, and operating temperature range.

The *concentration ratio* is defined as the aperture area divided by the receiver/absorber area of the collector.

7.3.2   Classifications of Solar Thermal Energy Collector/Solar Thermal Power Plant Systems

Figure 7.3 shows the general classification of solar conversion systems. The solar systems can be broadly classified into two categories, thermal heat production and electricity power generation. Brief descriptions of each collector are given below.

7.3.2.1   Evacuated-Tube Solar Collector

An evacuated-tube collector is essentially a swallow box filled with many glass tubes, usually double-walled tubes and reflectors, used to heat the fluid inside the tube. Each wall of the tubes is separated by a vacuum, where the

Table 7.1 Types of solar energy collectors

Motion	Collector type	Absorber type	Concentration ratio	Indicative temperature range ($^\circ$C)
Stationary	Flat plate collector (FPC)	Flat	1	30–80
	Evacuated tube collector (ETC)	Flat	1	50–200
	Compound parabolic collector (CPC)	Tubular	1–5	60–240
Single-axis tracking			5–15	60–300
	Fresnel lens collector (FLC)	Tubular	10–40	60–250
	Parabolic trough collector (PTC)	Tubular	15–45	60–300
	Cylindrical trough collector (CTC)	Tubular	10–50	60–300
Two-axes tracking	Parabolic dish reflector (PDR)	Point	100–1000	100–500
	Heliostat field collector (HFC)	Point	100–1500	150–2000

(After Kalogirou, 2003 [2].)

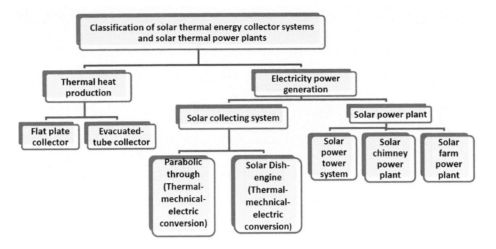

Figure 7.3
General classification of solar thermal energy collector/solar thermal power plants.

vacuum is used as an insulator to minimize heat loss.

7.3.2.2 Solar Thermal Collecting Systems

A few common designs of solar thermal collecting systems include:

Parabolic through system. Parabolic through systems concentrate the sun's energy through long, rectangular, U-shaped curved mirrors. The mirrors are tilted toward the sun, focusing sunlight on a pipe that runs down the center of the through. This heats the oil flowing through the pipe. The hot oil then is used to boil water to produce steam in order to feed a conventional steam generator to produce electricity.

Solar dish-engine system. This system uses a solar mirror dish to collect the solar radiation. The dish shape surface enables it to collect and concentrate the sun's thermal heat onto a solar receiver, which absorbs the heat and transfers the heat to the fluid. The fluid expands against a piston or turbine to drive the shaft and produce mechanical power. The mechanical power is then used to run an alternator or generator to produce electricity. Generally, the efficiency of solar disk-engine systems is low due to series conversion losses.

7.3.2.3 Solar Power Plants

In terms of solar power plant applications, there are a few common designs, which are described below.

Solar tower power plant (central receiver solar power plant). A solar power tower plant is similar to a heat engine in that it uses thousands of sun-tracking reflectors, the huge area of which is normally called a heliostat field, to direct and concentrate solar radiation onto a boiler located atop a tower (see Figure 7.4). The temperature in the boiler can rise to 500–7000°C, and the steam generated is used to drive a steam turbine for producing electricity.

Distributed parabolic collector power plant. Distributed parabolic collector system power plants are also called solar farm power plants due to their huge size. They contain a number of solar modules consisting of parabolic through systems, which are interconnected to increase the power output desired. Figure 7.5 shows the schematic diagram of a typical parabolic through power plant. The working fluid is heated in collectors and collected in hot storage tank. The hot thermo oil is used in the boiler to raise steam for the steam power plant. For backup purposes, the boiler is provided with a backup unit, which can be driven either by diesel or natural gas. The cooler oil is pumped back to the collector. According to the National Renewable Energy Laboratory (NREL), United States, the use of thermal energy storage (TES) allows parabolic trough power plants to achieve higher annual capacity factors, from 25% without thermal storage up to 70% or more with TES.

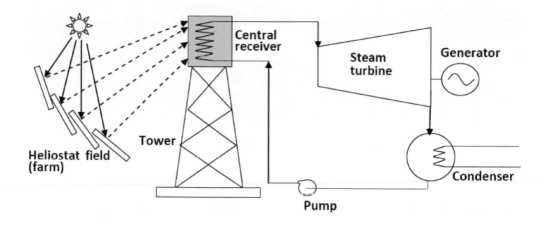

Figure 7.4
Schematic of a standard solar tower power plant (central receiver solar power plant).

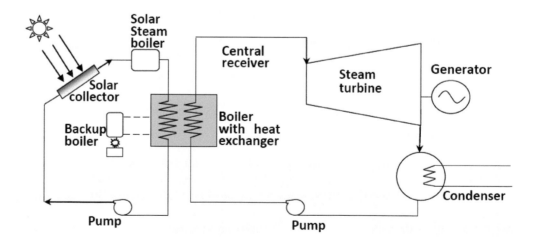

Figure 7.5
Schematic of a distributed parabolic through solar power plant.

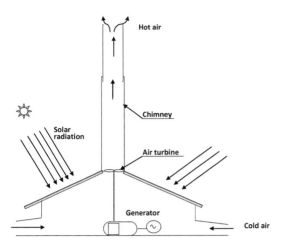

Figure 7.6
Schematic of a solar chimney power plant (SCPP).

Solar chimney power plant (SCPP). As shown in Figure 7.6, the air flows streams are heated by the solar radiation through the transparent cover. The hot air thus generated flows through the tall chimney structure due to pressure drop. The hot air stream flows through an air turbine, which drives the generator to produce electricity.

The chimney efficiency can be calculated by (Mullett, 1987 [3]):

$$\eta_{chimney} = \frac{gH}{c_p T_0} \tag{7.1}$$

Where,
g is the gravity
H is the chimney height (m)
c_p is the air heat capacity (J/kg·K); and
T_0 is the ambient temperature (K).

Example 7.1. For a chimney height of 1000 m operating at standard conditions for temperature and pressure, the chimney efficiency achieves the maximal value of 3%. Considering a collector efficiency (η_c) of 65% and a turbine efficiency (η_{tur}) of 78%, calculate the overall system efficiency.

Solution:

$$\eta_{overall} = \eta_{chimney} \cdot \eta_{collector} \cdot \eta_{turbine} = 0.03 \times 0.65 \times 0.78 = 0.014 = 1.52\% \; \# \tag{7.2}$$

The SCPP has notable advantages in comparison with other power production technologies, namely (Schlaich, 1995 [4]):

1. The collector uses both direct and diffuse radiation;

2. The ground provides a natural heat storage;

3. The low number of rotating parts ensure its reliability;

4. No cooling water is necessary for its operation;

5. Simple materials and known technologies are used in its construction; and

6. Non-OECD countries are able to implement such technology without costly technological efforts.

Table 7.2 Industrial demand and applications of solar energy

Industry	Process	Temperature (°C)
Dairy	Pressurization	60–80
	Sterilization	100–120
	Drying	120–180
	Concentrates	60–80
	Boiler feed water	60–90
Tinned food	Sterilization	110–120
	Pasteurization	60–80
	Cooking	60–90
	Bleaching	60–90
Textile	Bleaching, dyeing	60–90
	Drying, degreasing	100–130
	Dyeing	70–90
	Fixing	160–180
	Pressing	80–100
Paper	Cooking, drying	60–80
	Boiler feed water	60–90
	Bleaching	130–150
Chemical	Soaps	200–260
	Synthetic rubber	150–200
	Processing heat	120–180
	Pre-heating water	60–90
Meat	Washing, sterilization	60–90
	Cooking	90–100
Beverages	Washing, sterilization	60–80
	Pasteurization	60–70
Flours and by-products	Sterilization	60–80
Timber by-products	Thermo diffusion beams	80–100
	Drying	60–100
	Pre-heating water	60–90
	Preparation pulp	120–170
Bricks and blocks	Curing	60–140
Plastics	Preparation	120–140
	Distillation	140–150
	Separation	200–220
	Extension	140–160
	Drying	180–200
	Blending	120–140

(After Kalogirou, 2003 [2].)

7.3.3 Application Advantages of Solar Thermal Energy Conversion Systems

1. Solar thermal energy is low cost and has been used for generations.

2. Solar thermal energy is basically freely available at many locations of the globe.

3. Solar thermal energy conversion systems are suited for a wide range of industrial applications. Table 7.2 illustrates some of these.

7.3.4 Application Disadvantages and Technical Concerns of Solar Thermal Energy

Solar energy is basically freely available at many locations of the globe, but three main questions remain. The technical concerns are:

- How to collect it efficiently?

- How to store it?

- How to minimize energy losses from the series conversions, i.e., solar thermal →mechanical→ electrical energy?

7.3.5 How to Estimate Solar Energy Output

7.3.5.1 Example 7.2

Suppose the solar chimney power plant in Figure 7.7 has a $100\,m^2$ roof surface with incidence angle of 60° at about 9:00 in the morning. Given that the solar radiation arrives on the surface at power density approximately 0.9 kW/m²:

(a) Calculate power collected by the surface.

(b) Calculate power collected by the surface at 12:30 pm, but now with an incidence angle of 30°.

7.3.5.2 Solution:

(a) The power normal to the earth's surface P_N can be calculated as,

$$P_N = I_N \times A \tag{7.3}$$

where,
P_N = Power (Watts);
I_N = power density (W/m^3);
A = Area (m^3).
N denotes normal to the surface.

$$P_N = \{(0.9 \times 10^3)(W/m^3) \times cos(60^o)\} \times 100\,m^3$$

$$P_N = 0.045\,\text{MW}\#$$

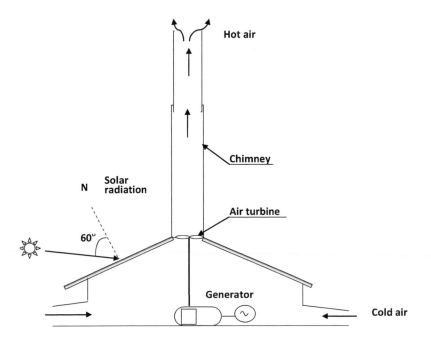

Figure 7.7
Example of solar chimney plant power output.

(b) Substituting the values in the equation, we obtain,

$$P_N = \{(0.9 \times 10^3)(W/m^3) \times cos(30^o)\} \times 100\, m^3 = 77,942\,\text{kW} \approx 78\,\text{MW} \#$$

We can notice that as the incidence angle becomes smaller and near to the normal line of the surface, more power can be expected due to higher solar intensity that can be absorbed by the surface.

7.3.6 *Case Study 1.* Solar Energy Application: A Simple Solar Clothes Dryer

As solar energy is available around the globe and is cheap, this form of energy can be utilized for our drying requirements as well as other daily activities. The solar drying systems can be classified according to their heating modes and how the solar heat is utilized, i.e., passive (natural-circulation) or active or forced-convection (hybrid solar dryer).

A solar clothes dryer has been designed, constructed, and experimentally tested by Abdullah et al. [5] to show the application of solar thermal energy for clothes drying. Figure 7.8 shows the processes involved, i.e., solar radiation, airflow, and ventilation in the solar clothes dryer. Solar radiation through the transparent sheet causes temperature to rise internally due to the greenhouse effect. A ventilation process either in the form of natural ventilation or forced ventilation helps in removal of the humid dense air through the holes and fan located at the bottom of the structure. Drying occurs when relatively low-humidity air passes over the wet cloth surface, picking up moisture which is then carried away through ventilation. The buildup of internal heat flux helps in the separation of the moisture from the cloth surface through enhancement processes of evaporation, convection, and diffusion.

Basic design consideration   Direct solar gain and ventilation gain are two important parameters for design consideration of the enclosure of the clothes dryer. Direct solar gain through transparent elements is estimated by the following equation:

$$Qs = G \times A \times Sgf \tag{7.4}$$

where

Q_s = total direct solar gain, Watts
G = the total solar radiation on the dryer, W/m^2
Sgf = solar gain factor.

The solar gain factor, Sgf is a function of the type of material and represents the amount of direct radiation that actually makes it through the element and into the dryer.

Ventilation gain was estimated by:

$$Q_v = 20 \times N_a \times V \times \triangle T \tag{7.5}$$

Where,

Q_v = total ventilation gain, Watts;
N_a = the number of air changes per minute within the dryer;
V = the total internal volume of the dryer, m^3;
$\triangle T$ = temperature difference between inside and outside of dryer, °C.

The value 20 in Eq. (7.5) resulted from the fact that moist air is assumed to have a volumetric heat capacity of 1200 Jm−3 K and that the flow rate is given as a factor of air changes per minute. Since 1 W equals 1 J/s, then (1200 J/m^3 K)/(60 min) = 20 J/m^3K. min.

Simulation.   For the simulation part, airflow modeling is described by the classic Navier–Stokes equations:

$$\frac{\partial}{\partial t}(\rho\varphi) + \nabla\left(\rho\upsilon\varphi - \psi^\varphi\nabla\varphi\right) = S^\varphi \tag{7.6}$$

[transient] + [convection] − [diffusion] = [heat source]

Heat flux due to solar radiation is

$$Q_k = \varepsilon_k A_k \sum_{j=1}^{n} G_{jk}\sigma(T_j^4 - T_k^4) \tag{7.7}$$

where

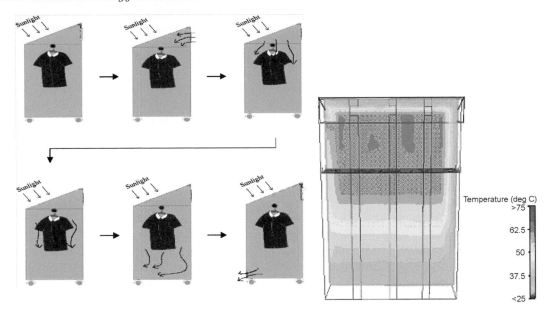

Figure 7.8
The clothes dryer and solar thermal heat application. (*Left*): The flow pattern of air inside the solar clothes dryer. (*Right*): The temperature buildup in the clothes; temperature profile developed at time $t = 180$ minutes. (From Abdullah et al. *International Journal of Thermal Sciences*, 45(2006)1027–1034. With permission.)

$Q_k=$ the net heat-flux arriving at the surface k for a general system of N radiant surfaces;

$G_{jk}=$ the proportion of radiative energy emitted by surface k that makes its way to surface i by all possible routes.

The simultaneous equations thus formed are solved iteratively to produce a solution that satisfies the conservation law of mass, momentum, and energy.

The drying performance result, from both experimental works and simulation study, shows that a solar dryer supplemented with forced-ventilation has a higher drying rate of 0.43 kg/h and shorter completion drying time of 3 hours compared to a solar dryer with natural ventilation, which has a drying rate of 0.25 kg/h and drying time of more than 4 hours for a typical day. Consistent with the experiments, the preliminary computational results based on the Navier–Stokes thermal transient simulation technique are able to demonstrate the temperature rise prevailing in the solar *cum* forced-ventilation system associated with the internal heat flux due to solar radiation and moisture removal. Good agreement is found between the numerical simulation and the corresponding experimental measurements. Parametric study by simulation was conducted subsequently to investigate the feasibility of employing higher capacity fans, i.e., higher forced-ventilation rates, on energy stored if the current system operated under the same ambient weather conditions. Initial comparison studies based on both experimental works (natural ventilation versus forced ventilation) and simulation (temperature rise versus forced ventilation rates) have shown that the application of higher ventilation rates would possibly further improve the dryer's performance and cause an increase of local heat flux and temperature rise, especially in the early stage of drying.

7.3.7 Recent Advancement and Future Applications of Solar Thermal Power

Solar thermal power can be harnessed for hot water heating, space heating and drying, and other commercial and industrial uses. The applications include the following:

1. Water heating for domestic, commercial and industrial use

2. Space heating and drying

3. Solar cooling through absorption and adsorption

4. Solar water pumping

5. Solar power generation

6. Solar CHP applications

The summary of recent advancements and future applications of solar thermal power are as follows:

Solar thermal for the textile industries.   Solar thermal is also used in the textile industry for heating water at temperatures close to 100 °C for bleaching, dyeing, and washing purposes. This can replace, in part or in whole, the fossil fuels used in most of the current fuel-powered textile industry. For instance, built-in solar storage water heaters were introduced in Pakistani textile plants for fuel saving and to improve overall stability of the solar heating systems [6].

Solar thermal for the food and agriculture industries.   At low grade thermal energy applications, the conventional solar dryers can be used extensively in the food and agriculture industries to improve both quality and quantity of production while reducing the wastes and minimize environmental problems. Industries that involve a drying process usually use hot air or gas with a temperature range between 140 °C and 220 °C. For higher thermal power applications, solar thermal systems must be integrated with other energy supplies to meet the system requirements [2]. Also, the improved heat storage capability of the system coupled with intelligent optimization control is expected in the near future.

SCPP.   In 1981, an SCPP plant was constructed in Manzanares, Spain with a collector diameter of 240 m, and a chimney diameter and chimney height of 10 m and 195 m, respectively. The plant successfully produced 50 kW of electricity.

SCPP for rural villages was studied in 2006 by Onyango and Ochieng [7], who emphasized some features for power generating. They disclosed that for a temperature ratio = 2.9 (i.e., the difference between the collector surface temperature and the temperature at the turbine to the difference between the air mass temperature under the roof and the collector surface temperature), 1000 W of electric power can be generated. The minimum dimension of a practical by a reliable SCPP to assist approximately fifty households in a typical rural setting has been determined to be chimney length = 150 m, height above the collector = chimney radius = 1.5 m. Ming et al. (2007) [8] developed a few mathematical models for heat transfer of the SCPP collector, chimney, and turbine, and validated them with the Manzanares SCPP prototype. They concluded that an output power of SCPP can exceed 10 MW. Peng et al. (2007) [9] developed a new mathematical model based on the relative static pressure concept, with optimized geometric dimensions of the collector outlet and the chimney inlet. As a result, the air flow velocity was enhanced (temperature profiles becoming more uniform, the overall energy losses reduced, and the associated relative static pressure decreased about 50%), resulting in a 14% increase in SCPP energy conversion performance. Ming et al. (2008a) [10] continued their work, carrying out numerical simulations analyzing characteristics of heat transfer and air flow in the solar chimney power plant system with an energy storage layer including the solar radiation and the heat storage on the ground. They concluded that the ground heat storage depends on the solar radiation incidence. Higher temperature gradients also increase the energy loss from the ground (Ming et al., 2008b [11] and Ming et al., 2008c [12]). Subsequently, Tingzhen et al. (2008) [13] included a three-blade turbine in the model simulation and validated the model, power output, and turbine; efficiency of 10 MW and 50%, respectively were reported. The SCPP energy/energy balance studies were carried out by Petela (2009) [14], for the energy distribution in the components for an input of 36.81 MW energy of solar radiation with an equivalent input of 32.41 MW of radiation energy, including a sensibility analysis.

From all the SCPP studies cited above, we can see that the performance of SCCP is very low, and that there is a possibility of further advancement of SCPP systems with improvement of control strategy and air flow optimization.

Evacuated tubes and Organic Rankine Cycle (ORC) turbines   Evacuated tubes are widely used in Europe, and in Japan, China, and other Asian countries. The evacuated tubes are mostly suited for solar hot water production, as well as high temperature production applications up to 185 °C. At such temperatures, the thermal heat generated is suitable for use to power organic Rankine cycle (ORC) turbines, with an overall efficiency of 10–13%.

Solar chimney technology   Solar chimneys have a capacity factor of about 10% and efficiency of about 0.53% to 1.3%. The overall efficiency of the solar chimney thermal power plants increases with plant size.

Hybrid solar thermal and other waste heat   Also, hybrid solar chimney waste heat from industrial flue gas sources (see, e.g., Romero, 2007 [15]) may be a future application advantage.

7.3.7.1   Application Examples of Solar Thermal Power Plants and PV-Powered Generators around Our World

7.3.7.1.1   *Solar Thermal Power Plants*

1. The 10 MWe SOLAR ONE pilot demonstration plant was operated in California from 1982 to 1988 (with steam as the heat transfer medium).

2. The 10 MWe SOLAR TWO plant, rebuilt from the 10 MWe SOLAR ONE pilot demonstration plant. The 10 MWe SOLAR TWO plant was a grid-connected system, which was successfully operated with a molten salt-in-tube receiver system and a two-tank molten salt storage system from 1997 to 1999. (> 15 % annual solar-to-electric plant efficiency, and > 90 % annual plant availability.) The plant has more than 936,000 mirrors, covering an area of more than 6.5 km^2.

3. Solar Energy Generating Systems power plant, California, USA, 354 MW.

4. Solnova Solar thermal power station, Sanlúcar la Mayor, Spain, 150 MW. The overall plant was completely built in 2010.

5. The Plataforma Solar de Almería (PSA) in Spain. The PSA was built in 2010 in the province of Almería. The plant has more than 20,000 m^2 of mirrors, installed on a 0.4 km^2 site.

7.4   Introduction to Electrochemical Energy Conversion

In this section, we will consider electrochemical energy conversion. Of particular importance is that the electrochemical or electro-mecha-chemical conversion systems are of two types, i.e., battery and fuel cell.

7.5   Battery

Battery — a key clean and fuel-efficient *enabling generator* for many applications, in particular portable energy systems and backup applications for renewable energy systems such as solar, wind, hybrid electric vehicles, and fuel cell electric vehicles.

7.5.1   What Is a Battery?

A battery is an energy device that converts the chemical energy it stored directly into electric energy via an electrochemical process that involves oxidation and reduction reactions. Figure 7.9 shows three types of common batteries for our everyday use.

7.5.1.1   What Are the Basic Components of a Battery as an Energy Generator?

A battery can be made up of many basic electrochemical units called "battery cells" or just "cells." The cells can be connected and arranged in series or parallel according to the required output voltage and current, thus the battery capacity.

Each cell consists of three major components:

1. *The anode (negative electrode).* It is the reducing electrode that gives up electrons to the external electric circuit and is oxidized during the electrochemical reaction; thus the anode electrodes is also called a "fuel electrode."

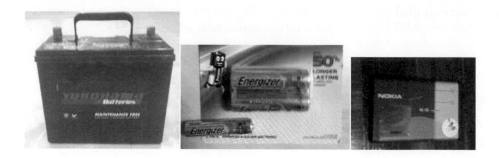

Figure 7.9

Three types of common batteries (from left to right): A typical car battery, a conventional touch light/toy battery, and a typical mobile phone battery. (Courtesy of author.)

2. *The cathode (positive electrode).* The cathode is the oxidizing electrode that accepts electrons from the external circuit and is reduced during the electrochemical reaction.

3. *The electrolyte (electrode medium).* The electrolyte is the medium or separator in between the anode and cathode, which serves as an ionic conductor for transfer of ions or charges between the two electrodes. It is usually a liquid with acids or alkalis to impart ionic conductivity. We must note that some batteries use solid electrolytes, thus we usually called them "dry cell" or "dry batteries."

A photo of a typical dry cell is shown in Figure 7.10. The various components of a typical dry cell are illustrated schematically in Figure 7.10b. Usually zinc is used as an anode and graphite (carbon) rod as a cathode, surrounded by moist electrolyte (chemical mix).

7.5.2 Classifications of Batteries

Generally, batteries can be classified into primary battery, secondary battery, and fuel cell (Figure 7.11). Primary batteries cannot be recharged, while secondary batteries are rechargeable. Unlike primary and secondary batteries, fuel cells, on the other hand, are a class of battery that operate with a continuous external fuel supply. Therefore, the amount of energy for a primary battery is limited to the reactants available in it; a secondary battery can operate in an intermittent mode, i.e., it can be rechargeable upon reaching low-charge level. A fuel cell essentially has the greatest amount of energy content, as the fuel, usually hydrogen, can be continuously supplied to the cells, thus "fuel" cell.

7.5.2.1 Dry cell battery

Dry cell batteries are of two types: the LeClanché cell and the "alkaline" cell battery. As the electrolyte used is somewhat "dry" or in paste/gel form, a dry battery is suitable for portable power applications, such as flashlights,

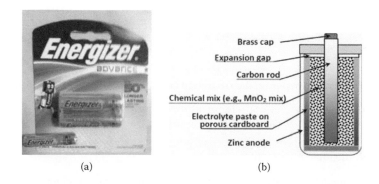

(a)   (b)

Figure 7.10

The "dry" cell: (a) Typical dry cell; (b) the schematic of a typical dry cell.

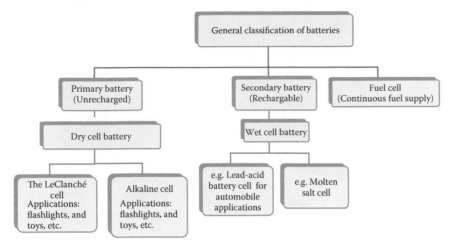

Figure 7.11
General classification of batteries.

toys, emergency power applications, etc.

The LeClanché cell. The LeClanché cell was invented by Georges Leclanché in 1866. The electrochemical reactions are,

$$Zn \rightarrow Zn^{2+} + 2e^- \tag{7.8}$$

$$2MnO_2 + 2H + +2e^- \rightarrow Mn_2O_3 + H_2O \tag{7.9}$$

"Alkaline" battery. The alkaline zinc–manganese dioxide or "alkaline" battery offers much higher energy density and hence capacity than either the carbon–zinc or zinc-chloride type. It is also capable of higher discharge current levels. The manganese dioxide (MnO_2) and carbon form the positive electrode, while the zinc is in powdered form as the negative electrode (anode), which is actually mixed to form a gel/paste with potassium hydroxide (KOH) as the electrolyte. Despite the fact that the alkaline battery is very expensive and somewhat heavier, it is commonly referred to as the "premium primary battery," due to the fact that it is superior to the carbon–zinc or zinc chloride types. Also, alkaline batteries are known to be long lasting due to their ability to avoid the corrosive effects due to acidic ammonium ion on the zinc. Alkaline batteries are especially suitable for applications that involve comparatively high-discharge current levels.

Lithium battery. The lithium–manganese dioxide primary battery is a relatively recent development, taking advantage of the high electrode potential and energy density of lithium metal. It offers considerably greater energy density and capacity than an "alkaline" battery, for a relatively small increase in cost. The lithium is in the form of very thin foil and is pressed inside a stainless steel can, to form the negative electrode. The positive electrode is manganese dioxide, mixed with carbon to improve its conductivity, and the electrolyte is lithium perchlorate dissolved in propylene carbonate. The nominal terminal voltage of the lithium cell is 3.0 V, twice that of "alkaline" and other primary batteries. It also has a very low self-discharge rate, giving it a very long shelf life. The internal resistance is also quite low, and remains so during the working life. The lithium battery performs well at low temperatures, usually below 60 °C, and advanced types are used in communications satellites, space vehicles, military and medical applications. Medical applications that require long-life critical devices, such as artificial pacemakers and other implantable electronic medical devices, use specialized lithium-iodide batteries that can run for years. Lithium batteries are also suitable for less critical applications for operating toys, clocks, and cameras. Although lithium batteries are more expensive in cost, they will provide longer life than "alkaline" batteries and minimize battery replacements.

Other common dry batteries include zinc–chloride batteries, mercury batteries, silver oxide batteries, zinc–air batteries, etc.

Figure 7.12 (SEE COLOR INSERT)

A few new and used automobile lead acid batteries. (Courtesy of author.)

7.5.2.2 Wet Cell Battery

A wet cell battery, such as lead-acid battery cell, has a liquid electrolyte. It is usually used in automobile batteries.

Lead-acid battery. The lead-acid battery was invented by Gaston Planté in 1859 and is widely used for automobile applications. Figure 7.12 shows a photograph of a few car lead-acid batteries. In a lead-acid battery, the positive electrode is lead dioxide, while the negative electrode is metallic lead. Lead acid batteries usually employ sulfuric acid as the electrolyte. As the cell discharges, the acid electrolyte is consumed, producing water, and both electrodes change into lead sulphate. When the cell is recharged, the chemical reaction reverses.

The anode reaction is,

$$Pb + HSO_4^- \rightarrow PbSO_4 + H^+ + 2e^- \qquad (7.10)$$

The metal lead (Pb) in the anode reacts with the ionized sulfuric acid (HSO_4^-) to produce lead sulphate ($PbSO_4$), hydrogen ions (H^+) in solution, and two electrons (e^-).

At the cathode, the reaction is,

$$PbO_2 + HSO_4^- + 3H^+ + 2e^- \rightarrow PbSO_4 + 2H_2O \qquad (7.11)$$

in which the lead dioxide (PbO_2) reacts with the ionized sulfuric acid (HSO_4^-) and the available hydrogen ions, and some excess electrons from the anode via the external electric circuit (thus generating electric current), and to produce lead sulphate ($PbSO_4$) and water.

The overall cell reaction is,

$$\text{Pb(s)} + \text{PbO}_2\text{(s)} + 2\text{H}_2\text{SO}_4\text{(aq)} \rightarrow 2\text{PbSO}_4\text{(s)} + 2\text{H}_2\text{O} \qquad (7.12)$$

A dry cell has a paste-type electrolyte. It is suitable for portable electric device applications. An example is a zinc-carbon battery.

A molten salt battery uses molten salt as its electrolyte. Molten salt batteries have high energy density, making them suitable for electrical vehicle applications.

7.5.3 Cell Performance

Primary cells   The performance (effective energy capacity) of primary cells depends on a number of factors, such as:

- The amount of electrochemical energy source stored in the cell

- Type of materials used

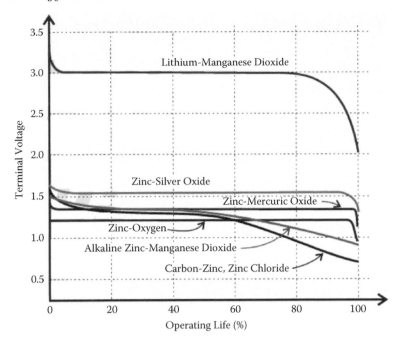

Figure 7.13
Discharge curves for some common primary batteries. (Courtesy of Jaycar Electronics.)

- The quantities of materials used

- The operating temperature

- The discharge rate, i.e., the rate at which the energy is drawn from the battery

We can see that different types of batteries deliver different levels or amounts of energy. Larger batteries are designed to deliver more energy than the smaller ones. That explains why many of the different types of primary batteries are constructed and packaged in different standard sizes: "D" cells (34 mm$\phi\times$61 mm), "C" cells (26 mm$\phi\times$50 mm), "AA" (14.5 mm$\phi\times$50.5 mm), "AAA" cells (10.5 mm$\phi\times$44.5 mm), 9V battery (26.2 mm$\times$17.5 mm$\times$45 mm), and button cells with different sizings, etc.

Now, in an ideal battery, the terminal voltage will be constant and plateau over the entire discharge period, i.e., until the battery is finally fully discharged, the voltage would drop sharply; in practice, however, the terminal voltage decreases as the charge reduces. That is the reason why, unlike secondary batteries, primary batteries are generally not given a capacity specification in either ampere-hours or milliamp-hours by most manufacturers; instead, only a maximum discharge current is usually given. Figure 7.13 shows typical discharge curves for some common primary batteries.

Example 7.3   Based on the discharge curves given in Figure 7.13:
(A) What is the nominal terminal voltage of a lithium battery ("alkaline" battery)?
(B) What are the advantages of lithium batteries as compared to the "alkaline" batteries or other common dry batteries?

Answer: (A) From the plots, the nominal terminal voltage of the lithium battery and the "alkaline" battery are 3.0 V and 1.5 V, respectively.
(B) Lithium batteries offer considerably greater energy density and capacity than "alkaline" batteries and other primary batteries; they deliver higher (about twice) terminal voltage compared to other primary cells, and the terminal voltage remains almost flat during the entire discharge life.

7.5.4 Application Advantages and Disadvantages of Batteries

1. As an efficient backup energy generator

2. Compact and handy for everday use, e.g battery for mobile phone, and so on.

Disadvantages:

1. Many batteries do not hold a lot of power.

2. Waste batteries contribute to landfill and the chemical/acid inside may cause pollution.

3. Many batteries contain highly corrosive electrolytes like potassium hydroxide or caustic potash.

4. Some batteries contain toxic materials such as mercuric oxide or highly reactive materials like lithium, which can explode on contact with water.

7.5.5 How to Estimate Batteries' Energy Output

The power in a battery can be estimated as follows. Let us consider a car battery in the following example:

Example 7.4 A typical car battery is rated at 45 amp/hours and 12 volts.
(A) Estimate for how long the battery can deliver an amperage of 2.25 amps.
(B) In actual practice, what is the most likely start-up voltage, ending voltage, and average voltage if one measures the corresponding voltage with a multimeter?
(C) What is the total power of the battery?

Answer: (A) The battery can deliver 2.25 amps for 20 hours.
(B) We shall notice that the battery will have a start-up voltage range of about 12.9 to 13 volts (7.5%–8% surge from the rated voltage) and drop to ~11 volts (around 8% reduction of rated voltage), with an average of about 12 volts during the running period.
(C) The power delivered by the battery in 1 hour is,

$$Power = ampere \times average\,voltage = 2.25 \times 12 = 27\,Watts$$

Therefore, total power for 20 hours is,

$$27\,watts/hour \times 20\,hours = 540\,Watts = 0.54kW\#$$

7.6 Fuel Cell

An energy conversion technology that has proven highest efficiency!

7.6.1 What Is a Fuel Cell?

A fuel cell is an electrochemical cell similar to a primary battery or a secondary battery that converts chemical energy from a fuel directly into electric energy; but unlike the batteries, the energy conversion process is continuous in a fuel cell due to a continuous external fuel supply.

7.6.2 Fuel Cell Classifications

7.6.2.1 Fuel Cell Classification I: Based on Electrolyte

Figure 7.14 shows a classification of 6 main fuel cell types based on electrolyte used.

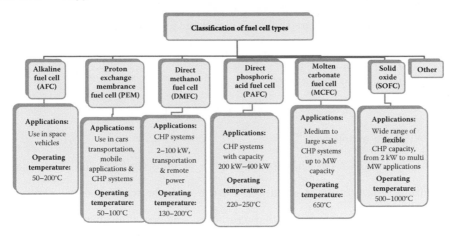

Figure 7.14
Fuel cell classification I: based on electrolyte.

Alkaline fuel cell (AFC). AFC was used for space applications, e.g., on the Apollo and Shuttle Orbiter craft. AFC must use pure hydrogen and oxygen, and be free from CO_2.

The general reaction for AFC is as follows:

At the anode, alkali hydroxyl OH^- reacts with hydrogen to form water, and releases electrons, i.e., electric energy,

$$2H_2 + 4OH^- \rightarrow 4H_2O + 4e^- \tag{7.13}$$

And at the cathode, oxygen reacts with electrons and water in the electrolyte to form new OH^- ions,

$$O_2 + 4e^- + 2H_2O \rightarrow 4OH^- \tag{7.14}$$

The advantages of AFC are:

1. Works at low temperature and with simple operations.

2. Fast startup.

3. High efficiency, and low operating cost.

4. It uses very little amount of catalyst, thus lowers costs.

5. It has low weight and volume, thus compact in size.

The disadvantages of AFC are:

1. It is extremely intolerant of CO_2 (up to 350 ppm) and shows a certain intolerance to CO. This limits the type of both oxidant and fuel.

2. Oxidant must be pure oxygen or air free of CO_2. The fuel must be pure hydrogen.

3. AFC requires use of corrosive liquid electrolyte, e.g., KOH.

4. It has a relatively short lifetime.

Phosphoric acid fuel cell (PAFC). PAFC was the first type of fuel cell that reached the commercialization stage. For instance, some PAFC systems around 200 kW have been installed by American and Japanese companies. PAFC solves the hydrogen fuel problems by employing alternative fuel sources via natural gas reforming, i.e., methane gas and CH_4.

The PAFC make use of highly concentrated phosphoric acid, $H_4P_2O_7 \geq 95\%$, as the electrolyte. PAFC uses gas diffusion electrodes with Pt supported on a porous carbon at the anode and cathode. The general reaction for PAFC is:

At the anode, gas hydrogen is oxidized, and releases electrons, i.e., electric energy,

$$2H_2 \rightarrow 4H^+ + 4e^-$$ (7.15)

At the cathode, oxygen is reduced to form water,

$$O_2 + 4H^+ + 4e^- \rightarrow 2H_2O$$ (7.16)

The advantages of PAFC are as follows:

1. PAFC can tolerate up to 30% CO_2, therefore it can use air directly from the atmosphere.

2. Operating at medium temperature, it is suitable for use with the waste heat for cogeneration.

3. PAFC can use an electrolyte with stable characteristics, and low volatility even for temperatures above 200 °C.

The disadvantages of PFAC are:

1. It has a maximum tolerance of 2% CO.

2. It utilizes liquid electrolyte that is corrosive at average temperatures, which involves handling and safety problems.

3. PAFC cannot auto-reform fuel.

4. PAFC is generally slow in start-up, i.e., it needs to reach a certain temperature before starting to work.

5. Generally, it is big in size, bulky, and heavy.

Direct methanol fuel cell (DMFC). A direct methanol fuel cell (DMFC) directly converts the chemical energy stored in methanol to electricity [17]. DMFC employs steam reforming of methanol at a somewhat low temperature of 200 C. This is an advantage as methanol is in liquid form at room temperature, thus providing easy fuel handling, transportation, and storage. This is especially useful for transportation or rural mobile power applications.

DMFC can be subdivided into two types, i.e., active and passive DMFC.

The DMFC is more competitive with conventional battery technologies, a passive DMFC (see Figure 7.15), has no auxiliary liquid pump, gas blower or compressor, but relies on diffusion and natural convection to supply the fuel and oxygen [17].

The general related reactions are as follows:

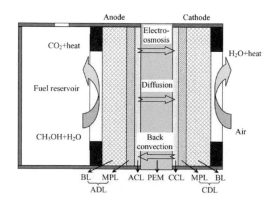

Figure 7.15
Schematic of MEA and heat/mass transport in passive DMFC. (From Zhao et al. 2009 [17].)

Figure 2.3
An offshore drilling platform. An offshore structure where typical drilling and associated production activities are carried out in the oil and gas industries. (From NOAA, public domain.)

Figure 2.4
Electrical transmission and distribution. (a) A typical electrical power substation, a subsidiary station for electrical power transmission and distribution at which required voltage is stepped up or stepped down via a transformer. (b) A typical electrical transmission tower. (Courtesy of author.)

Figure 2.13
The Melaka refinery. The 926-acre refinery complex has a maximum production rate of 265,000 barrels per day (bpd), supplying about 30% of the Malaysian petroleum product needs. (Courtesy of Petronas.)

Figure 2.14
Typical cement plant. The plant has an installed annual Portland cement production capacity of 1 million metric tonnes. (Courtesy of author.)

Figure 2.15
Cloth produced from the textile industry. Depending on cloth types and requirements, about 40–90 % of thermal energy and 10–60 % of electricity energy are needed for textile production (also see Table 2.2). (Courtesy of author.)

Figure 2.21
A typical modern building under construction. Photo taken in March 2011, the building construction completed in September 2011. Depending on the building, the total quantity of energy consumed in a building during its lifetime is usually many times that consumed in its construction. (Courtesy of author.)

Figure 2.25
Typical car — a product of the automotive industry. This 2.0 liter ~100 kW car can accommodate 8 people, with a fuel consumption rate of around 11 km per liter of petrol. (Courtesy of author.)

Figure 2.28
Ferry transport. This typical diesel-driven ferry can accommodate 24 vehicles per trip for river crossing transportation applications. The ferry employs four Cummins engines, each with norminal capacity of 261kW @ 1800 rpm. (*Left*): during loading; (*Right*): river transportation. (Courtesy of author.)

Figure 2.31
A typical affordable-cost airboat constructed/assembled by students. The airboat or other energy systems can be designed, assembled, and constructed in-house. (After Abdullah et al. 2010 [26].)

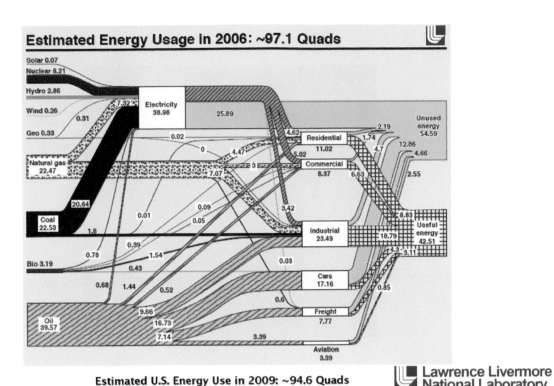

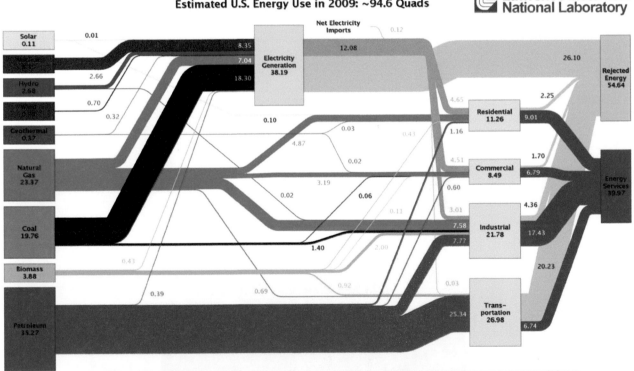

Figure 3.28
Estimated energy usage in (*top*) 2006 and (*bottom*) 2009. [From LLNL 2008 data based on DOE/EIA-0384 (2006) and LLNL 2010 data based on DOE/EIA-0384 (2009), respectively. Credit is given to the Lawrence Livermore National Laboratory and the Department of Energy, under whose auspices the work was performed.]

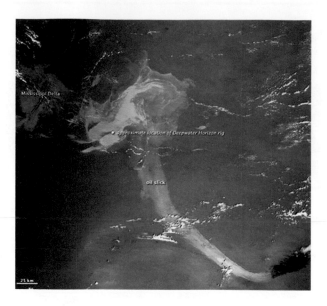

Figure 4.10
An example of a massive oil spill. As seen in the photo, a hundreds of kilometers long oil slick propagated from the deep water oil well drilling rig, Gulf of Mexico. (From NASA.)

Figure 4.11
Pollution from a typical SME-related factory. (Courtesy of author.)

(a)   (b)

Figure 4.12
A thermal power plant using sea water for effective water rejection. This power plant has two cooling ponds whose water is continuously recycled from the nearby South China Sea. (a) The power plant. (Courtesy of Sejingkat Power Corporation.) (b) One of the cooling water ponds. (Courtesy of author.)

Figure 4.18
Typical POME retention pond. (Courtesy of author.)

Figure 5.12
A typical ammonia storage tank. This typical tank can store about 5,000 metric tons of ammonia. (After M.O. Abdullah et al. [51].)

Figure 5.13
Typical dry cask storage containers. (Courtesy of the U.S. Nuclear Regulatory Commission.)

Figure 5.17
A typical capacitor. This capacitor is designed with a capacity of 35μF at 450 VAC, suitable for air-conditioning unit application. (Courtesy of author.)

Figure 7.12
A few new and used automobile lead acid batteries. (Courtesy of author.)

Figure 8.14
A typical wave power generator — the Pelamis Wave Power Device. (From Wind & Hydropower Technologies Program, U.S. Department of Energy, Energy Efficiency and Renewable Energy, public domain.)

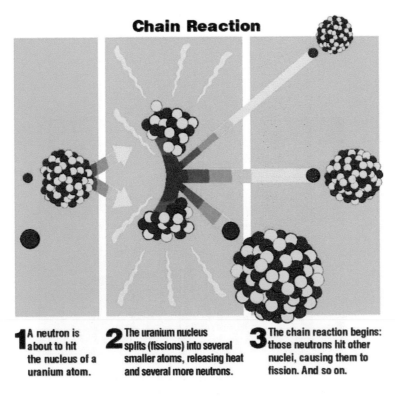

Chain Reaction

1 A neutron is about to hit the nucleus of a uranium atom.

2 The uranium nucleus splits (fissions) into several smaller atoms, releasing heat and several more neutrons.

3 The chain reaction begins: those neutrons hit other nuclei, causing them to fission. And so on.

Figure 9.2
Nuclear reaction — fission, energy, and heat production. (From United States Nuclear Regulatory Commission, public domain.)

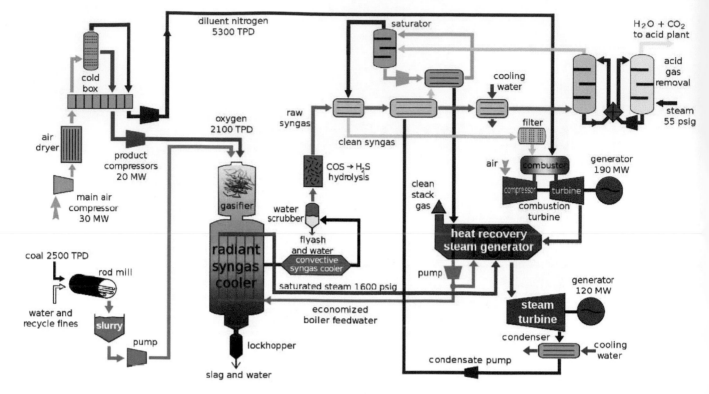

Figure 10.6
Schematic of IGCC plant. (From Wikimedia Commons, public domain.)

Figure 10.15
Hybrid solar thermoelectric-adsorption cooling system. (From Abdullah et al. 2009 [32].)

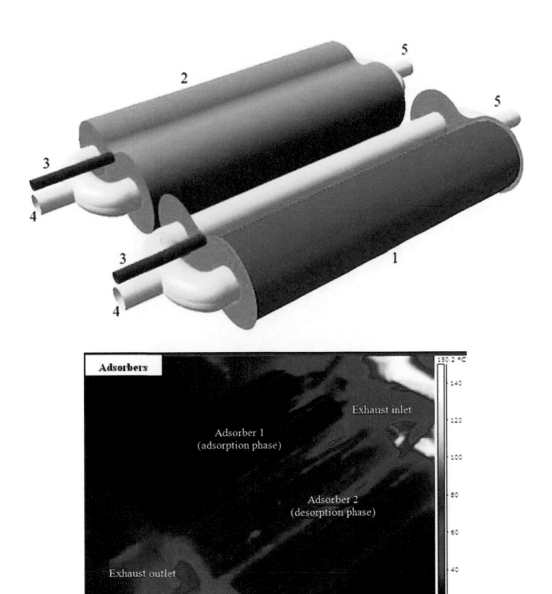

Figure 11.8
A typical adsorption system. (*Top*): the "chemical" compressor, 3-D diagram view. (*Bottom*): the thermograph photo of the chemical adsorber system showing temperature variation of the adsorbers during operation. The system works intermittently where a pair of the adsorbers is cold while the other pair is heated, and vice versa. (From Leo, S.L., and Abdullah, M.O., HVAC&R Research 16 (2) (2010), pp. 221–231. With permission.)

Figure 11.9

SEM image of a typical activated carbon used in an adsorption system. Activated carbon properties: type = granular; average size = around 3.0 mm; density = 0.431 g/cm^3; heat of adsorption = 1800 kJ/kg; surface area = 1000-1100 m^2/g. (From Leo and Abdullah, HVAC&R Research 16 (2) (2010). With permission.)

At the anode, part of the methanol is oxidized to generate electrons, protons, CO_2, and heat,

$$CH_3OH + H_2O \rightarrow CO_2 + 6H^+ + 6e^- + heat \tag{7.17}$$

And, at the cathode, part of the oxygen reacts with the protons that are conducted through the membrane from the anode and the electrons that come from the external circuit to form water and heat,

$$6H^+ + 6e^- + \frac{3}{2}O_2 \rightarrow 3H_2O + heat \tag{7.18}$$

while the remaining oxygen electrochemically reacts with the permeated methanol to produce CO_2, water, and heat. Therefore, the overall reaction in the DMFC is,

$$CH_3OH + \frac{3}{2}O_2 \rightarrow CO_2 + 2H_2O + heat \tag{7.19}$$

Passive DMFC is expected to gain space in the market because they have a higher lifetime compared to the lithium ion battery and can be recharged by simply changing the cartridge of fuel. These types of fuel cells are being developed by Samsung (Korea), Toshiba, Hitachi, NEC, and Sanyo (Japan). Like PEMFC, these fuel cells use a polymer electrolyte membrane; however, in DMFC the anode catalyst extracts hydrogen from liquid methanol, eliminating the need for a fuel reformer. They show efficiencies around 40% and work at temperatures around 130 °C (Andújar and Segura 2009 [18]).

Advantages of DMFC are:

1. It uses liquid fuel. The size of the deposits is smaller and can take advantage of the existing infrastructure provision.

2. It does not need any reforming process.

3. Its electrolyte is a proton exchange membrane, similar to the PEMFC fuel cell type.

Disadvantages of DMFC:

1. The drawbacks of DMFC are the need for concentrated toxic methanol to achieve beneficial energy densities and the problem of methanol cross-over.

2. It has low efficiency with respect to the hydrogen cells.

3. It needs large amounts of catalyst for electro-oxidation of methanol at the anode.

Proton exchange membrane fuel cell (PEMFC). The electrolyte used is of the solid polymer type, the mobile ion is proton, H^+. This is one of the most simple types of fuel cells.

The general reaction for PEM is similar to that of PAFC:

At the anode, gas hydrogen, H^2, shown schematically in Figure 7.16, is oxidized, and releases electrons, i.e., electric energy,

$$2H_2 \rightarrow 4H^+ + 4e^- \tag{7.20}$$

At the cathode, oxygen is reduced to form water,

$$O_2 + 4H^+ + 4e^- \rightarrow 2H_2O \tag{7.21}$$

The advantages of a PEM fuel cell are:

1. PEM fuel cells use solid polymer film that is non-corrosive electrolyte. And, unlike acid-based fuel cells, the PEM fuel cell eliminates the need to handle corrosive acid.

2. It operates at relatively low temperatures and low pressure (1 or 2 bars); therefore, system handling and assembly structure are less complex than in most other types of fuel cells.

3. PEM fuel cells can tolerate CO_2 (but not CO and impurities of hydrogen), thus they can use the oxygen from atmospheric air.

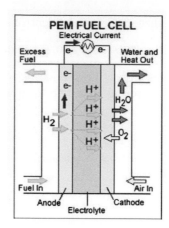

Figure 7.16
Schematic of a PEM fuel cell. (Courtesy of DOE-EERE [19].)

4. PEM fuel cells employ a solid electrolyte, so they eliminate the handling of liquids and the problems of resupply.

5. PEM has high voltage, current, and power density.

6. It is simple, compact, and robust, and suitable for portable power applications.

The disadvantages of a PEM fuel cell are:

1. It uses an expensive catalyst: platinum and membrane from solid polymer.

2. PEM fuel cells are very sensitive to impurities of hydrogen.

3. PEM fuel cells do not tolerate more than 50 ppm of CO and have a low tolerance to sulfur particles.

4. PEM fuel cells need humidification units of reactive gases. If water is used for humidification of gases, the operating temperature of the fuel cell must be less than the boiling water, restricting the potential for cogeneration.

Molten carbonate fuel cell (MCFC). MCFC operating at high temperature enables it to have a better reaction rate. As such, it can use relatively lower-cost catalyst material, i.e., nickel. It could also power by methane (CH_4) and coal gas (CO) directly.

Advantages of MCFC are:

1. It allows spontaneous internal reforming fuel

2. It generates a lot of heat

3. It has high-speed reactions

4. High efficiency

5. Noble metal catalyst, such as platinum, is not required, thus minimizing overall cost.

The disadvantages of MCFC are:

1. It has corrosion problems and needs dimensionally stable and resistant materiel construction.

2. The catalyst of nickel oxide cathode can be dissolved in the electrolyte, causing a malfunction. Dimensional instability can cause distortion, changing the active area of the electrodes.

3. It has high intolerance to sulfur. In particular, the anode does not tolerate more than 1.5 ppm of sulfur particles in the fuel. Otherwise, the fuel cell will suffer deterioration.

4. It has liquid electrolyte, thus the corresponding handling problems.

5. It requires preheating before starting to work.

Solid oxide fuel cell (SOFC). The operating temperature of SOFC ranges from 500–1000 °C. It can have a high reaction rate due to high operating temperature; thus an inexpensive catalyst such as nickel, and alternative gas such as natural gas can be used directly via a process normally called *internal reform*. SOFC requires an air and fuel pre-heater for high temperature requirements.

The advantages of SOFC are as follows:

1. It allows spontaneous internal reforming of fuel. Because the oxide ions travel through the electrolyte, this fuel cell can be used to oxidize any combustible gas. SOFC can operate on high-energy density hydrocarbon fuels such as propane and butane, compared to PEMFC, which requires pure hydrogen as the fuel source.

2. It generates a lot of heat, and thus is suitable for high temperature CHP applications.

3. The chemical reactions are very fast.

4. It has high efficiency.

5. It can work at current densities higher than molten carbonate fuel cells.

6. The electrolyte is solid, which avoids the problems of liquid handling.

7. No need of noble metal catalysts.

The disadvantages of SOFC are:

1. It is difficult to find suitable materials that have sufficient conductivity and high resistance, and which remain solid at temperatures of operation and chemically compatible with other components of the cell.

2. It is moderately intolerant to sulfur (50 ppm).

3. It has slow start-up time, thus less useful for mobile applications.

7.6.2.2 Other Fuel Cell Types

Other fuel cell types include microbial fuel cell (MFC), the Redox fuel cell, etc.

The Redox fuel cell Charging of the Redox fuel cell can be accomplished with hydrogen and oxygen. The advantage of the Redox system is that, unlike other types of fuel cells, it can make use of simple and inexpensive (non-porous) electrode materials. For example, with titanium Redox, the reactions are:

$$Ti(OH)^{3+} + H^+ + e^- \rightleftharpoons Ti^{3+} + H_2O \tag{7.22}$$

At the negative compartment,

$$VO^{2+} + 2H^+ + e^- \rightleftharpoons VO^{2+} + H_2O \tag{7.23}$$

The titanium redox couple can be regenerated with hydrogen at 60 °C over $Pt - Al_2O_3$,

$$2Ti(OH)^{3+} + H_2 \rightarrow 2Ti^{3+} + 2H_2O \tag{7.24}$$

The vanadium redox couple can be regenerated with oxygen at around 75 °C in concentrated nitric acid,

$$2VO^{2+} + \frac{1}{2}O_2 + H_2O \rightarrow 2VO^{2+} + 2H^+ \tag{7.25}$$

Microbial fuel cell (MFC) A microbial fuel cell (MFC) is a special kind of fuel cell in that it uses microorganisms to convert chemical energy into electricity.

The current (I) through the electrical circuit was calculated based on the measured voltage (V) and the resistance according to I = V/Re; therefore, power density (W/m^3) can be calculated as,

$$P = \frac{V^2}{Re \times \nu} \tag{7.26}$$

where P is power density (W/m^3); V is measured voltage (V); ν is the volume of the anode chamber.

The coulombic efficiency was calculated as (Rabaey et al., 2003 [20]),

$$E_c = \frac{\sum\limits_{i=1}^{n} V_i T_i}{ReFb \cdot \triangle s \cdot V_{An}} M \times 100\,\%$$

(7.27)

where E_c is the coulombic efficiency(%), Vi is the measured voltage (V), Ti is the time duration (sec), F is Faraday constant (96,485 C/mol), b is the mole of electrons produced per mole of substrate (b = 4 mole/mol for glucose in this study), DS (g/L) is the overall removal efficiency of glucose in TOC, M is the molecular weight of TOC (12 g/mol), VAn is the volume of the anodic solution.

The advantages of MFC include:

1. It has mild reaction conditions, possible beneficial companion events (such as carbon fixation), and ability in producing sustaining power in a natural environment, making it easier to apply and operate over other kinds of fuel cells (Logan, 2008 [21]).

2. MFC has been considered as an alternative power source that is more environment friendly than fossil fuels, and conventional fuel cells that use fossil fuels.

A typical recent bio-energy generation study using membrane-free baffled MFC is as shown in Figure 7.17 for power generation and power density, respectively.

7.6.2.3   Fuel Cell Classification II: Based on Power Size

Figure 7.18 shows a classification of fuel cell types based on power capacity.

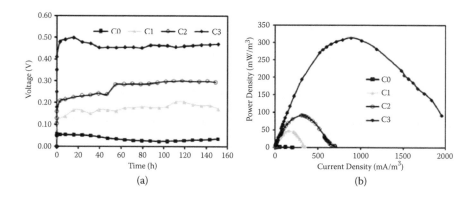

(a)   (b)

Figure 7.17
Power generation of the MFCs at different initial Cu^{2+} concentrations. (a) Cell voltage; (b) power density. C0 = zero (control), C1 = 200 mg/L, C2 = 500 mg/L, C3 = 6400 mg/L. (After Taoet al., 2011 [22].)

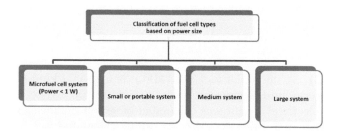

Figure 7.18
Fuel cell classification II: based on power size.

Figure 7.19
VTT micro fuel cells. (Courtesy of VTT [23].)

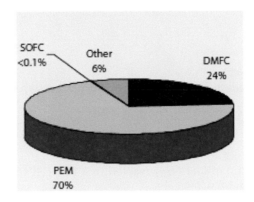

Figure 7.20
Portable fuel cells by electrolyte type. (After Butler, 2009 [23].)

7.6.3 Example Fuel Cell Systems

1. *Microfuel cell system (Power < 1 W).* This class of fuel cells is still generally at the pre-commercial or R&D stage. In early 2009, the world's smallest working fuel cell (a metal hydride-based 3 mm by 3mm by 1mm unit needing limited external fueling) was developed by the University of Illinois at Urbana-Champaign. The unit is capable of generating 0.1-1 mA for 30 hours until the metal hydride is depleted, and it uses surface tension rather than pumps. Although largely still at the R&D stage, these units could be used to power medical devices that require only short power duration. Similarly, researchers at the University of British Columbia announced in April 2009 that they had developed a micro fuel cell that uses yeast feeding on the sugar in human blood to generate electricity. This microbial fuel cell produces around 40 nW of power, and when used with capacitors, may be sufficient to power intraspinal electrodes for treating paralysis or devices such as heart pacemakers. The fuel cell is effectively a living source of power that is able to regenerate itself and eliminate the need for regular operations to replace batteries. Another novel application currently under development, and arguably more commercially ready, is an enzyme-catalyzed printed micro fuel cell being developed by VVT of Finland (see Figure 7.19). The water-activated microbial fuel cell is printed on packaging paper that contains embedded sensors for monitoring, for example, the condition of food in "intelligent packaging." The technology has other applications; for example, in biomedical monitoring when integrated into a sticking plaster, and has the benefit of being both cheap and capable of being mass produced (Butler, 2009 [23]).

2. *Small or portable system.* Figure 7.20 shows the distribution of portable fuel cell systems classified in accordance with the electrolyte type. From a typical survey conducted, it has shown that about 70% of the portable systems commercially available are of PEM type (Butler, 2009) [23]. The portable systems are mainly fuel cell toys, military applications, and for remote (battery charger and mobile phone) applications.

3. *Medium and large system.* Fuel cell engines for tranport applications, especially in vehicles such as buses and cars, have been reported; see, e.g., [29]; [30]. A typical example is the 260 kW PEM fuel cell system, designed and manufactured by Ballard Power Systems. The PEM fuel cell stacks are arranged as two sub-units connected in parallel, each sub-unit consisting of 10 stacks, each with power 13 kW. Collectively, the maximum electrical

power is thus 260 kW. The operating voltage of the fuel cell stack is about 450–750 Volts. The operating temperature is only around 90 °C. Therefore, compared with a normal internal combustion engine, the system has advantages of being environmentally friendly and having low thermal heat emissions. Currently, however, fuel cell operating vehicles are still very costly and the overall systems are still under development.*

7.6.4   Fuel Cell Voltage and Fuel Cell Efficiency

Fuel cells can easily obtain high efficiency, of between 60% and 85%.

The fuel cell theoretical efficiency can be defined by the cell voltage to the maximum (reversible) fuel cell voltage E_{emf}:

$$Fuel\,cell\,theoretical\,efficiency\,\eta = \frac{V_c}{E_{emf}} \times 100\,\% \tag{7.28}$$

where,

η=fuel cell efficiency, %,

V_c= fuel cell voltage, V,

E_{emf}= reversible open circuit voltage.

The actual fuel cell efficiency, considering fuel utilization effect, is:

$$Fuel\,cell\,efficiency\,\eta = \frac{V_c}{E_{emf}} \times \mu_{fuel} \times 100\,\% \tag{7.29}$$

where,

μ_{fuel}= fuel utilization factor, = ratio of fuel (H_2) reacted in a fuel cell to the total amount of fuel (H_2) input to the fuel cell, the value normally around 0.9–0.95.

The electrical work done to move a z Coulombs of electron,

$$The\,electrical\,work\,done = charge \times cell\,voltage = -z\,F\,E_{emf} \tag{7.30}$$

where,

E_{emf} = the voltage of a fuel cell, V.

For a reversible fuel cells, i.e., no losses, the electrical work done is equal to the Gibbs free energy released $\triangle \bar{g}_f$ such that the reversible open circuit E_{emf} can be estimated by

$$E_{emf} = \frac{-\triangle \bar{g}_f}{zF} \tag{7.31}$$

The amount of Gibbs free energy released ($\triangle \bar{g}_f$) depends on the operating temperature.

For a hydrogen fuel cell, the following can be assumed: E_{emf} =1.48 V (at HHV); or = 1.25 (at LHV).

7.6.5   Fuel Cell Energy Output Estimation

Let us consider a PEMFC, from Eq. 7.20. For each mol of H_2, gas hydrogen is oxidized, and releases 2 electrons e^- as follows:

$$H_2 \rightarrow 2H^+ + 2e^- \tag{7.32}$$

Knowing the following constants and conversions:

1A= 1 coulomb / sec,

F= Faraday constant, the charge on 1 mole of electron = 96485 Coulombs/mol,

we can show that the flow rate of the fuel is:

$$V_{gmol} = 5.182 \times 10^{-6} g\,mol\,H_2/(A.sec) \tag{7.33}$$

and, the corresponding mass flow rate of fuel $\dot{m}_{H_2}$ is:

$$\dot{m}_{H_2} = 1.045 \times 10^{-8} kg\,H_2/(kA{\cdot}sec) \approx 10^{-8} kg\,H_2/(kA{\cdot}sec) \tag{7.34}$$

---

*Various hybrid fuel cell systems are also under research and development worldwide.

where,

1 g mol H_2 = 2. 0158 g;

1 kg=1000 g.

Equation 7.34 is an easy to remember figure, and can be conveniently used to estimate the hydrogen fuel (H_2) requirement in a fuel cell.

The power of a fuel cell stack, for a stack of n cells can be estimated by Eq. 7.35

$$Power,\ P_e = V_c \times I \tag{7.35}$$

where,

P_e = power, watt,

V_c = voltage in each cell in the fuel cell stack, volts,

I = current, ampere (A),

n = total number of cells in a fuel cell stack.

The air usage can be calculated by Eq. 7.36,

$$Air\ usage,\ \dot{m}_{air} = 3.57 \times 10^{-7} \times \lambda \times \left(\frac{P_e}{V_c}\right) \tag{7.36}$$

where,

$\dot{m}_{air}$ = air usage, kg/sec,

λ = stoichiometry,

P_e = Power (Watt),

V_c = voltage in each cell in the fuel cell stack, volts.

Water production is given by Eq. 7.37,

$$Water\ production,\ \dot{m}_{water}(moles/sec) = \frac{P_e}{2V_c F} \tag{7.37}$$

where,

$\dot{m}_{water}$ = water production, $moles/sec$,

P_e = power, Watt,

V_c = voltage in each cell in the fuel cell stack, volts,

F = Faraday constant = 96485 Coulombs,

Or, expressed in Kg/s,

$$Water\ production,\ \dot{m}_{water}(kg/sec) = 9.34 \times 10^{-8} \left(\frac{P_e}{V_c}\right) \tag{7.38}$$

To compare different fuel cell systems, *power density* and *specific power* are two commonly used parameters,

$$Power\ density\ = \frac{Power}{Volume} \tag{7.39}$$

where power density is in kW/m^3; power in kW; and volume in m^3.

$$Specific\ power\ = \frac{Power}{Mass} \tag{7.40}$$

where specific power is expressed in W/kg or kW/kg, and mass in kg.

7.6.6 Application Advantages of Fuel Cell as Energy Generator

Fuel cell generators are used in particular for combined heat and power (CHP) applications. Fuel cells also can be used for remote and mobile power applications. This includes portable computers, mobile telephones, and military communication equipment. Fuel-cell powered cars are still in the research stage. The main advantages that fuel cells can offer are as follows:

1. Environmentally friendly, since the fuel source hydrogen and oxygen are used.

2. High system efficiency, more than other power systems.

3. No mechanical moving parts except the fan.

4. Silence in operation.

7.6.7   Application Disadvantages or Limitations of Fuel Cells

Some of the general limitations of fuel cells for energy production applications are:

1. *Slow reaction rate of fuel cells for electricity production.* To improve the solid oxide fuel cell's (SOFC) reaction rate to some extent, three suggestions have been normally made, i.e. (a) to use a catalyst such as platinum, (b) raising the operating temperature, (c) increase of electrode area; and (d) increase the number of cells in the fuel cell stack.

2. *Fuel availability problem*: The main fuel, i.e., hydrogen is not yet readily available in the market.

3. *High overall cost.* The cost of platinum and hydrogen fuel for fuel cell energy production is relatively high compared to other energy production methods.

Now, for vehicle power applications, until today, the two most important challenges for fuel cells for HV and HEV applications were *cost and durability*. The cost for conventional automotive internal combustion engines (ICEs) power plants is about \$25–35 per kW (Chalk and Miller, 2006 [24]) . The DOE target cost was \$45 per kW in 2010 and \$30 per kW in 2015 for transportation applications. The cost is \$61 per kW in 2009 for transportation fuel cells (Yun et al., 2011 [25]). Major contributors to the cost are the electrocatalyst, the membrane, and the bipolar plates.

For stationary power plant applications, such as combined heat and power (CHP), PEMFC, PAFC, MCFC, and SOFC are normally used. However, in general, the limitations reported are related to high overall cost, high corrosion problems (gradual dissolution of nickel oxide from the oxygen electrode, anode creep and corrosion of metal parts), high operating temperature (except PEMFC) and intolerance to sulfur.

7.6.8   *Case Study 2.* Parametric Study of an Alkaline Fuel Cell (AFC)

Fuel cells produce electricity without involving combustion processes. They generate no noise, vibration, or air pollution and are therefore suitable for use in many vibration-free, power-generating applications. In this case study ([26]), a mini alkaline fuel cell signal detector system has been designed, constructed, and tested. The initial results have shown the applicability of such systems for use as an indicator of signal disturbance from cellular phones. For instance, a small disturbance even at 4mV/cm, corresponding to an amplitude of 12–18 mG in terms of electromagnetic field, can be well detected by such a device.

The number of anecdotal reports of symptoms experienced by hand-phone users due to electromagnetic effects and the low-level radio frequency fields around the world is increasing. These symptoms include headaches, dizziness, warmth or tingling around the ears and face, and difficulties in concentrating. It is believed that if such undesired disturbances could be detected, identified, and measured, this could provide a better understanding of the hazards. Also, disturbance identification is useful in providing safety control measures, especially at electrical signal-sensitive areas. In the study, an affordable, non-noble material for the construction of AFC electrodes of the AFC has been used. The study does not aim to improve the AFC design or to increase the power density and efficiency, but to investigate its feasibility for small micro-electronic and mechanical signal applications, particularly for mobile phone signal detection. To further the experimental work, a parametric study based on thermodynamics and sensitivity analysis has been conducted to provide many useful generic design data for the AFC.

The overall system, as shown in Figure 7.21, consists of four main sections, i.e., the AFC, signal induction, signal convertor and detection, and recording section. The mini AFC produced the desired direct current. The voltage generated before and after signal disturbance was then sensed by a Picooscilloscope. The results were plotted by a computer that used Pico Technology software.

The mini AFC is of the dissolved alkaline fuel cell type design. The anode was made from nickel, rather than from the usual, relatively expensive, platinum catalyst. The cathode was carbon since it does not take part in any chemical reaction, but allows electrons to flow through it. The fuel used was a mixture of potassium of potassium hydroxide

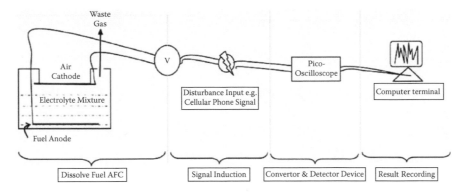

Figure 7.21
Schematic diagram of the dissolved fuel AFC signal detector. (After Abdullah and Gan, 2006 [26].)

(KOH) and methanol solution, in a 75%:25% weight ratio. Reduction of oxygen at the cathode and oxidation of hydrogen at the anode produce electrical energy according to the following reaction:

$$2H_2 + O_2 \rightarrow 2H_2O + electrical\,energy \tag{7.41}$$

The reduction of oxygen in an alkaline solution is given by:

$$O_2 + 2H_2O + 4e^- \rightarrow 4OH^- \tag{7.42}$$

and the anode reaction, i.e., oxidation of hydrogen, is given by:

$$CH_3OH + 6OH^- \rightarrow 5H_2O + CO_2 + 6e^- \tag{7.43}$$

Gibbs free energy, entropy-temperature, and pressure relations   The Gibbs free energy available from a fuel cell can be described either in terms of electrical work (1) or the change in enthalpy and entropy as follows:

$$G = -nFE = \triangle H - T\triangle S \tag{7.44}$$

so that the cell voltage, E, can be evaluated by:

$$E = -\frac{\triangle H - T\triangle S}{nF} = -\frac{\triangle H}{nF} + T\left(\frac{\triangle S}{nF}\right) \tag{7.45}$$

Differentiating the above equation with respect to T:

$$\left(\frac{\partial E}{\partial T}\right)_p = \frac{\triangle S}{nF} \tag{7.46}$$

Now, for a practical fuel cell, a correction factor C_f is introduced to account for the overall cell gains or losses, so that Eq. 7.46 becomes:

$$\left(\frac{\partial E}{\partial T}\right)_p = C_f \frac{\triangle S}{nF} \tag{7.47}$$

and so:

$$\triangle S = \frac{nF}{C_f}\left(\frac{\partial E}{\partial T}\right)_p \tag{7.48}$$

The temperature–entropy (T–S) and temperature–enthalpy (T–H) diagram for H_2 in the region of 130–300K can be found from literature derived from earlier experimental works; The data extracted from these studies were employed and established the following entropies–temperature relationships at different pressure levels:

$$\left[S_1 bar = -1.256E^{-04}T_2 + 5.985E^{-02}T + 60.468\right] \tag{7.49}$$

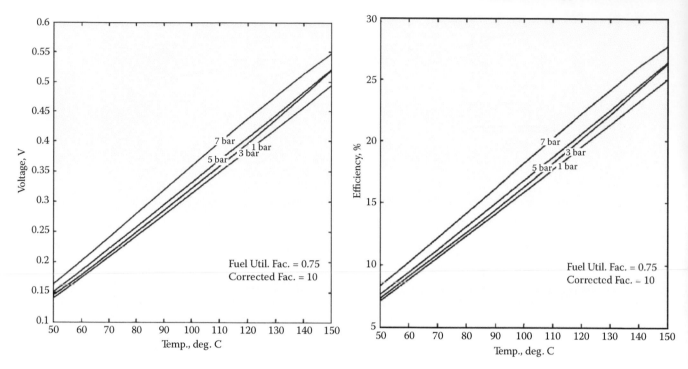

Figure 7.22
Plots of (a) theoretical voltage and (b) theoretical efficiency at various temperatures and pressures ($C_f = 10$).

$$\left[S_3bar = 2.930E^{-04}T_2 - 4.186E^{-04}T + 64.277\right] \tag{7.50}$$

$$\left[S_5bar = -2.093E^{-05}T_2 + 4.44E^{-02}T + 65.013\right] \tag{7.51}$$

$$\left[S_7bar = -8.371E^{-06}T_3 + 2.093E^{-03}T_2 - 0.125T + 74.385\right] \tag{7.52}$$

where,

$S =$ entropy, $J/gm.K$

$T =$ temperature, °C.

The fuel cell efficiency is calculated by Eq. (7.53) and Eq. (7.54):

$$Cell\ efficiency, \eta_{actual} = \mu_f \frac{E}{EMF} \times 100\,\% \tag{7.53}$$

where μ_f is the fuel utilization coefficient.

$$EMF = \frac{-\triangle \bar{h}_f}{nF} = 1.48\,V \text{ (at high heating value, HHV)} \tag{7.54}$$

where, $\bar{h}_f$ is the enthalpy of formation per mole.

Parametric study. The parametric study was carried out to investigate the thermodynamic perspective as well as the performance of the fuel cell at various operating conditons. The effect of temperature and pressure on fuel cell performance and at different correction factors, C_f, is presented in Figure 7.22, Figure 7.23, and Figure 7.24, for C_f values of 10, 15, and 20, respectively. It is apparent that the theoretical voltage, and hence the efficiency, increases with increasing pressure.

If we consider the well-known Nernst equation, which relates the potential generated by an electrochemical cell, E, to the activities a of the chemical species in the cell reaction, $\alpha A + \beta B \leftrightarrow \gamma C + \delta D$, and to the standard potential, E_0, i.e.,

$$E = E_0 + \frac{RT}{nF} ln \frac{{}^a C^{\gamma a} D^{\delta}}{{}^a A^{\alpha a} B^{\beta}} \tag{7.55}$$

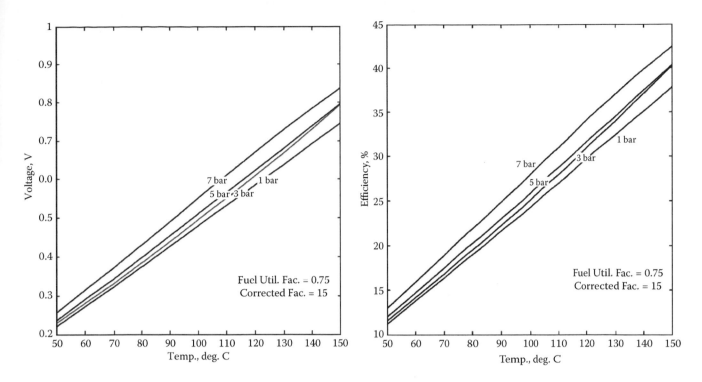

Figure 7.23
Plots of (a) theoretical voltage and (b) theoretical efficiency at various temperatures and pressures ($C_f = 15$).

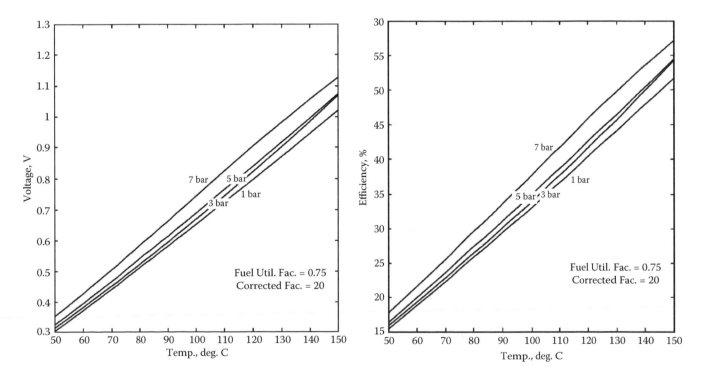

Figure 7.24
Plots of (a) theoretical voltage and (b) theoretical efficiency at various temperatures and pressures ($C_f = 20$).

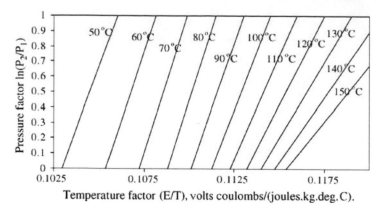

Figure 7.25
Performance curves for AFC at various temperatures and pressures.

a simplified equation can be written:

$$E = E_0 + \frac{RT}{nF} \ln\left(\frac{P_{H_2}}{P_{H_2O}}\right) + \frac{RT}{nF} \ln\left(P_{O_2}\right)^{\frac{1}{2}} \qquad (7.56)$$

where PH_2, PO_2 and PH_2O are the partial pressures for hydrogen, oxygen, and water, respectively.

Considering only the hydrogen partial pressure and that it changes from P_1 to P_2, Eq. (7.56) can be reduced to:

$$E \propto \frac{RT}{nF} \ln\left(\frac{P_2}{P_1}\right) \qquad (7.57)$$

by rearranging,

$$\frac{nF}{R}\left[\frac{E}{T}\right] \cong \ln\left(\frac{P_2}{P_1}\right) \qquad (7.58)$$

where,
$n = 2$
$F = 96,485 \ C/mol$
$R = 8314.3 \ J/kg.mol.$

The $\left[\frac{E}{T}\right]$ and $\left(\frac{P_2}{P_1}\right)$ terms are the temperature factor and pressure factor, respectively. Eq. (7.58) is a partial-dimensionless equation because the temperature factor is not dimensionless.

A plot of pressure factor $ln\left(\frac{P_2}{P_1}\right)$ versus temperature factor $\left[\frac{E}{T}\right]$ is shown in Figure 7.25. The temperature factor $\left[\frac{E}{T}\right]$ increases with increasing temperature T. Therefore, this further proves that the fuel cell will have higher performance at higher temperature. Also, an increasing pressure factor at constant temperature increases the performance indicator, i.e., the temperature factor.

Two of the general conclusions from the theoretical study are:

- The alkaline fuel cell performance increases with temperature, pressure, and correction factor, C_f.

- The temperature factor $\left(\frac{E}{T}\right)$ increases with increasing temperature and pressure factor.

7.7   Summary

In this chapter, we have covered various solar energy and electrochemical energy conversions, together with their respective application aspects:

Solar thermal is an attractive source for high grade (high-thermal energy buildup) thermal energy, to provide high-power energy applications (can power up to MWs level) such as electricity generation and district space cooling/heating as well as industrial chemical processing. The lower grade thermal applications are suitable for food, non-metallic, textile, building, chemical, and SMEs related industries (such as food drying). On the other hand, solar PV electricity is widely applied in the telecommunications, agricultural, water desalination, and building industries, in particular, to operate low grade electrification applications, notably lights, fans, water heaters, refrigerators, small pumps, and engines.

As to the electrochemical–mechanical conversions, batteries are used for effective storage and backup for most energy applications — both in residential, commercial, and industrial applications. The fuel cell is both an established as well as an emerging technology that has the highest system efficiency. Hybrid fuel cell technology has great potential for transport and clean power generation applications.

7.8   *Application Projects, Questions, and Problems*

Application Project

Linking knowledge with applied energy design and practice

1. *Mini solar chimney project.* Your group is required to design and build a small solar chimney called a mini SCPP. Subsequently, you have to conduct a few experiments. Determine (a) the maximum efficiency of your energy product; (b) explain ways that you can improve the total design of your product. Optional: employ a CFD visualization programming and conduct visualization of the heat flux flow profiles to support your findings. Finally, submit a report on your product. Maximum length, 10 pages.

2. *Fuel cell study using a fuel cell simulator.* Your group is required to conduct a series of experiments using a fuel cell simulator. Determine the following:

   (a) The fuels usage

   (b) The electricity production characteristics of the fuel cell

   (c) The efficiency of the fuel cell system

   (d) Discuss how to improve the fuel cell systems.

   Note: Submit a group report on the overall study. Maximum length, 10 pages.

Questions and Problems

Multiple choice questions

1. Which one of the following statements is *not* correct?

   (A) The solar energy source is unlimited, available globally, and low cost compared to other energy systems.

   (B) We can define solar photovoltaic as the generation of an electromotive force due to the absorption of ionizing radiation in converting solar energy to electrical energy, the process of which is usually called *photovoltaic effect.*

   (C) A solar PV cell is an energy conversion device employed to produce the photovoltaic effect, i.e., via conversion of sunlight to electric energy.

(D) Solar PV module or solar array refers to the combination of many individual solar PV cells into a modular form in order to increase the total electrical power output.

(E) Solar energy can be generally classified into two categories: solar PV and solar thermal system.

(F) Output of PV is not consistent, i.e., it varies with the day and sun radiation's availability.

2. The following are the application advantages of solar energy systems, *except*,

(A) Environmentally friendly

(B) High system efficiency

(C) No moving parts

(D) The energy source is unlimited and available globally

(E) PV systems can be fabricated easily

3. The performance, i.e., effective energy capacity of primary battery cells depends on a number of factors, *except*:

I The amount of electrochemical energy source stored in the cell

II Type of materials used

III The quantities of material used

IV The operating temperature

V The discharge rate

(A) I only

(B) I and II only

(C) I, II, and III only

(D) I, II, III, and IV

(E) All of the above

4. The overall methanol fuel cell reaction is

$$2CH_3OH + 3O_2 \rightarrow 4H_2O + 2CO_2$$

If the free Gibbs energy released is -698.2 kJ/mole, what is the the reversible open circuit E_{emf}?

(A) 0.60

(B) 0.65

(C) 0.70

(D) 0.75

(E) 1.21

(F) 1.30

5. The following are the application advantages of fuel cells, *except*,

(A) Environmentally friendly

(B) High system efficiency

(C) Fuel cells can be used for remote and mobile power applications.

(D) Fuel cells can be used for combined heat and power systems.

(E) Fuel cell technology is a clean and economical technology.

Theoretical questions

1. With the aid of a schematic diagram, describe briefly how a solar tower power plant (central receiver solar power plant) works.

2. Briefly discuss how a solar chimney power plant works.

3. Discuss *any one* of the following solar power plants:

   (a) Solar tower power plant (central receiver solar power plant)
   (b) Distributed parabolic collector system power plant

4. With the aid of a schematic diagram, describe briefly the various basic components of a battery cell.

5. List and briefly discuss any three types of fuel cells.

6. What is an alkaline fuel cell (AFC)? Briefly discuss the chemical reactions in the cathodes of an AFC.

7. Discuss the advantages and disadvantages of the following fuel cell types.

   (a) MCFC
   (b) PEM
   (c) SOFC

8. With the aid of a schematic diagram, discuss the concept of a passive DMFC. Also, provide the chemical reactions that occur in a DMFC.

9. Draw a schematic diagram of a dry battery, then label and briefly discuss the various components.

10. Define a wet cell battery, and discuss the basic electrochemical reactions in a lead-acid battery.

Homework problems

1. *Fuel cell application.* A hydrogen fuel cell has the reaction $H_2 + \frac{1}{2}O_2 \rightarrow H_2O$ with the Gibbs free energy released $\Delta \bar{g}_f$ given as in Table 7.3. If the fuel cell is operating at 200 °C with operating voltage at 0.7 V, determine:

   (a) The reversible open circuit E_{emf} of the cell.
   (b) The fuel cell theoretical efficiency.
   (c) The fuel cell efficiency, assuming fuel utilization factor $= 0.95$.
   (Solution: 1.14 *Volts*; 61.4%; 58.3%*)*

Table 7.3 Gibbs free energy as a function of temperature

Temperature (°C)	Δg_f (kJ/mole)	Water production
25	-237.2	Liquid
80	-228.2	Liquid
80	-226.1	Gas
100	-225.2	Gas
200	-220.4	Gas
400	-210.3	Gas
600	-199.6	Gas

2. *Fuel cell.* Given the following constants and conversions of units: 1A= 1 coulomb/sec; $F=$ Faraday constant, the charge on 1 mole of electron $= 96485$ Coulombs/mol; 1 g mol $H_2 = 2.0158$ g. Show that:

   (a) the flow rate of the fuel in a fuel cell is $V_{gmol} = 5.182 \times 10^{-6}$ g mol H_2/(A.sec);

   (b) the corresponding mass flow rate of fuel is $\dot{m}_{H_2} = 1.045 \times 10^{-8}$ kg H_2/(kA·sec).

3. A 1.0 MW fuel cell stack is operated with a cell voltage of 700 mV on pure hydrogen with a fuel utilization, U_f, of 80%.

   (a) How much hydrogen will be consumed in kg/s?

   (b) What is the required fuel flow rate?

   (c) What is the required air flow rate for a 25% oxidant utilization, U_{ox}?

   (Solution: $1.5 \times 10^{-2} kg/sec$; $1.86 \times 10^{-5} kg/sec$; $3.7 \times 10^{-4} kg\,mol/sec$)

Answers to multiple choice questions:

1. (A) Solar energy has *high* initial (capital) cost

2. (B) Solar energy possesses low energy density due to diffused radiation — this causes overall low system conversion efficiency

3. (E) All the above

4. (E) is correct.

   $E_{emf} = \dfrac{-\Delta \bar{g}_f}{zF} = \dfrac{698.2 \times 10^3}{(6)(96485)} = 1.21$ Volts#. Since electron for each mole of methanol is passing from anode to cathode $z = 6$ in the reaction.

5. (E) Generally fuel cell technology is a clean but generally it does not have the economic advantage at the moment due to few factors including high capital H_2 fuel cost.

Bibliography

[1] Willson, R. C., Mordvinov, A. V. (2003), Secular total solar irradiance trend during solar cycles 21–23, *Geophys. Res. Lett.*, 30(5), 1199.

[2] Kalogirou, S. (2003), The potential of solar industrial process heat applications, *Applied Energy*, 76 (December (4)), 337–61.

[3] Mullett, L. B. (1987), Solar chimney — overall efficiency, design and performance, *International Journal of Ambient Energy*, 8, 35–40.

[4] Schlaich, J. (1995), *The Solar Chimney: Electricity from the Sun*, Edition Axel Menges, Stuttgart.

[5] Abdullah, M. O., Mikie, F. A., Lam, C. Y., (2006), Drying performance and thermal transient study with solar radiation supplemented by forced-ventilation, *International Journal of Thermal Sciences*, 45, 1027–1034.

[6] Muneer, T., Maubleu, S., Asif, M., Prospects of solar water heating for textile industry in Pakistan. *Renewable and Sustainable Energy Reviews*, 10 (February (1)), 1–23.

[7] Onyango, F. N., Ochieng, R. M., (2006), The potential of solar chimney for application in rural areas of developing countries, *Fuel*, 85, 2561–2566.

[8] Ming, T. Z., et al. (2007), Numerical simulation of the solar chimney power plant systems with turbine, Zhongguo Dianji Gongcheng Xuebao/Proceedings of the Chinese Society of Electrical Engineering, 27, 84–89.

[9] Peng, W., et al. (2007), Research of the optimization on the geometric dimensions of the solar chimney power plant systems, Huazhong Keji Daxue Xuebao (Ziran Kexue Ban)/*Journal of Huazhong University of Science and Technology* (Natural Science Edition), 35, 80-82.

[10] Ming, T., et al. (2008a), Numerical analysis of flow and heat transfer characteristics in solar chimney power plants with energy storage layer, *Energy Conversion and Management*, 49, 2872–2879.

[11] Ming, T., et al., (2008b), Numerical analysis of heat transfer and flow in the solar chimney power generation system, Taiyangneng Xuebao/*Acta Energiae Solaris Sinica*, 29, 433–439.

[12] Ming, T. Z., et al. (2008c), Experimental simulation of heat transfer and flow in the solar chimney system, Kung Cheng Je Wu Li Hsueh Pao/*Journal of Engineering Thermophysics*, 29, 681–684.

[13] Tingzhen, M., et al. (2008), Numerical simulation of the solar chimney power plant systems coupled with turbine, *Renewable Energy*, 33, 897–905.

[14] Petela, R. (2009), Thermodynamic study of a simplified model of the solar chimney power plant, *Solar Energy*, 83, 94–107.

[15] Romero, M., (2007), Waste heat recovery and air pollution control, AIChE Chicago symposium.

[16] Chalk, S. G., Miller, J. F., (2006), Key challenges and recent progress in batteries, fuel cells, and hydrogen storage for clean energy systems, *Journal of Power Sources* 159, 73–80.

[17] Zhao, T. S., Chen, R., Yang, W. W., Xu, C., (2009), *Journal of Power Sources* 191, 185–202.

[18] Andújar, J. M., Segura, F., (2009), Fuel cells: History and updating. A walk along two centuries, *Renewable and Sustainable Energy Reviews* 13, 2309–2322.

[19] DOE-EERE, (2009), FCT fuel cells: types of fuel cells; <https:// www1.eere.energy.gov/hydrogenandfuelcells/fuelcells/fc_types.html> accessed April 30, 2011.

[20] Rabaey, K., Lissens, G., Siciliano, S. D., Verstraete, W., (2003), A microbial fuel cell capable of converting glucose to electricity at high rate and efficiency. *Biotechnol. Lett.* 25, 1531–1535.

[21] Logan, B. E., (2008), *Microbial Fuel Cells*. John Wiley & Sons, Inc., New York.

[22] Tao, H.-C., Li, W., Liang, M., Xu, N., Ni, J.-R., Wu, W.-M., (2011), *Bioresource Technology* 102, 4774–4778.

[23] Butler, J., (2009), Portable Fuel Cell Survey, available at http://www.fuelcelltoday.com, accessed April 13, 2011.

[24] Chalk, S. G., Miller, J. F., (2006), Key challenges and recent progress in batteries, fuel cells, and hydrogen storage for clean energy systems, *Journal of Power Sources* 159, 73–80.

[25] Wanga, Y., Chen, K. S., Mishler, J., Chan Cho, S., Cordobes Adroher, X., (2011), A review of polymer electrolyte membrane fuel cells: Technology, applications, and needs on fundamental research, *Applied Energy* 88, 981–1007.

[26] Abdullah, M. O., Gan, Y. K., (2006), Feasibility study of a mini fuel cell to detect interference from a cellular phone, *Journal of Power Sources* 155, 311–318.

[27] Raugei, M., Frankl, P., (2009), Life cycle impacts and costs of photovoltaic systems: Current state of the art and future outlooks, *Energy*, 34(3), 392–399.

[28] Castro, M., Delgado, A., Argul, F. J., Colmenar, A., Yeves, F., Peire, J., (2005), Grid-connected PV buildings: analysis of future scenarios with an example of Southern Spain, *Solar Energy* 79, 86–95.

[29] Koppel, T., (1999), Powering the Future - the Ballard Fuel Cell and the Race to Change the World, John Wiley & Sons Canada Ltd., Ontario, Canada.

[30] Spiegel, R.J., Gilchrist, T., House, D.E., (1999), Fuel cell bus operated at high altitude, *Procedings of the Institution of Mechanical Engineeers*, 213, Part A, 57–68.

8

Hydro, Wind, and Geothermal Energy

In this chapter, we will cover matters relating to energy applications that can be harnessed from three of the Earth's resources:

- Water (usually called hydro energy)

- Air (Wind energy)

- Earth (geothermal energy)

8.1 Hydro Energy

Water energy from streams, rivers, and waterfalls can all be harnessed by the most traditional energy generator, the "hydropower turbine," to produce clean electrical energy for our applications.

8.1.1 Introduction to Hydro Energy

The hydro energy application is the most successful type of alternative energy usage to date, in terms of total energy production and applications. Worldwide, hydro power plants provide around 15% of the total power of the world (see, e.g., [18]), and the total hydro potential of our World is about 5,000 GW. It is expected that worldwide energy demand will increase by about 30% between 2010 and 2030, with hydro and other renewables having the highest growth rates [18]; [20].

The power developed from a hydro-electric power plant can be approximately calculated as,

$$Power = w \times Q \times h \times \eta_{overall} \tag{8.1}$$

where
Power = watts;
w = weight density of water, N/m^3 ($w = \rho g = 1000 \times 9.81 = $ around $9810\,\text{N/m}^3$);
Q = water volumetric flow rate, m^3/sec;
h = height of waterfall or "head";
$\eta_{overall}$ = overall conversion efficiency.

Example 8.1   A local SME manufacturing company considers designing and setting up a micro-hydro power system for an electrical power enterprise. The water source is from the waste condenser water from a power plant, delivered at a constant rate. The water turbine average efficiency is 75%.

(a) What is the expected power drawn from the system in kW from a water height of 2 m and volume flow rate of $3\,m^3/sec$?

(b) What is the possible energy saving per year, if the management decided to change to a new turbine of the same capacity but with higher efficiency of 82%? (assume operating hours = 3,920 hours/year).

Solution:
(a) The power that can be drawn is,

$$Power = w \times Q \times h \times \eta_{overall} = 9810 \times 3 \times 2 \times (0.75)$$

$$= 44,145\,W = 44.145 kW\#$$

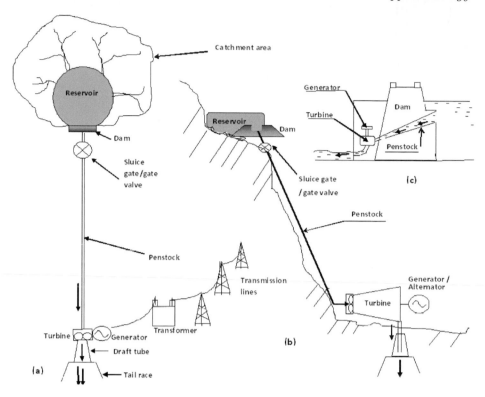

Figure 8.1
Schematic of a hydro-electric power plant together with its basic components: (a) top view of the power plant; (b) side view of the power plant; and (c) close view of a typical dam, penstock turbine, and generator arrangement.

(b) The power expected from the new improved turbine,

$$Power = w \times Q \times h \times \eta_{overall} = 9810 \times 3 \times 2 \times (0.82)$$

$$= 48.265 \, kW$$

Therefore, energy saving is,

$$Energy \, saving = (48.265 - 44.145)kW = 4.12 kW$$

And, the total energy saving per year is,

$$Total \, annual \, saving = 4.12 \, kW \times 3920 \, hours/year$$

$$= 16,150 \, kWh \, per \, year\#$$

8.1.2   The Basic Components of a Hydro-Electric Power Plant

The basic components of a hydro-electric power plant (see Figure 8.1) are:

Catchment area   A general geography term referring to a suitable place drained by rivers or other bodies of water for dams and hydropower plants.

Dam   A dam is a barrier used to store water and create water head.

Reservoir   The reservoir is the water storage, and can be natural or man-made.

Penstock   The penstock carries water to the turbine.

Power house, turbine, and generator   The hydroturbine is rotated by the force of water exerted on the blades of the turbine, and the rotation turns the generator, which provides electricity. All hydro equipment is kept in a power house.

Transformer   The step-up transformer is used to step up the electricity from the generator/alternator and supply the electricity through the transmission lines.

Transmission line   Transmission lines conduct electricity, and ultimately transfer the electricity to residential or industrial areas.

8.1.3   Site Selection for a Hydro-Electric Power Plant

There are a few factors that we have to take into consideration for efficiency and sustainability of a hydro-electric plant:

Water flow rate and water availability.   Water volume flow rate and energy availability throughout the year are one of the most important parameters required by a hydroelectric power plant. The information includes the average, maximum, and minimum quantity of water throughout the year; all the information of water flow rate is recorded systematically throughout the year and usually referred to as *hydrographs*. This detailed information is important to:

1. determine the total capacity of the hydro plant;

2. optimize design for base load and peak load requirements;

3. provide adequate spillways or gate relief for flood contingency plan;

4. set up peak load plant(s) for hybrid hydro energy applications, such as combination with a steam turbine, diesel turbine, gas turbine, or nuclear power plant, and so forth.

2. Water head.   Sufficient *water head* should be available. The higher the head available, the lower the requirement of water volume flow rate in accordance with Eq. 8.1.

3. Water storage and reservoir.   In order to ensure continuous functionality of the turbine power plant, water storage in the reservoir must be enough as a means for backup and to supply continuous power production.

4. Other considerations.   Other considerations include site accessibility, location of plant, land structure, surrounding conditions, and so on.

8.1.4   Hydro-Electric Turbine Classifications

All hydroelectric turbines convert (potential and kinetic) energy from water into mechanical energy through the turbine, which in turn generates electricial energy by means of a generator. Hydroelectric turbines can be classified in many ways.

1. Figure 8.2 shows the classification based on size or capacity of the hydropower output. Definitions of size or capacity vary in literature. However, the following is the average or general capacity sizing:

    (I)   Large hydropower: > 30MW — large power plants supply electricity to many customers for both residential and industrial applications.

    (II)   Small: 100 kW to 30 MW — small power plants for residential or small industrial applications.

    (III)   Micro: 5 kW to 100 kW — A micro hydropower system can provide electricity for a community applications.

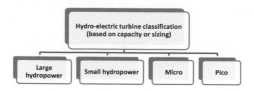

Figure 8.2
Hydro-electric turbine classification (based on capacity sizing).

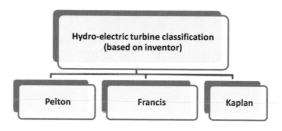

Figure 8.3
Hydro-electric turbine classifications (based on inventor).

(V)   Pico hydro: less than 5 kW. Generally Pico hydro is suitable to power a stand-alone house.

2. Figure 8.3 shows the classification based on inventor. The Pelton turbine, Francis turbine, and Kaplan turbine were named after Lester Allen Pelton, James Bichens Francis, and Victor Kaplan, respectively.

3. Figure 8.4 depicts a simple classification based on turbine shaft inclination, i.e., either vertical or horizontal. The Pelton turbine, for instance, is horizontal in turbine shaft inclination to allow the buckets to tap the maximum impulse force from the jetting effects of water.

4. Figure 8.5 shows the popular classification according to the water action on the rotating turbine blades, i.e. either impulse type due to jetting effect or the reaction type. Usually impulse turbines are cheaper and simpler in design requirements compared to reaction turbines, which require specialist pressure casing and carefully designed engineered clearance.

5. Table 8.1 and Figure 8.6 present the general classification of turbines based on the dimensionless parameter *specific speed* (see Eq. 8.2).

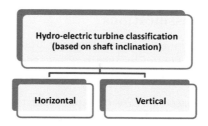

Figure 8.4
Hydro-electric turbine classifications (based on turbine shaft inclination).

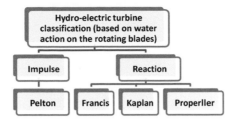

Figure 8.5

Hydro-electric turbine classifications (based on water action on the turbine blades).

Table 8.1 General hydro turbine application range — specific speed range for various types of turbines and runners

Turbine type	Runner type	Specific speed N_s	Head
Pelton impulse	Slow	10–20	200 m
	Normal	20–30	to
	Fast	30–40	2000 m
Turgo impulse		50–100	200 to 2000m
Francis	Slow	80–120	15 m
	Normal	120–180	to
	Fast	180–300	800 m
Kaplan/Propeller		300–1000	5–30 m
Tube, bulb, cross-flow	Normal	20–100	2–30 m

8.1.4.1 What Is *Specific Speed* of Turbine?

Specific speed is a standard theoretical feature of hydro-electric turbines defined as the number of revolutions per minute (rpm) at which a given runner would produce a power of 1 kW under a water head of 1 m (see Eq. 8.2).

$$\text{Specific speed of turbine, } N_s = \frac{N\sqrt{P}}{(H)^{5/4}} \tag{8.2}$$

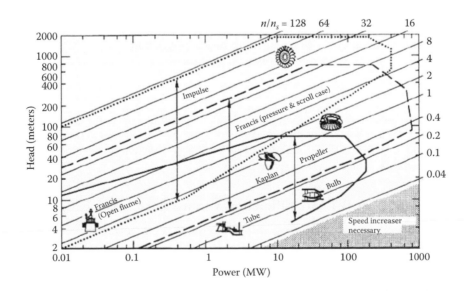

Figure 8.6

Hydro turbine application chart. (From Arndt, R.E.A., 1991. In *Hydropower Engineering Handbook*, J.S. Gulliver and R.E.A. Arndt, eds., McGraw-Hill, New York. pp. 4.1–4.67. With permission.)

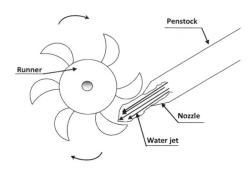

Figure 8.7
Schematic of a Pelton turbine.

where
    N_s = working speed in rpm;
    P = power output of the turbine, kW;
    H = Net head (m).

We shall note that Eq. 8.2 is developed based on similitude or dimensional analysis. The specific speed expressed in dimensionless parameters allows us to conveniently select a suitable type of turbine for hydro-electric generation applications.

Some of the conventional specific speed ranges for various types of runners are given in Table 8.1. And, a general hydro turbine application chart is presented in Figure 8.6.

8.1.4.2   What Are the Common Prime-Movers or Turbines?

8.1.4.2.1   Pelton Turbine   The Pelton turbine was first invented by Pelton in 1880. It is an axial flow impulse turbine type generally mounted on a horizontal shaft. A number of buckets are mounted a round the periphery of the wheel. The flow of water is directed towards the wheel through a nozzle or nozzles (Figure 8.7). The hydro-electricity efficiency of the Pelton wheel lies between 85–95%. Peltons are used specifically for *high head* applications.

8.1.4.2.2   Francis Turbine   The Francis turbine is named after J.B. Francis. It is a reaction type of turbine.

8.1.4.2.3   Kaplan Turbine   Propeller and Kaplan turbines make use of a large volume of available water with low head, and are designed to provide huge flow area while allowing the turbine machine to run on low speeds. It is an axial flow reaction turbine that is suitable for medium head applications, generally between 5–70 m, and specific speed of 300–1000. A typical cut-away view of Kaplan turbine together with the generator is given in Figure 8.8.

8.1.4.3   Dams Versus Weirs

Hydroelectric plants around the globe are ranging in size-from large dams (with many turbines) to small/micro-hydro systems using weirs. Big dams are generally classified (see Figure 8.9) based on their structure and materials used: Timber dams, mansory-made dams, and embankment (fill) dams (either earth-filler or rock fill) are concrete-made. Weirs are basically "small dams" placed across rivers or streams.

8.1.5   Turbine Efficiency: Full Load Efficiency versus Part Load Efficiency

Depending on the target capacity, water resources, and load requirements, a turbine can operate in various operating modes.

Full load efficiency is the maximum efficiency a turbine can develop. It depends on various parameters such as the type of turbine design and the runner use. Unlike full load efficiency, *part load efficiency* is the efficiency obtained when the system is operating below the maximum efficiency.

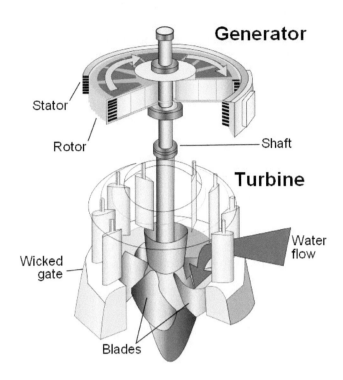

Figure 8.8
Kaplan turbine and electrical generator, cut-away view. (Image edited by Mikhail Ryazanov. From U.S. Army Corps of Engineers, public domain.)

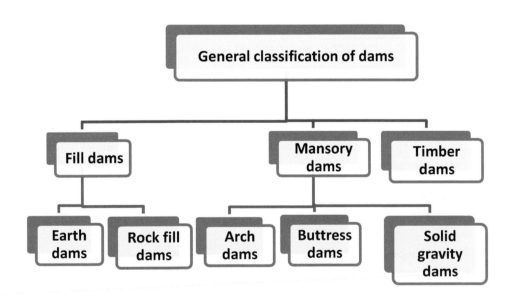

Figure 8.9
General classification of dams.

8.1.6 Micro-Hydro Applications

In the preceding sections, we had discussed hydro energy with big plant capacity up to MWs power level (see Subsection 8.1.4); in the next section, we will consider micro-hydro applications the power range of which is 5–100 kW level, a level that is ideal for SME and rural communities' applications.

Unlike the large hydropower application, which required a large plant with huge generator-turbine facilities, micro-hydro requires minimal requirements.

8.1.6.1 Plant Factor

Plant factor (PF) is defined as the ratio of total energy use to total energy available,

$$\text{Plant factor, PF} = \frac{\text{Total energy use}}{\text{Total energy available}} \tag{8.3}$$

Plant factor is a parameter to assess the feasibility or power availability. It is useful to indicate hydro schemes whether or not it is likely to be successful. The ideal PF factor for a factory would perhaps be between 0.4–0.7; however, it is desirable that the plant factor is to be designed as high as possible for rural electrification applications due to unforeseen future power requirements.*

8.1.6.2 Example 8.2

A typical SME food factory is using 15 kW for food processing applications from 9 am to 9 pm. What is the plant factor if it is designed at 20 kW capacity and for continuous use?

$$\text{Plant factor} = \frac{\text{total energy use, kW}}{\text{total energy available, kW}} = \frac{15\,\text{kW} \times 12\,\text{hours}}{20\,\text{kW} \times 24\,\text{hours}} = 0.375\#$$

8.1.6.3 Load Factor

Load factor (LF) is a term to define the ratio of total energy consumed in a particular period of application to total energy available at a particular period of application consideration.

$$\text{Load factor} = \frac{\text{total energy consumed in specific period, kWh}}{\text{total energy available in specific period, kWh}} \tag{8.4}$$

The *load* usually refers to the utilities, that is, electrical consumption, with units in kWh.

8.1.6.4 Example 8.3

Consider a typical SME food factory using 3 kW for food processing applications from 9 am to 9 pm (12 hours). What is the plant factor if the maximum load demand is 3 kW load capacity and for continuous 24-hour use?

$$\text{Load factor} = \frac{\text{total energy consumed in specific period, kWh}}{\text{total energy available in specific period, kWh}} = \frac{3\,\text{kW} \times 12\,\text{hours}}{3\,\text{kW} \times 24\,\text{hours}} = \frac{36\,\text{kWh}}{72\,\text{kWh}} = 0.5\#$$

8.1.6.5 Unit Energy Cost

Unit energy cost is a financial term used to define the total annual cost of energy consumed per year.

$$\text{Unit energy cost} = \frac{\text{total annual cost}}{\text{total useful energy consumed per year}} \tag{8.5}$$

Express Eq. 8.5 mathematically,

$$\text{Unit energy cost} = \frac{C_{\text{annual}} + (O + M)}{P_{\text{installed}} \times (365\,\text{days} \times 24\,\text{hours/day}) \times \text{PF}} \tag{8.6}$$

---

*Rural communities would be likely to increase future energy demand, both in the number of houses and the associated electrical appliances.

where,

Unit energy cost is given in currency per kWh, e.g. \$/kWh;

C_{annual} is the annual cost of the capital spent on the installation of the hydro machinery and facilities, \$;

(O+M) is the annual operational and maintenance cost, \$;

$p_{installed}$ is the installed power capacity;

PF is the plant factor.

8.1.6.6 Example 8.4

Now let us consider again the same SME food factory in Example 8.2. Suppose the annual cost of the capital spent (from a loan) is \$4000 per year, annual operational and maintenance cost around \$200. What is its unit energy cost?

$$\text{Unit energy cost} = \frac{C_{annual} + (O + M)}{p_{installed} \times (365 \text{days} \times 24 \text{hours/day}) \times \text{PF}}$$

$$= \frac{4000 + 200}{20kW \times (365 \times 24) hours \times 0.375} = 0.06 \, \$/kW\#$$

Unit energy cost is useful as an economic indicator for various energy resource comparisons.

For further treatment of the details of micro-hydro designs including hydrology survey and financial aspects, the reader can refer to, for example, Harvey's micro-hydro manual (1993) [1].

8.1.7 Application Advantages of Hydro Energy and Hydro Power Plant

The application advantages of hydro energy are:

1. Zero fuel requirement. Unlike fossil-fuel-power plants, hydro energy requires zero fuel charges.

2. Low operational and running-cost.

3. Environmentally friendly, no carbon or other unwanted emissions.

4. Long operating life — can be 100 years or more.

5. Hydro energy is one of the most sustainable energies.

The application disadvantages of hydro energy are:

1. High capital costs.

2. Hydro plants usually require long transmission lines for power delivery, thus loss of energy. This is because a hydro plant and power house are usually constructed near a hilly site, far away from the station center.

3. Construction of dams and hydro power plants usually faces opposition from many environmentalists and people around the land where they are to be constructed.

4. The amount of electricity generated by hydro power in OECD countries has remained at the same levels over the past few years. This assumption reflects the increasing difficulty of locating sites for large-scale hydroelectric projects in OECD countries (IEA, 2002, [11, 18]).

8.1.8 Examples of Hydro Energy Power Plants around the World

1. The Three Gorges Dam, China, provides the world's largest hydro-electric power station, with total capacity of 22,500 MW. The total length of the dam axis is 2309.47m, with a crest elevation of 185m and the maximum dam height at 181m. The normal reservoir storage water level is 175 m, and the total reservoir storage capacity is 39.3 billion m^3.

2. The Itaipu Hydroelectricity Power Plant, Brazil, is the second largest hydroelectric power plant in the world. With around 14,000 MW of installed capacity from 20 generator units, the dam was completed in 2003.

3. The Guri Dam, Venezuela is the third biggest hydroelectricity power plant in the world, with a capacity of 10.2 GW. The Hydroelectric Power Station Guri at Necuima Canyon supplies about 73% of Venezuela's electricity.

4. The Grand Coulee Dam Hydroelectric Power Plant, USA, consists of three power plants with total capacity of around 7,000 MW.

5. The Sayano–Shushenskaya Hydroelectric Power Plant, Russia, was built in 1978. It has a total installed capacity of 6,400 MW, and average annual production is 23.5 TWh.

6. The Robert-Bourassa Hydroelectric Power Plant, Québec, Canada, has two plants built on the Le Grande River with total capacity of 5600 MW.

8.2   Ocean Energy (Tidal, Wave, Thermal)

Oceans cover $\geq 70\%$ of the Earth's surface, making them potentially the world's largest energy collectors. Our ocean can produce three types of primary energy, i.e., ocean wave energy, ocean tidal energy, and ocean thermal energy.

8.2.1   Tidal Energy

The restless tides around the thousands of coastlines, straits, estuaries, and river deltas around the globe provide huge energy sources — literally inexhaustible and sustainable as long as our Moon continues to function!

8.2.1.1   What Is Tidal Energy?

Tidal energy is the energy inherited from the rise and fall of water levels usually called *tidal cycles*, due to the gravitational force of the Sun and Moon on the Earth and the centrifugal force produced by the rotation of the Earth and Moon about each other [25]; [26].

8.2.1.2   What Are the Application Advantages of Tidal Energy?

The large difference between low tide and high tide, usually called *head* or *tidal range*, is the main requirement for harnessing tidal energy. Tidal ranges of ≥ 5 m are generally regarded as economical for power operation.
   Advantages of tidal power generation:

1. *Tidal power is sustainable.* The tidal energy is inherited from the rise and fall of water level due to the action of the Moon, which is independent of levels of precipitation.

2. Tidal power is environmentally friendly, being free of carbon emission and pollution.

Disadvantages and limitations of tidal power generation:

1. Tidal power is not uniform, due to variations in tidal head or tidal range.

2. Tidal power is subject to corrosion related problems. Some tidal devices may require mooring systems that are subject to biofouling and corrosion, affecting the survivability of the system [24].

3. Tidal power plants are relatively expensive in operation. For instance, electricity transmission is a cost-related issue and in some cases transmission to shore over longer distances may be required [24].

4. Tidal current turbine technology is currently not economically viable on a large scale, as it is still in an early stage of development [24].

5. Possibility of sedimentation and erosion problems.

8.2.1.3 How Tidal Energy Is Used for Electric Energy Production?

The difference between the water level elevations at high tide and low tide of a tidal cycle result in a *differential head* that can be utilized for potential energy → electrical energy conversion. Quite similar to hydro-electric, tidal energy application can be realized by having a very long dam built across the basin, and employing two sets of turbines underneath the dam. As the tide comes in, water flows into the basin of one set of turbines; likewise, at low tide, the water flows out of the basin operating under another set of turbine.

The potential energy stored in the full basin is the work done in lifting the mass of water above the ocean surface, given by:

$$E_{potential} = \rho g A \int x.dx \tag{8.7}$$

where,

$E =$ energy;
$\rho =$ sea water density;
$h =$ tidal range (m);
$A =$ area of water stored in basin (m^2).

$$E_{potential} = \frac{1}{2}\rho g A h^2 \tag{8.8}$$

Average potential power can also be estimated from,

$$\therefore P_{average} = \frac{\frac{1}{2}\rho g A h^2}{\left(\frac{T}{2}\right)} = \rho g A h^2 \left(\frac{A}{T}\right) \tag{8.9}$$

where,

$T =$ amplitude of the tide.

Now, the kinetic energy of the water mass due to velocity is given by,

$$E_{kinetic} = \frac{1}{2}mV^2 \tag{8.10}$$

Therefore, the total tide energy equals the sum of its potential and kinetic energy components, i.e.,

$$E_{total} = \frac{1}{2}\left(\rho g A h^2 + mV^2\right) \tag{8.11}$$

8.2.1.4 Classification of Tidal Power Plants and Water Turbines

Tidal power plants can be classified into two main categories based on their basin arrangements, i.e., either single basin arrangement or double basin arrangement (Figure 8.10).

In a single basin arrangement, only one basin interacts with the sea such that power can be generated only intermittently due to high and wide tides. The basin and the sea are separated by a dam, and flow between them is through sluiceways along the dam. In a double basin scheme, the system consists of 2 basin, each designed at different elevation (water head) connected through turbine, thus providing continuous water flow and provide constant power generation.

Water turbines can be complex due to the many proposed designs available. However, water turbines can be generally classified into two categories in terms of their physical arrangements and energy conversion mechanisms: horizontal axis turbine, and vertical or cross-flow axis turbine, where the axis of rotation is perpendicular to the flow direction (see Figure 8.11).

8.2.1.5 Examples of Tidal Applications around the World

- La Rance, St Mao, France, tide range 13.5 m, basin area = 22 km^2, installed power= 240 MW, built in 1967 and one of the oldest in the world.

- Kislogubsk, Russia, one of the oldest in the world.

- Canada Fundy Bay Tidal Power; 16.2 m (highest tide range)

- San Jose (South America): 10.7m, 777 km^2, 19,900 MW

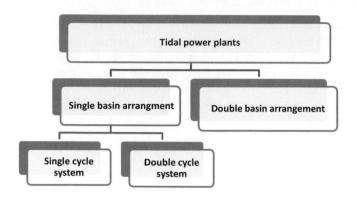

Figure 8.10
Classification of tidal power plants.

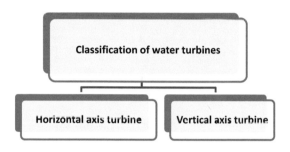

Figure 8.11
Classification of water turbines.

- Server (UK): 9.8 m, 70 km^2, 8,000 MW

- Passanaquoddy Bay (North America): 5.5 m, 262 km^2, 1,800 MW

- Seven Estuary, England: 14.5 m

- JiangXia, China, 5.08 m (mean tide), 1.4 km^2, 3.9 MW

8.2.1.6   Some Recent Advancements of Tidal Technologies

According to Gorlov (2011), traditional river power plants have a poor ecological reputation because the dams block fish migration, destroying their population, and damage the environment. Moreover, the relatively low water head in tidal power plants creates a difficult technical problem for designers. Even the very efficient, mostly propeller type hydraulic turbines developed for high river dams are inefficient, complicated, and very expensive for low-head tidal power application. These environmental and economic factors have forced scientists and engineers to look for a new approach to exploitation of tidal energy that does not require massive ocean dams and the creation of high water heads. The key component of such an approach is using new unconventional turbines, which can efficiently extract the kinetic energy from a free unconstrained tidal current without any dams. One such turbine is the Helical Turbine. The turbine (Figure 8.12), consists of one or more long helical blades that run along a cylindrical surface like a screw thread, having a so-called airfoil or "airplane wing" profile. The blades provide a reaction thrust that can rotate the turbine faster than the water flow. The turbine shaft (axis of rotation) must be perpendicular to the water current, and the turbine can be positioned either horizontally or vertically. Due to its axial symmetry, the turbine always develops unidirectional rotation, even in reversible tidal currents. This is a very important advantage, which simplifies design and allows exploitation of the double-action tidal power plants. A pictorial view of a floating tidal power plant with a number of vertically aligned triple-helix turbines is shown in Figure 8.12.

Figure 8.12
Helical turbine and environmentally friendly helical tidal system. (*Left*): the helical turbine. (*Right*): the floating tidal power plant with vertical triple-helix turbines for Uldolmok Strait in Korea. Electric generators are seen sitting above the water. (After Gorlov, 2011 [2].)

Figure 8.13
Underwater wave power generation: Tuscanit. (Wikimedia Commons author, GNU Free Documentation License, public domain.)

Development of new, efficient, low-cost and environmentally friendly hydraulic energy convertors suited to free-flow waters, such as triple-helix turbines, can make tidal energy available worldwide. This type of machine, moreover, can be used not only for multi-megawatt tidal power farms but also for mini-power stations with turbines generating a few kilowatts. Such power stations can provide clean energy to small communities or even individual households located near continental shorelines, straits, or on remote islands with strong tidal currents (Gorlov, 2011 [2]).

We will consider wave energy in the next section.

8.2.2 Wave Energy

Wave energy is a result of the interaction of wind energy and the surfaces of the oceans, caused by the wind blowing over the surface of the ocean. It has high power densities and is available in vast areas of our oceans.

The west coasts of the United States and Europe and the coasts of Japan and New Zealand are good sites for harnessing wave energy.

8.2.2.1 How Is Wave Energy Used for Electric Energy Production?

One way to harness wave energy is to divert and focus the waves into a narrow channel, thus increasing wave size and associated power intensity. The waves can then be channeled into a catch basin or used directly to spin turbines.

Many more ways to capture wave energy are currently under development. Some of these devices being developed are placed underwater (see Figure 8.13), anchored to the ocean floor, while others ride on top of the waves. The world's first commercial wave farm using one such technology opened in 2008 at the Aguçadora Wave Park in Portugal (see Figure 8.14).

Figure 8.14 (SEE COLOR INSERT)
A typical wave power generator — the Pelamis Wave Power Device. (From Wind & Hydropower Technologies Program, U.S. Department of Energy, Energy Efficiency and Renewable Energy, public domain.)

8.2.2.2 What Are the Advantages and Disadvantages of Wave Energy Applications?

Advantages of ocean wave energy:

1. Ocean wave energy is pollution free.

2. Ocean wave energy is sustainable and free.

3. Ocean wave power systems do not require large land areas.

Disadvantages of ocean wave energy:

1. The variation of wave frequency and amplitude makes it an unsteady energy source.

2. Similar to tidal energy, an ocean wave power system is subject to corrosion related problems.

8.2.2.3 Example of Wave Energy Applications around the World

Next, we will consider another energy resource from our ocean.

- WaveGen wave energy plant, Isle of Islay, Scotland, 0.5 MW.

- OWC Pico Power plant, Pico Island, Azores, Portugal (complete in 2006); 2.25 MW.

- Aegir-Schetland Wafe Farm, off Shetland Island (complete in development); 10 MW.

8.2.3 Ocean Thermal Energy

It is possible to capture ocean heat from thermal gradients for energy production and applications.

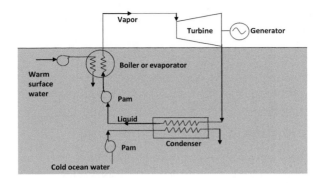

Figure 8.15
Schematic of OTEC closed cycle Rankine plant.

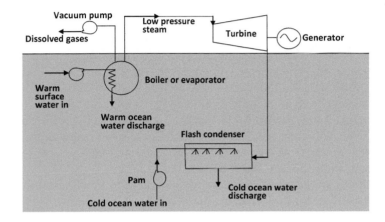

Figure 8.16
Schematic of an OTEC open cycle (Claude) plant. In a Claude open cycle plant, the seawater itself acts as (i) a heat source (surface warm water), (ii) working fluid (instead of using ammonia), (iii) coolant for heat rejection, and heat sink.

8.2.3.1 What Is Ocean Thermal Energy?

Our oceans cover about 70% of the total Earth surface, the vast area of which becomes a gigantic energy absorber. Warm water from the tropical ocean's surface flows from the tropics towards the poles, while the cold water from the poles circulates at the ocean bottom from the poles to the tropics, resulting in a continuous cycle and a heat difference of about 20 ºC, with about 25 ºC–27 ºC at the surface and 5 ºC–10 ºC at the ocean bottom.

8.2.3.2 Ocean Thermal Energy Conversion Plant (OTEC)

In 1992, Barjot proposed a closed cycle OTEC concept, but the concept was developed by Anderson in 1992, so that it is also called an Anderson Cycle OTEC Plant.

Figure 8.15 shows a schematic diagram of a typical OTEC, which works based on the closed Rankine cycle. Warm surface water is used to vaporize a working fluid that has a low boiling point, such as ammonia. The vapor expands and turns the turbine, which in turn rotates a generator to produce electricity.

As to the open cycle (Claude) OTEC system (see Figure 8.16), it functions by boiling the sea water inside an evaporator or vaporizer at low pressures. The low pressure in the evaporator is accomplished by employing a vacuum pump that also removes dissolved non-condensable gases that form in the evaporator. This produces continuous low pressure steam that powers the turbine and generator. The first OTEC plant was constructed in Cuba by Georges Claude, a Frenchman, back in 1929.

8.3   Wind Energy

Wind is basically the air motion caused by heat gradients on our Earth's uneven surfaces, and due to the sun and the rotation action of our Earth. Since air moves around from place to place, it carries primarily kinetic energy.

8.3.1   Introduction to Wind Energy and Its Applications

In the past, the wind has been used as a source of power for pumping water and grinding grains for more than a thousand years. It is estimated that before the industrial revolution, there were some 10,000 windmills in England, but these fell into disuse with the introduction of reliable steam engines, and electrical pumps following the extensive rural electrification programs of the 1940s. Dr C.F. Brush built a 17-m diameter, 12 kW direct current (DC), multi-bladed wind turbine in Cleveland, Ohio, in 1888, while Prof. P. LaCour conducted important experiments that led to the construction of several hundred wind generators in the range 5–25 kW in Denmark in the early 1900s [19].

Large wind turbine development with a capacity of ≥100 kW began to be erected in 1930 and gradually increased to many parts of the world in the 1950s. In 1957, for example, Dr U. Hutter in Germany constructed an advanced 100 kW lightweight machine. Around this time, individual wind turbines of around 100 kW were also constructed in France and the United Kingdom, but the low price of oil led to limited commercial interest in wind energy [19].

However, once again, in the early 1970s the world price of oil tripled, stimulating wind energy research and development programs in a number of countries including the United States, United Kingdom, Germany, and Sweden. In general, these programs supported the development of large, technically advanced wind turbines by major aerospace manufacturing companies. Typical examples included the Mod 5B (97.5-m diameter, 3.2 MW) machine in the United States, and the LS-1 (60-m diameter, 3 MW) machine in the United Kingdom. These measures provided a market for manufacturers to supply much smaller, simpler wind turbines.

Initially a wide variety of designs were used, including vertical axis wind turbines, but over time the so-called 'Danish' concept of a three-bladed, upwind, stall regulated, fixed-speed wind turbine became dominant. Initially these turbines were small, sometimes rated at only 30 kW, but were developed over the next 15 years to around 40-m diameter, 800–1,000 kW. However, by the mid-1990s it was becoming clear that for larger wind turbines it would be necessary to move away from this simple architecture and to use a number of the advanced concepts (e.g., variable-speed operation, pitch regulation, advanced materials) that had been investigated in the earlier government funded research programs [19].

Large wind turbines up to 100 m in diameter, with capacity rated at 3–4 MW, are now being developed using the concepts of the large prototypes of the 1980s but building upon the experience gained from over twenty years of commercial operation of smaller machines [19].

8.3.1.1   Some Applications of Windmills

The main usage of windmills, in particular, big wind farms, is for electrical power generation. Some other energy applications, in particular, small windmills, are designed specially for isolated electrification usages, such as:

1. Rural electrifications

2. Farming and post-harvest applications such as water pumping, drainage, food processing, etc.

3. SMEs applications such as powering various small factories for grinding grams, saw milling, food processing, etc.

8.3.2   How to Estimate Wind Energy Output

The well-known formula for estimating the power output from a wind turbine is:

$$Power, \ P = \frac{1}{2} C_p \rho A V^3 \tag{8.12}$$

where,

P = power available, W;

$\rho = air\,density\,(1.225\,kg/m^3)$;

A = rotor swept area $= \frac{\pi D^2}{4}$ where D = is rotor diameter);

V = wind speed;

C_p=p Power coefficient, ranging from 0.25 to 0.45 (theoretical maximum, i.e., the Betz limit = 0.59 which denotes 59%).

Example 8.5   An offshore turbine, rated at 6 MW in 15 m/s winds, has a rotor blade diameter of 130 meters. Given that the air density is $1.23\,kg/m^3$, calculate the power coefficient C_p if the actual measured power is at 5.93 MW.

Solution: Rotors sweep area, A, $=>\pi(\frac{D}{2})^2 = \pi(\frac{130\,m}{2})^2 = 13,273\,m^2$

Wind power $P = \frac{1}{2}C_p\rho AV^3 \Rightarrow \frac{1}{2} \times C_p \times 1.23\,kg/m^3 \times 13,273\,m^2 \times 15^3 = 5,930,000$

$$\therefore C_p = 0.21\,or\,21\%\#$$

As we can see from Eq. 8.12, the rotor swept area, A, is important as it is the part of the turbine that captures the wind energy. So, the larger the rotor and the larger the sweep area A, the more energy it can capture. The air density, ρ, changes slightly with air temperature and with elevation.

We shall notice that the wind speed, V, has an exponent of 3, which implies that even a small increase in wind speed will result in a large increase in power. That explains why for land wind turbines, a taller tower will increase the productivity of wind turbine plant.

Now, since the rotor swept area A has circular diameter D, then

$$A = \pi(\frac{D}{2})^2 = \frac{\pi D^2}{4} \tag{8.13}$$

Substituting Eq. 8.13 into Eq. 8.12, the wind power is:

$$P = \frac{1}{2}C_p\rho(\frac{\pi D^2}{4})V^3 = \frac{1}{8}\rho\pi D^2 V^3 \cdot C_p \tag{8.14}$$

Now, we shall also note that more complicated or accurate energy output estimation for wind turbines has been dealt with by a number of researchers who considered a perturbation from mean wind speed and variance.

In 2000, Kainkwa [3] suggested the following formula,

$$P_a = \frac{1}{2}\rho[\overline{V}^3 + 3\overline{V}\sigma^2] \tag{8.15}$$

where,

ρ is the air density,

$\overline{V}$ represents mean wind speed; and

σ = the variance of wind speed.

Paul Gipe [4] introduced the "swept area's method." It consists of determining the wind power and then estimating the potential production of energy E_a, simply by knowing the area swept by the rotor A:

$$E_a = 8760\,h/year \times 1kW/1000W \times \left[\frac{1}{2}\rho AV^3 F\eta\right] \tag{8.16}$$

where

E_a = potential production of energy;

A = area swept by the rotor;

F = the Rayleigh distribution factor;

η = the overall efficiency of the wind conversion system.

All the above-mentioned methods can give an approximate estimation of wind turbine energy; to account for all critical factors that affect the amount of wind energy, other research and references focus on more "complicated" methods based on wind speed distribution models, in particular, the Weibull wind speed distribution (see for instance, Dorvlo 2002 [5] for further detail). A typical global wind power estimation for wind class at 10 m height is presented in Figure 8.17, showing the wind power potential and estimated distribution.

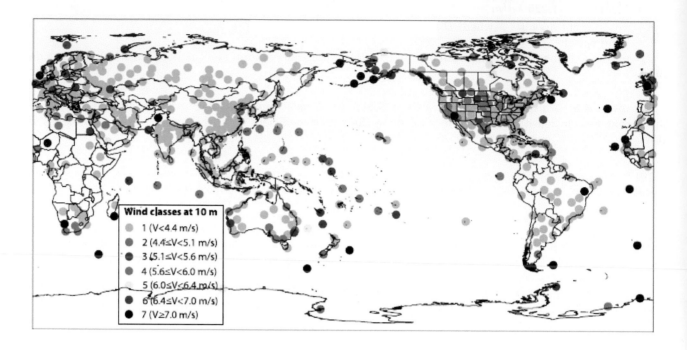

Figure 8.17
Global wind power estimation. (After Dorvlo, 2002.)

8.3.3   Wind Energy Extraction and Aerodynamic Effects on Windmill Power Output

The power output of a windmill turbine depends on the turbine design and its aerodynamic features. Some of the aerodynamic parameters are explained as follows:

8.3.3.1   *Tip Speed Ratio*

Tip speed ratio is the ratio of wind velocity to rotational tip velocity of rotor

$$\text{Tip speed ratio, } \mu = \frac{\text{Tip speed of blade}}{\text{Wind speed}} = \frac{\pi DN}{V} \tag{8.17}$$

where,
μ= tip speed ratio.
D= diameter of rotor, m.
N= revolution per seconds, RPS. ($= \text{RPM} \times \frac{1}{60}[\frac{min}{sec}]$)
V= wind speed, m/s.

8.3.3.2   *Solidity*

The term *solidity* can be defined as the ratio of blade area to disk area, which indicate to percentage of the area of motor containing solid material rather than air,

$$\text{Solidity, } \sigma = \frac{n}{\pi \times D} \times 100\% \tag{8.18}$$

where,
σ = solidity
n = no. of blades
D = diameter of rotor, m

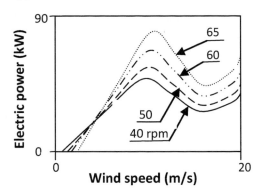

Figure 8.18
Effect of rotational speed effect on windmill power output.

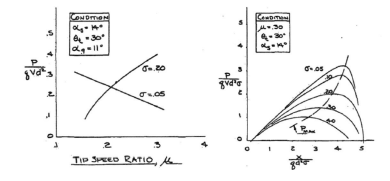

Figure 8.19
Effect of (*left*): tip speed ratio and (*right*): solidity on windmill power output. (After Wiesner, [7].)

The power output of a windmill turbine running at constant speed is strongly governed by operational rotational speed (Figure 8.18). For a typical aerodynamic design, at high wind speed > 8 m/s, a great deal of wind power can be extracted, with higher electrical power production proportional to the speed rotation. Conversely, at low available wind speeds of ≤ 8 m/s, there is a reduction of power with increasing rotational speed. Therefore, it is wise to operate at an average optimum level, as maximize energy capture above higher average level will result in rather high wind cut-in speed.*

Figure 8.19 shows that power output is dependent on tip speed ratio and that for each operating tip speed ratio there will be an optimum solidity to produce maximum power output. Generally, the widely accepted optimum tip speed ratio for modern wind turbines is between 6 and 8 (see, e.g., Yurdusev et al., 2006 [6]).

As a rule of thumb, for maximum power output, the optimum tip speed ratio occurs at

$$\text{Optimum tip speed ratio} = \mu_{optimum} = \frac{4\pi}{n} \tag{8.19}$$

where,
n = number of blades.

It is to be noted that the calculation of the ideal tip speed ratio demands detailed information of the turbine design, best to be evaluated by detailed computational works such as artificial neural networks, e.g., Yurdusev et al., 2006 [6].

Figure 8.19 (right) shows that power output normalized to $(\frac{P}{qVd^2\sigma})$ decreases with solidity, and that Wiesner's further study shows that power output is maximum at a solidity between 0.20 and 0.40 for the baseline conditions studied.

---

*The lowest wind speed where useful power generation is possible.

Table 8.2 Distribution of offshore wind farms by country as of August 2008

Country	Capacity (MW)	Number of wind farms	% capacity
Denmark	425.2	8	28.57
Netherlands	241	4	16.19
Sweden	133	4	8.94
U.K.	598	9	40.18
Ireland	25	1	1.68
Germany	12	3	0.81
Belgium	30	1	2.02
Finland	24	1	1.61
Italy	0.08	1	0.01
Total	1488.28	32	100

(After Snyder and Kaiser 2009 [9].)

8.3.4 Offshore Windmills

Offshore windmills are windmills that are installed in the ocean. Unlike onshore windmills, offshore windmills can avoid land-use disputes, and the usual onshore noise and visual impact problems, as well as taking advantage of stronger and more constant winds. The offshore windmills have recently become more popular. In Europe, the total installed capacity of wind farms is now close to 1,500 MWs (see Table 8.2).

The advantages of offshore windmills are:

1. *Globally, offshore windmills are more promising than land windmills due to the higher available wind velocity.* As reported by Archer and Jacobson in 2005 [8], based on both measurement and calculations, the global average 10-m wind speed over the ocean from measurements was 6.64 m/s (class 6); that over land is 3.28 m/s (class 1). The calculated 80-m values are 8.60 m/s (class 6) and 4.54 m/s (class 1) over ocean and land, respectively.

2. *Vast continuous areas availability.* Offshore wind farms suitable for MWs scale energy production can be installed due to availability of big, open, and continuous ocean areas.

The disadvantages of offshore windmills are:

1. *Higher overall capital cost.* Overall cost of installation (including difficulties with sensitive marine foundations), operation, and maintenance are much higher than those of onshore windmills. Long-term corrosion problems may occur due to salt water.

2. *Complex in electrical integration.* Generally, offshore windmills are more complex in the integration of electrical network requirements due to weak coastal grids and remote offshore areas for transmission.

8.3.5 Some Examples of WindMill Plants

1. In 2004, Nissan installed a 6 turbine (4 MW) wind farm at the Nissan Car Factory near Sunderland, United Kingdom, which is the first wind farm within the global Nissan group that could generate around 5% of the factory's annual electricity requirement, or around £800,000 in savings annually [22].

2. Ulfborg, Denmark (2 MW) wind turbine (Dornberg, 1979 [10]). Currently, Denmark gets about 20% of its electricity from land and offshore wind farms. (see Figure 8.20.)

3. Thanet Offshore Wind Farm, 100 turbines, Ramsgate, United Kingdom.

Figure 8.20
The world's oldest 2 MW windmill in Denmark. Seen in the photo are workers installing the rotor section of the windmill atop a 53.3 m tower. The epoxy-fiberglass made blades are 27.1 feet long. (Courtesy of Popular Science.)

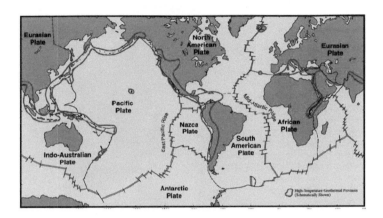

Figure 8.21
World geothermal provinces. (Courtesy of U.S. Department of Energy.)

8.4 Geothermal Energy

"Heat from our Earth": Geothermal energy makes use of the energy from high temperatures underground. And, ocean thermal energy is essentially one of the group of geothermal energy sources.

Geothermal is used commercially in more than 20 countries and generates more than a quarter of the total electricity used in the Philippines and Iceland. The world's main geothermal provinces are given in Figure 8.21

Figure 8.23 shows worldwide installed capacity and energy production for geothermal energy applications. Globally, direct-use heating applications supply the majority of energy, while electric generating capacity exceeds 10,000 megawatts (MW). And, the worldwide growth of installed geothermal generating capacity is on the increase, as displayed in Figure 8.22. Installed capacity and energy production of geothermal among the continents are given in Figure 8.2.

From an application standpoint, more than 90% of geothermal production is from power plants in the United States. These plants, with aggregate capacity of more than 3,100 MW, accounted for 5% of the U.S. total renewable energy and almost 13% of non-hydro renewable generation in 2008. (EIA, 2009 [11]; EPRI, 2010 [12]).

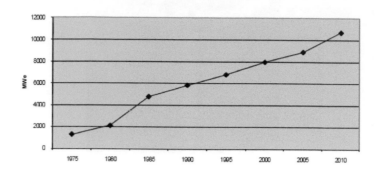

Figure 8.22
Worldwide growth of installed geothermal generating capacity. (From International Geothermal Association.)

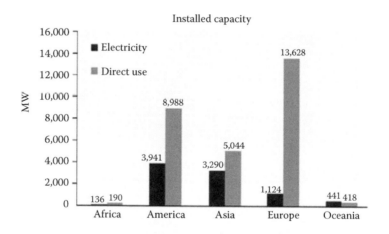

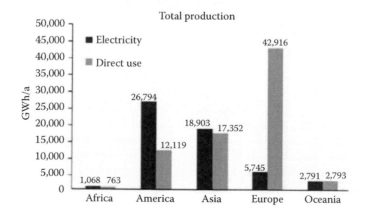

Figure 8.23
Installed capacity (top) and energy production (bottom) for geothermal electricity generation and direct use by continent. Most geothermal energy is used directly for heating applications, while the United States and Asia are at the top in electric generating capacity and production. (From World Energy Council, Survey of Energy Resources, 2007.)

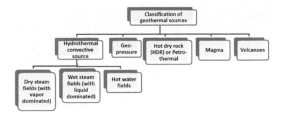

Figure 8.24
Classification of geothermal sources.

8.4.1   What Is Geothermal Energy?

Geothermal energy is the heat energy derived from our Earth. Geothermal energy is an enormous, underused heat and power resource that is clean and reliable. The temperature gradient of our earth is around 0.03 °C/m such that the temperature increases with depth. The center of our Earth is estimated at temperatures up to 10,000 K due to decay of radioactive isotopes. The total steady geothermal energy flow towards the Earth's surface is 4.2×10^{10} kW, but the average flow energy is only $0.063 W/m^2$.

8.4.2   Classification of Geothermal Energy Resources and Applications

The geothermal sources can be classified into five main types (Figure 8.24).

Geothermal resources can also be conveniently classified based on temperature range, as shown in Table 8.3 (White and Williams [13]), and this classification is useful from an energy application standpoint. In general, resources above 150 °C are used for electric power generation. Resources below 150 °C are usually for direct use such as for heating and cooling. Ambient temperatures in the 5–30 °C range can be used with geothermal (ground-source) heat pumps that provide both heating and cooling.

As discussed in the preceding section, the core of our Earth is very hot and there are many hot spots deep under the Earth's surface where we can obtain this geothermal energy, i.e., via volcanoes, hot springs, geysers, and methane under water in the oceans. Ground water, when it comes into contact with these hot spots, can from either dry or wet steam.

We now briefly explain the various steam and HDR resources [23]:

8.4.2.1   Vapor-Dominated Steam ("Dry Steam")

Vapor-dominated steam is steam produced from boiling of deep, saline waters in low permeability rocks. These reservoirs, few in number, The Geysers in northern California, Larderello in Italy, and Matsukawa in Japan and are being exploited to produce electric energy.

Table 8.3 Geothermal resource types

Resource Type	Temperature range °(C)
Convective hydrothermal resources	
Vapor dominated	≈240°
Hot-water dominated	20°–350°
Other hydrothermal resources	
Sedimentary basin	20°–150°
Geopressured	90°–200°
Radiogenic	30°–150°
Hot rock resources	
Solidified (hot dry rock)	90°–650°
Part still molten (magma) >600°	

(After White and Williams [13].)

8.4.2.2   Water-Dominated Steam ("Wet Steam")

Water-dominated steam is produced by ground water circulating to depth and ascending from buoyancy in permeable reservoirs that have a uniform temperature over large volumes. There is typically an upflow zone at the center of each convection cell, an outflow zone or plume of heated water moving laterally away from the center of the system, and a downflow zone where recharge is taking place. Surface manifestations include hot springs, fumaroles, geysers, travertine deposits, chemically altered rocks, or sometimes, no surface manifestations in which case it is called a blind resource.

8.4.2.3   Hot Dry Rock (HDR) Resources.

Hot dry rock resources are defined as heat stored in rocks within about 10 km of the surface, from which energy cannot be economically extracted by natural hot water or steam. These hot rocks have few pore spaces or fractures, and therefore, contain little water and have little or no interconnected permeability. In order to extract the heat, experimental projects have artificially fractured the rock by hydraulic pressure, followed by circulating cold water down one well to extract the heat from the rocks and then producing from a second well in a closed system.

By drilling to these locations, we can tap the hot water or steam for power generation and for heating purposes such as space heating.

Generally, there are three ways to harness geothermal energy:

1. If the hot water reaches the surface as steam (which is very rare), it can be harnessed to drive an electric turbine. This is called a "dry steam" system.

2. If the hot water remains a liquid at a high enough temperature, it can be "flashed" into steam and then used to drive the electric turbine. This is called a "flash steam" system.

3. If the hot water is not hot enough to be flashed directly into steam, it can be used to flash another liquid with a lower boiling point—such as ammonia—and the resulting steam is then captured in turbines. This is called a "binary" system.

8.4.3   How to Estimate Geothermal Energy Output

We first consider a typical geothermal plant using the Stirling cycle as the heat engine for energy generation (Figure 8.25). Heat is supplied from the geothermal source (process 3-4), and rejected to an external sink (process 1-2) from the condenser and cooling tower. By considering heat transfer of the system, we can estimate the Stirling efficiency.

$$\eta_{Stirling} = \frac{Net\,work\,output}{Net\,thermal\,energy\,input} = \frac{Q_{out}}{Q_{in}} \tag{8.20}$$

The net work output, Q_{out},

$$Q_{out} = (C_v.\Delta T + \Delta s.T_{max}) - (C_v.\Delta T + \Delta s.T_{min}) = \Delta s\,(T_{max} - T_{min}) \tag{8.21}$$

The net work input from geothermal heat,

$$Q_{in} = \Delta s.T_{max} \tag{8.22}$$

Therefore, from Eq. 8.20 the Stirling efficiency,

$$\eta_{Stirling} = \frac{Q_{out}}{Q_{in}} = \frac{\Delta s\,(T_{max} - T_{min})}{\Delta s.T_{max}} = \frac{(T_{max} - T_{min})}{T_{max}} \tag{8.23}$$

Example 8.6:   A typical geothermal plant is operating based on the Stirling engine cycle. Evaluate the following:

(a) Heat supply

(b) Power available

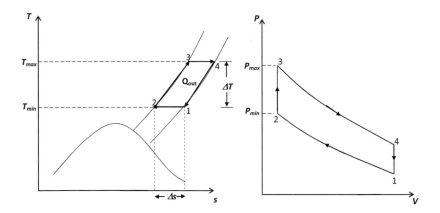

Figure 8.25
T-s and P-V diagram of a Stirling cycle for geothermal energy production.

(c) Electricity production, if total working hours = 700 hours per month

Given: Geothermal water delivery $\dot{m}_w$ is 250,000 kg/h; outlet temperature from well = 95 °C; average plant operating temperature = 24 °C; $Cp = 1.163 \times 10^{-3}$.

Solution

(a) Temperature difference $\triangle t = 95 - 24 = 71$ °C

$$\text{Heat supplied } Q_{supply} = \dot{m}_w \, Cp \, \Delta t$$

$$= (250,000 \, \text{kg/h}) \times (1.163 \times 10^{-3}) \times (71°C) = 20,639.5 \, \text{kW}_t \#$$

(b) Since Carnot efficiency $\eta_C = \dfrac{\Delta t}{T_{max}} = \dfrac{71\,°R}{95 + 273\,°R} = 0.193 = 19.3\%$

$$\therefore Power, \; P = Q \times \eta_c = 20,639.5 \, \text{kW}_t \times 0.193 = 3983.4 \, \text{kW}_e = 3.984 \, \text{MW}_e \#$$

(c) Electricity production is

$$E = P \times z = 3983.4 \, \text{kW} \times 700 \; hours/month = 2,788,380 \; kWh\#$$

We now turn the subject to the potential thermal energy available in *hard dry rock (HDR)*, also known as *petrothermal*:

The thermal energy or heat content of the rock can be estimated by [13],

$$Q = \rho C_p V \Delta T \tag{8.24}$$

where,
Q = heat thermal content or heat capacity, joules;
ρ = rock density $(kg/m^3) = 2,550 \; kg/m^3$;
Cp = heat capacity of the rock $J/kg°K$;
V = reservoir volume (m^3).
We can define the recoverable fraction of thermal energy, Fr, [13] as in Eq. 8.25,

$$Fr = \frac{\rho V_{active} C_r (T_{r,i} - T_{r,a})}{\rho V_{total} C_r (T_{r,i} - T_0)} \tag{8.25}$$

$$F_r = \phi_v \frac{(T_{r,i} - T_{r,a})}{(T_{r,i} - T_0)} \tag{8.26}$$

where

Q_{rec} = recoverable thermal energy content of the reservoir;
Q_{total} = total thermal energy content of the reservoir;
ϕ_v = active reservoir volume/total reservoir volume;
ρ = rock density (kg/m^3);
V_{total} = total reservoir volume (m^3);
V_{active} = active or effective reservoir volume (m^3);
C_r = rock specific heat (J/kg °C);
$T_{r,i}$ = mean initial reservoir rock temperature (°C);
T_o = mean ambient surface temperature (°C);
$T_{r,a}$ = mean rock temperature at which reservoir is abandoned (°C).

With a recovery factor and an abandonment temperature specified, the recoverable heat can be determined from the total energy in place:

$$Q_{rec} = F_r \rho V_{total} C_r (T_{r,i} - T_o) \tag{8.27}$$

and, the average MWe of capacity could be estimated by

$$MW_e = \eta_{th} \times Q_{rec} \times 1MJ/1000kJ \times 1/t \tag{8.28}$$

where,

Q_{rec} = recoverable thermal energy (heat) in kWs (or kJ);
η_{th} = net cycle thermal efficiency (fraction);
t = duration of operation, seconds.

8.4.3.1   What Are the Application Advantages of Geothermal Energy?

Advantages of geothermal energy:

1. Geothermal energy is cheaper than many other types of energy.

2. Unlike solar or wind energy, geothermal is not weather-dependent. Therefore, geothermal energy can be considered a constant and sustainable energy as it can be obtained consistently from our Earth's interior geothermal heat.

3. It is clean and environmentally friendly. We can clearly observe that the comparison of CO_2 and sulfur emissions from geothermal fuel and fossil-based power plants has shown that geothermal plants emit far less CO_2 and sulfur (see Figure 8.26).

4. It is useful for rural applications.

Disadvantages of geothermal energy:

1. Low production efficiency of about 15–20%.

2. *Land use.* Like onshore oil well drilling, the drilling of suction and reinjection wells geothermal wells requires large areas during the drilling period. However, a comparison study by DiPippo in 1991 [15] on land uses for typical geothermal flash and binary plants versus those of coal and solar photovoltaic plants shows that a solar–thermal plant requires about 20 times more area than a geothermal flash or binary plant; and a solar photovoltaic plant (in the best insolation area in the United States) requires about 50 times more area than a flash or binary plant per MW. The ratios are similar on a per MWh basis. The coal plant, including 30 years of strip mining, requires between 30–35 times the surface area for a flash or binary plant, on either a per MW or MWh basis. A nuclear plant occupies about seven times the area of a flash or binary plant.

3. *Thermal pollution.* Although thermal pollution is currently not a specifically regulated quantity, it does represent an environmental impact for all power plants that rely on a heat source for their motivating force. Considering only thermal discharges at the plant site, a geothermal plant is two to three times worse than a nuclear power plant with respect to thermal pollution, and the size of the waste heat rejection system for a

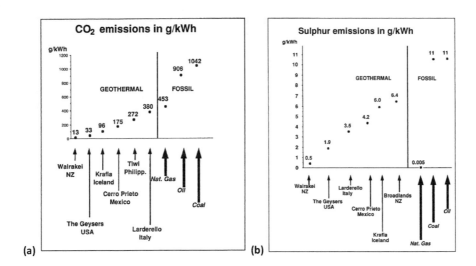

Figure 8.26
Comparison of CO2 and S emissions from geothermal fuel and fossil-based power plant. (After Barbier, 2002 [14].)

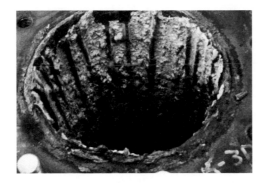

Figure 8.27
Chemical scale deposition and corrosion caused by impurities in hydrothermal fluids are major contributors to high operating and maintenance costs for geothermal power plants. [Courtesy of National Renewable Energy Laboratory (NREL)].

100 MW geothermal plant will be about the same as for a 500 MW gas turbine combined cycle [15]. Therefore, cooling towers or air-cooled condensers are much larger than those in conventional power plants of the same electric power rating.

4. Geothermal fluids often contain significant amounts of gases such as hydrogen sulphate and other dissolved chemicals, which can cause corrosion and scaling, leading to damage in the casing, pipelines, valves, and turbine blades (e.g., see Figure 8.27).

8.4.4 Type and Classifications of Geothermal Energy Applications

Generally, the geometry applications can be classified into two main types, i.e., geothermal power plants; and direct geothermal heat applications (Figure 8.28). We shall first discuss on geothermal power plant in Section 8.4.5; and follows by direct geothermal application in Section 8.4.6.

8.4.5 Geothermal Power Plant

Classification of geothermal power plants:

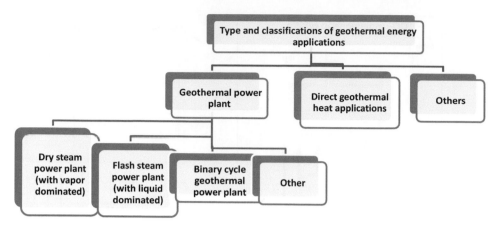

Figure 8.28
Type and classifications of geothermal energy applications.

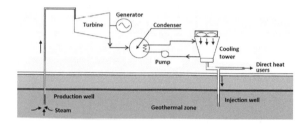

Figure 8.29
Schematic diagram of a typical dry steam geothermal power plant.

8.4.5.1   Dry Steam Power Plant (with Vapor Dominated)

Dry steam power plants derive energy from steam available from geothermal reservoirs underground (Figure 8.29). The steam is piped from the underground geothermal wells via the production well to the turbine inlet in the power plant to generate electricity. Turbine exhaust is run through a condenser, where the steam is turned back into liquid form to facilitate reinjection into the reservoir. Heat is rejected from the condenser via either wet or dry cooling towers. The leftover water and condensate (condensed steam) provide heat to the direct heat users, and the rest is reinjected back into the geothermal reservoir via the injection well.

8.4.5.2   Flash Steam Power Plant (with Liquid Dominated)

Flash steam power plants make use of hot water available from the geothermal reservoir, with temperatures of above approximately 180 °C (see Figure 8.30). This hot water flows up through the production well due to pressure difference. As it flows upward, the pressure decreases and some of the hot water boils into steam. This process is called *flash evaporation*. Flash evaporation is allowed to happen inside one or more flash tanks, which are large vessels allowing a portion of the liquid to expand to steam.

The steam is then separated from the water in the flash chamber and delivered to the turbine, which rotates and produces electricity. The leftover water and condensed steam provide heat to the direct heat users, and the rest are reinjected back into the geothermal reservoir via the injection well.

8.4.5.3   Binary Cycle Geothermal Power Plant

Figure 8.31 displays a binary cycle geothermal power plant, usually a suitable and cost-effective generation option for moderate-enthalpy hydrothermal fluids below 180 °C. Unlike the two open cycles just mentioned, the *binary* cycle power plant employs two cycles, i.e., the geothermal fluid and the organic working fluid (vapor) cycle as the

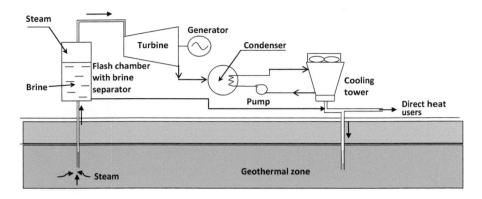

Figure 8.30
Schematic of flash steam geothermal power plant.

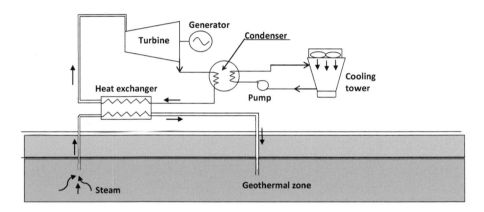

Figure 8.31
Binary cycle geothermal power plant.

secondary working fluid. The organic fluids suitable for such applications are isopentane or ammonia, both of which have a low boiling point and are easy to vaporize when subjected to high temperatures, and condense easily during cooling at the cooling tower. The hot reservoir fluid is passed through a heat exchanger, which heats the secondary fluid. The secondary fluid is recycled back into the heat exchanger and forms a closed loop, while the cold reservoir fluid is reinjected into the reservoir at the injection well. The water and working fluid are confined to distinct, closed loops during the whole process. Depending on the primary fluid available from the geothermal source, the binary power plants are normally between 7–12% in efficiency.

Present binary cycles can transform heat into power at hydrothermal fluid temperatures down to 105 °C (225 °F). In addition to their standalone applications, binary cycles may serve as bottoming cycles to generate power from brines at flash-steam units, providing an alternative to the use of a second or third flash stage for further increase in productivity. However, the improvement in hot-water rate must more than offset the capital cost of the additional cycles (EPRI, 2010 [12]).

We shall discuss the hybrid geothermal–fossil fuel energy plants in Chapter 10.

8.4.6 Direct Geothermal Heat Applications

There are many diverse uses of this energy, and the users are sometimes small and located in remote areas. The total installed capacity, reported at the end of 2009, for the world's geothermal direct utilization, is 50 583 MWt (World Energy Council, 2010). Figure 8.32 shows the various direct use of geothermal energy around the globe, with the geothermal heat pump as the most used candidate (49%), in terms of energy use of 210,000 TJ/year. Of course, bathing and swimming, in particular, hot springs, are also popular in usage (about 25% or a quarter). Other

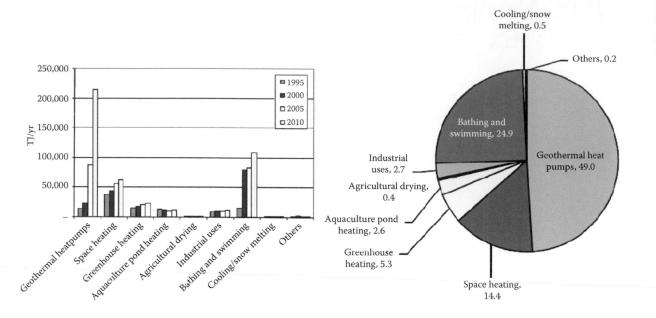

Figure 8.32
Worldwide geothermal energy direct use in (*left*): bar chart, and (*right*): pie chart. (From International Geothermal Association.)

applications are for space heating, greenhouse heating, industrial usages, etc.

Figure 8.33 and Figure 8.34 show the operation of a typical geothermal heat pump in cooling or heating mode, respectively. A desuperheater can be provided to use rejected heat in the summer and some input heat in the winter for domestic hot water heating (World Energy Council, 2010 [23]).

8.4.7   World Geothermal Energy Production

The top 24 countries active in geothermal energy application is given in Table 8.4. The top three countries are the United States, the Philippines and Indonesia, producing 29%, 18% and 11%, respectively. Thus, cumulatively these three countries produce close to 60% of world geothermal production.

Geothermal power plants operated in at least 24 countries in 2010, and geothermal energy was used directly for heat in at least 78 countries. According to Renewable 2011 Global status report by Ren21 [27], although power development slowed in 2010, with global capacity reaching just over 11 GW, a significant acceleration in the rate of deployment is expected as advanced technologies allow for development in new countries. Heat output from geothermal sources increased by an average rate of almost 9% annually over the past decade, due mainly to rapid growth in the use of ground-source heat pumps. Use of geothermal energy for combined heat and power is also on the rise. Significant geothermal power capacity (and CHP) was in project pipelines around the globe by year-end, with 46 countries forecast to have new geothermal capacity installed within the next five years.

Examples of common geothermal heat pump installations are given in Figure 8.35.

By the end of 2010, total global installations came to just over 11 GW, up an estimated 240 MW from 2009, and geothermal plants generated about 67.2 TWh of electricity during the year. Although geothermal developments slowed in 2010 relative to 2009, the lull was expected to be temporary. The lack of available drilling rigs (due to competition with the oil and gas industry) has hindered geothermal developers worldwide [27].

8.4.8   Some Examples of Geothermal Power Plants Around the World

1. First geothermal power plant, Laudarello, Italy.

2. Larderello geothermal steamfield, Tuscany, Italy. The installed geothermal electric capacity of the field was 547 MWe in year 2000.

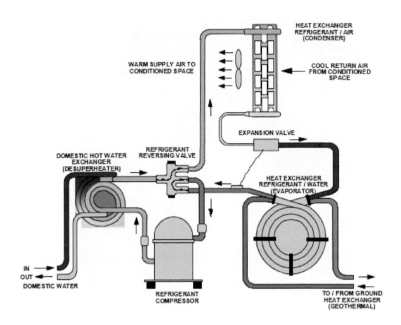

Figure 8.33
Typical geothermal heat pump (cooling cycle). (From World Energy Council, 2010.)

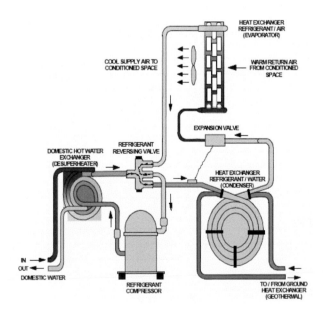

Figure 8.34
Typical geothermal heat pump (heating cycle). (From World Energy Council, 2010.)

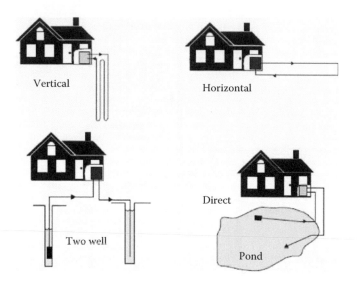

Figure 8.35
Examples of common geothermal heat pump installations. (From Lund, J. et al., GHC Bulletin, Sept. 2004. With permission.)

Table 8.4 Countries generating geothermal power

Country	Installed Capacity (MW)	Ranking in 2010
USA	3,086	1 (29%)
Philippines	1,904	2 (18%)
Indonesia	1,197	3 (11%)
Mexico	958	4
Italy	843	5
New Zealand	628	6
Iceland	575	7
Japan	536	8
El Salvador	204	9
Kenya	167	10
Costa Rica	166	11
Nicaragua	88	12
Russia	82	13
Turkey	82	14
Papua New Guinea	56	15
Guatemala	52	16
Portugal	29	17
China	24	18

(Adapted from Geothermal Energy Association.)

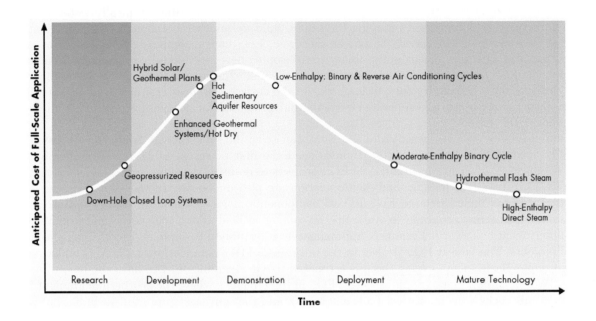

Figure 8.36
Grubb Curve displays geothermal power generation technologies at varying stages of development and commercial maturity.

3. The Geysers, Northern California, United States (dry steam power plant).

4. Ormat geothermal binary power plant exploiting hot water, 30 MWe, East Mesa, California. (Ormat, USA.)

5. Matsukawa geothermal power plant, the eldest plant in Japan, 23.5 MWe.

6. Wairakei geothermal power station, New Zealand, 181 MW (wet steam binary-cycle).

7. Kawerau geothermal power plant, Phillippines, 100 MW, completed in 2008.

8. Malitbog Power Station, Phillippines, 233 mW.

9. Hellisheidi geothermal power plant, Iceland, 133 MW (flash steam CHP combined cycle power plant), commissioned in 2010.

10. Cerro Prieto, geothermal power plant, Mexico, 720 MW (world's largest geothermal plant).

11. Wayang Windu geothermal power plant, Indonesia, 227 MW.

8.4.9   Future of Geothermal Technology and Applications

According to EPRI, geothermal power technologies are at different stages of development (see Figure 8.36), depending on resource type. Commercially mature direct- and flash-steam plants convert the highest-quality hydrothermal resources into electricity. Binary-cycle technologies for moderate-enthalpy hydrothermal resources are commercially available, and technologies for lower-enthalpy hydrothermal and hot dry rock (HDR) resources are emerging, with potential to vastly expand power production. Advanced and hybrid technologies are in the early development stages.

Enhanced or engineered geothermal systems (EGS) aim at using the heat of the Earth where no or insufficient steam or hot water exists and where permeability is low. EGS technology is centered on engineering and creating large heat exchange areas in hot rock. The process involves enhancing permeability by opening pre-existing fractures and/or creating new fractures. Heat is extracted by pumping a transfer medium, typically water, down a borehole into the hot fractured rock and then pumping the heated fluid up another borehole to a power plant, from which it is pumped back down (recirculated) to repeat the cycle. EGS can encompass everything from stimulation of already existing sites with insufficient permeability to developing new geothermal power plants in locations without geothermal fluids. EGS has been under development since the first experiments in the 1970s on very low permeability rocks, and is also known as hot dry rock technology. On the surface, the heat transfer medium (usually hot water) is used in a binary or flash plant to generate electricity and/or is used for heating purposes (Chandrasekharam and Chandrasekhar, 2010 [17]).

Among current EGS projects worldwide, the European scientific pilot site at Soultz-sous-Forêts, France, is in the most advanced stage and has recently commissioned the first power plant (1.5 MWe), thereby providing an invaluable database of information. In 2011, 20 EGS projects were under development or under discussion in several EU countries. EGS research, testing, and demonstration are also under way in the United States and Australia. The United States has included large EGS RD&D components in its recent clean energy initiatives as part of a revived national geothermal program. In Australia, 50 companies held about 400 geothermal exploration licenses in 2010. The government has awarded grants of approximately $205 million to support deep drilling and demonstration geothermal projects. The largest EGS project in the world, a 25 MWe demonstration plant, is under development in Australia's Cooper Basin. The Cooper Basin is estimated by Geodynamics Ltd to have the potential to generate 5 GWe to 10 GWe! In China, there are plans to test EGS in three regions where the geothermal gradient is high: in the northeast (volcanic rocks), the southwest (volcanic rocks), and the southeast (granite). In India, hot rock resources have been estimated to be abundantly available, because of a large volume of heat-generating granites throughout the country, but geothermal energy exploitation has yet to be initiated (Chandrasekharam and Chandrasekhar, 2010 [17].

8.5 *Application Projects, Assignments, Questions, and Problems*

Application Project

Linking knowledge with applied energy design and practice

1. *Mini Wind Turbine Project.* You are required to design, build, test and demonstrate a mini wind turbine system at the energy laboratory. It is understood that under different operating conditions, we would obtain different energy outputs i.e., power measured in Watts (W). Your group is required to conduct simple experiments at various operating conditions namely at five or six different blower speeds. What is the energy efficiency of your mini wind turbine system? Discuss your result obtained. The following are given: a multimeter with electrical wiring, an alternator (or small generator), timing pulleys, turbine shaft, and blades. Also, a multi-speed electrical wind blower is provided during commissioning and testing. (Group Project: 3-4 persons per group). (Adopted from FYP student individual project #5, 2006).

Assignments

1. *Small hydro project evaluation.* Suppose under the "green" energy for rural electrification application, you are asked to propose a micro-hydro scheme for a rural village, which requires electrification for two main purpose, i.e., (i) lighting; and (ii) powering a village community-based 5 kW fruit processing machine. The available stream gross head is 25 meters, and the hydrograph for a typical year is given (Figure 8.37). You are ask to evaluate the following:

   (a) Considering the water demand throughout a year and the shaded area of "dry season," calculate and pro-

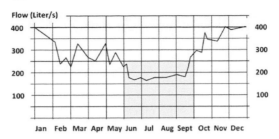

Figure 8.37
Hydrograph for 2009.

pose a design capacity for your hydro turbine. Explain your suggested capacity. Assume turbine efficiency = 65%.

(b) Calculate the plant factor for primary load requirements. Assuming lighting requirement = 15 kW operating from 6 pm to 6 am daily. Additionally, the food processing machine will be operating from 8 am to 4 pm daily (8 hours per day).

(c) Calculate actual plant factor, assuming total down-time is 1 month per year for operation and maintenance.

(d) Calculate unit energy cost. Assume capital cost = $4000, and allow 10% of the annual operating cost for the annual operating and maintenance cost.

Note: Individual project

Solution:

(a) Suggested capacity: $P = h \times Q \times 9.81 \times 0.65$

$$P = 25 \times (0.15) \times 9.81 \times 0.65 = 23.9\,kW = 20kW\,(suggested)\#$$

(b) Plant factor for primary load

$$= \frac{\text{total energy use, kW}}{\text{total energy available, kW}}$$

$$= \frac{(8\,\text{hours} \times 5\,\text{kW food processing}) + (12\,\text{hours} \times 15\,\text{kW lightings})}{20\,kW \times 24\,hours} = 0.5\#$$

(c) Actual Plant factor = Plant factor × actual time operating = $0.5 \times \frac{11}{12} = 0.46\#$

(d) Unit energy cost,

$$\text{Unit energy cost} = \frac{C_{\text{annual}} + (O + M)}{P_{\text{installed}} \times (365\text{days} \times 24\text{hours/day}) \times PF}$$

$$= \frac{(4000) + (\frac{10}{100} \times 4000)}{(4000) \times (365\text{days} \times 24\text{hours/day}) \times (046)} = \$0.0003/\text{kWh}\#$$

Questions and Problems

Multiple choice questions

1. A pelton turbine is classified as a _____ turbine, and it is usually designed in such a way that it is _____. Hence, it is best suited for _____ applications.

   (A) reaction; horizontal in turbine shaft inclination for maximum water volume intake; high head

   (B) impulse; horizontal in turbine shaft inclination for tapping maximum impulse force; high head

   (C) reaction; vertical in turbine shaft inclination for maximum water volume intake; low head

   (D) impulse; vertical in turbine shaft inclination allow for axial flow action; high head

(E) reaction; horizontal in turbine shaft inclination to allow for radial flow action, low water volumetric flow rate

2. Which of the following are the main differences between large hydro power plants and the micro-hydro power schemes?

I Large hydro power plants have higher system efficiencies.

II Large hydro plants come with big plant capacity up to MWs power level, while the micro-hydro applications range is usually between 5–100 kW.

III Large hydro power plants are designed for large centralized electrical power, while micropower plants are suited to small-scale isolated power generation.

IV Large hydro power plants provide virtually any large power requirements for big industries, towns and district applications, while micro-hydro is suited for SMEs and rural electrification applications.

(A) I and II only

(B) I and III only

(C) II and III only

(D) II, III, and IV only

(E) I, II, III and IV

3. What is the flow rate that must be supplied by hydro-power to a 500 kW capacity factory if the water head available is 20 meters high? Given: the estimated system component losses are: transmission losses = 7%, transformer-generator losses = 19%, turbine losses = 10%, and penstock losses = 12%.

(A) 612.2 m^3/s

(B) 61.2 m^3/s

(C) 0.00425 m^3/s

(D) 4.25 m^3/s

4. Which of the following statements *correctly* explain the advantages of ocean wave energy?

I Ocean wave energy is pollution free.

II Ocean wave power systems do not require large land areas.

III Ocean waves have consistent frequency and amplitude that make them a steady energy source.

IV The ocean waves have high power density.

V Ocean wave energy is subject to corrosion related problems.

(A) I and II only

(B) I and III only

(C) II and III only

(D) I, II, and IV only

(E) I, II, and III only

5. If a 15 m/s wind is blowing on a wind turbine and the tips of its blades are rotating at 80 m/s, what is the tip speed ratio?

(a) 4.0

(b) 5.0

(c) 5.3

(d) 0.2

(e) 0.5

6. A 0.5 m diameter rotor has 3 blades, each 0.15 meter wide. What is its solidity?

   (a) 24%

   (b) 27%

   (c) 29%

   (d) 16%

   (e) 30%

7. Estimate the thermal heat content of the rock of a typical geothermal site, given: rock density $= 2,550 \ kg/m^3$, heat capacity of the rock $= 0.00163 J/kg^{\circ}K$, temperature difference $= 80 \ ^{\circ}K$.

   (A) 3.3 MW

   (B) 0.33 MW

   (C) 519.6 kW

   (D) 0.5196 MW

   (E) 960 kW

8. What is the Carnot efficiency of a typical geothermal plant and its power capacity in MW? Average heat supply from the geothermal production well is around 1.92×10^7 kW. The average ambient temperature and maximum temperature are 23.7 °C and 80 °C, respectively.

   (A) 15.9 %; 3.05 MW

   (B) 39.5 %; 7.58 MW

   (C) 15.9 %; 7.85 MW

   (D) 25.9 %; 3.05 MW

   (E) 35.9 %; 5.87 MW

9. Offshore windmills are becoming more popular recently. Worldwide, the total installed capacity of wind farms is now close to 1,500 MWs. Which of the following statements *correctly* explain the advantages of ocean wave energy?

   I Offshore windmills can avoid land-use disputes, and the usual onshore noise and visual impact problems.

   II Offshore windmills have the advantage of stronger and more constant winds.

   III The operation and maintenance costs of offshore windmills are lower than those of onshore windmills.

   IV Offshore windmills can easily meet the integration of electrical network requirements.

   (A) I only

   (B) I and II only

   (C) I, II, and III only

   (D) I and IV only

   (E) III and IV only

Theoretical questions

1. With the aid of a schematic diagram, briefly discuss how a Pelton turbine works.

2. Discuss the differences between Pelton, Francis, and Kaplan turbines.

3. Discuss any two of the following for power production applications. Also list their application advantages and disadvantages.

   (a) tidal power

(b) wave energy

(c) ocean wave power

4. What is wind energy? Briefly discuss some applications of windmills.

5. What is geothermal energy? With the aid of a schematic diagram, describe how we can generate electricity using a geothermal energy-driven dry steam power plant.

6. List and briefly explain the advantages and disadvantages of geothermal energy applications.

Homework problems

1. *Windmill application.* A windmill generator has a 6-meter diameter rotor, rotating at 45 revolutions per minute (rpm) while the wind speed is at 7 m/s.

    (a) What is the total power generated if $C_p = 0.4$?

    (b) What is the tip speed ratio of the rotor?

    (c) What is the optimum tip speed ratio if the rotor is of the 4-bladed rotor type?
    (Answer: 2.376 kW; 2; 4)

2. *Micro-hydro applications.* You have been asked to design two micro-hydro schemes for the following applications:

    (a) A micro-hydro scheme to supply 75 kW to a small SME factory. There is a waterfall near the factory that is 20 m high. How much volume flow rate is required?

    (b) A micro-hydro scheme to power lighting for a rural village community. There is a small stream just adjacent to the village with water flow of 230 liters/s, and a head of 40 m. What is the estimated power output?

    Given: Net power, $Power_{net} = g \times Q \times h \times \eta_{overall}$; where, $Power_{net}$ in watts; Q in m^3/sec; $g = 9.81\,m/s^2$; $\eta_{overall}=60$ %.
    (Answer: 0.64 m^3/s; 54.2 kW)

3. *Hydro-electric power plant.* In a typical hydro-electric power plant, water is flowing at the available high volume flow rate of 505 m^3/s with an average head of 27 m. A certain manufacturer provides turbines that run at a speed of 150 rpm, with overall efficiency of 79%.

    (a) Calculate the specific power of the turbines if they have the maximum specific speed of 460.

    (b) Calculate the maximum power available from the turbines.

    (c) Find the number of turbines required for the plant.
    (Answer: 35, 624 kW; 105, 669.9 kW ;3)

4. Construct a parametric graph of power versus sweep area A, at wind speed V of 1, 10, 100, 1000, 5000 m^2, with swept area of from 1 to 5000 m^2, and at $C_p = 0.25, 0.30, 0.35, 0.40, 0.45$.

    Hint: for convenience of calculation, use spreadsheet software such as EXCEL, or write a computer program to generate the required output.

5. A typical small wind energy generator system has a 5-m diameter rotor that is rotating at 15 rpm.

    (a) Calculate the tip speed ratio of the rotor, at the following wind speeds: 3 m/s, 5 m/s, and 7 m/s

    (b) What is the relationship between tip speed ratio and available wind speeds?

6. *Ocean wind farm and power estimation.* A wind farm consisting of 100 individual wind turbines is constructed in a typical near-shore ocean. An average 10 m/s wind is blowing consistently at normal pressure and at average temperature of 15 °C. Compute the following:

    (a) The total power density from the wind

(b) The maximum obtainable power density

(c) The actual power density; assume $\eta = 45\%$

(d) The power of each of the turbines if the turbines' diameters are 100 m, 120 m, and 140 m, respectively

(e) What is the total power of the windfarm if the wind farm consists of 100 individual wind turbines, that is, 100 units of 100 m turbine diameter; or, 100 units of 120 m turbine diameter; or 100 units of 140 m turbine diameter?

(f) What can you infer from (e) above in relation to the turbine diameter and the total power it is possible produce?

(**Ans.:** $0.613kW/m^3$; $0.363kW/m^3$; $0.276kW/m^3$; $2,168\,kW$, $3,121kW$, $4,249kW$; $21.68MW$, $31.21\,MW$, $42.49\,MW$)

7. *Geothermal power plant and electric production.* A typical geothermal power plant is operating based on Stirling engine cycle. Evaluate the following:

(a) Heat supply

(b) Power available

(c) Electricity production, if total working hours $= 700$ hours per month

Data given: Geothermal water delivery $\dot{m}_w$ is 250,000 kg/h; outlet temperature from well $= 95$ °C; average plant operating temperature $= 24$ °C; $Cp = 1.163 \times 10^{-3}$.

Answers to multiple choice questions:

1. (B)

2. (D) II, III, and IV only.

3. (A) $612.2\,\mathrm{m^3/s}$

   Overall efficiency $\eta = 0.93 \times 0.81 \times 0.9 \times 0.88 = 0.6$

   Since *Power* $= w \times Q \times h \times \eta_{overall}$

$$\Rightarrow Q = \frac{Power}{w \times h \times \eta_{overall}} = \frac{500\,kW \times 1000\frac{W}{kW}}{(9.8 \times 1000) \times 20 \times 0.6} = 612.2\,\mathrm{m^3/s}\,\#$$

4. (D) I, II and IV only.

5. (C) The tip speed ratio is: tip speed ratio, $\mu = \frac{\text{Tip speed of blade}}{\text{Wind speed}} = 80/15 = 5.3\#$

6. (C) $Solidity = \frac{No\,of\,blade \times width}{\pi \times dia\,of\,rotor} \times 100\% = \frac{3 \times 0.15m}{\pi \times 0.5m} = 29\%$

7. (A). $Q = \rho C_p V \Delta T = 2550\,\frac{kg}{m^3} \times 0.00163\frac{J}{KgK} \times 10,000m^3 \times 80\,K = 3.3\,\mathrm{MW}\#$

8. (A).

   Carnot efficiency $\eta_c = \frac{\Delta T}{T_{max}} = \frac{(80+273)-(23.7+273)K}{(80+273)K} = \mathbf{15.9\ \%}\#$

   Power capacity $= Q \times \eta_c = 1.92 \times 10^7 \times 0.159 = 3.05\,MW\#$

9. (B). I and II only

Bibliography

[1] Harvey, A., Brown, A., (1993), *Micro-hydro design manual, a guide to small scale power schemes*, Intermediate Technology Publications: London.

[2] Gorlov, A. M., 2011, *Tidal energy*, 2955–2960, In Encyclopedia of Ocean Sciences, Academic Press, London.

[3] Kainkwa, R.M.R., Wind speed pattern and the available wind power at Basotu,Tanzania, *renewable Energy* 21(2000) 289–295.

[4] Gipe, P., 2004, *Wind power: renewable energy for home, farm, and business*, White River Junction, Chelsea Green Pub. Co.

[5] Atsu S. S. Dorvlo, Estimating wind speed distribution, *Energy Conversion and Management*, Volume 43, Issue 17, November 2002, pp. 2311–2318.

[6] Yurdusev, M.A., Ata, R., Çetin, N.S., Assessment of optimum tip speed ratio in wind turbines using artificial neural networks, *Energy*, Volume 31, Issue 12, September 2006, pp. 2153–2161.

[7] Wiesner, W., *The effect of aerodynamic parameters on power output of windmills*, The Boeing Vertol Company, Philadelphia, pp. 89–95.

[8] Archer, C. L., Jacobson, M. Z., *Journal of Geophysical Research*, Vol. 110, D12110, doi:10.1029/2004JD005462, 2005.

[9] Snyder, B., Kaiser, M. J., A comparison of offshore wind power development in Europe and the U.S.: Patterns and drivers of development, *Applied Energy*, Volume 86, Issue 10, October 2009, Pages 1845-1856.

[10] Dornberg, J., Danish amateur builds the world's biggest windmill. *Popular Science*, January 1979, pp. 81–84.

[11] EIA, Renewable Energy Consumption and Electricity Preliminary Statistics 2008. July 2009, DOE/EIA, Washington.

[12] Electric Power Research Institute (EPRI), Geothermal Power: Issues, Technologies, and Opportunities for Research, Development, Demonstration, and Deployment, February 2010.

[13] White, D.E., Williams, D.L., Eds., 1975. *Assessment of Geothermal Resources of the United States – 1975, U.S. Geological Survey Circular 727*, U.S. Government Printing Office, 155 p.

[14] Barbier, E., Geothermal energy technology and current status: an overview. *Renewable and Sustainable Energy Reviews* 2002, 6, 3–65.

[15] DiPippo, R., 1991, Geothermal Energy: Electricity Production and Environmental Impact, A Worldwide Perspective, *Energy and Environment in the 21st Century*, pp. 741–754, MIT Press, Cambridge.

[16] Geothermal Energy Association, *Geothermal Energy: International Market Update*, May 2010.

[17] Chandrasekharam, D., Chandrasekhar, V., (2010), Hot Dry Rock Potential in India: Future Road Map to Make India Energy Independent, proceedings at World Geothermal Congress 2010, Bali, Indonesia, April 25–29, 2010.

[18] Renewables 2010 Global Status Report. Renewable Energy Policy Network for the 21st Century, Paris.

[19] Fox, B., Flynn, D., Bryans, L., Jenkins, N., Milborrow, D., O'Malley, M., Watson, R., Anaya-Lara, O., 2007, *Wind Power Integration: Connection and System Operational Aspects*. London: The Institution of Engineering and Technology.

[20] BP Energy Outlook 2030, London, January 2011.

[21] Arndt, R.E.A. 1991. In |em Hydropower Engineering Handbook, J. S. Gulliver and R.E.A. Arndt, eds., pp. 4.1–4.67. McGraw-Hill, New York.

[22] http://www.theenergyworkshop.co.uk/case-studies.asp accessed September 25, 2011.

[23] World Energy council, 2010 Survey of Energy Resources, ISBN: 978 0 946121 021, London, United Kingdom.

[24] Rourke, F. O., Boyle, F., Reynolds, A., Tidal energy update 2009, Applied Energy, Volume 87, Issue 2, February 2010, Pages 398-409.

[25] Owen, A., Trevor, M.L., Tidal current energy: origins and challenges. In: *Future energy*. Oxford: Elsevier; 2008. pp. 111–28.

[26] Mazumder, R., Arima, M., Tidal rhythmites and their implications. *Earth–Science Rev*, 69 1–2 (2005), pp. 79–95.

[27] Ren21, Renewable 2011 Global Status Report, Paris, 2011.

[28] Lund, J., Sanner, B., Rybach, L., Curtis, R., Hellström, G., Geothermal ground source heat pump, *GHC Bulletin*, Sept 2004.

9

Nuclear Energy and Energy from Biomass

In this chapter, we will consider nuclear energy in the first part, and energy from biomass in the second part.

- Nuclear

- Biomass

9.1 Nuclear Energy and Nuclear Power Plant

Nuclear energy can be harnessed in a nuclear reactor, similar to energy from our Sun and the countless stars in some ways, for electrical and thermal energy production applications.

Nuclear power is economically feasible and meets $> 20\%$ of the world demand for electricity, generated by some 440 nuclear power reactors in 32 countries. The extraordinary high energy density of nuclear fuel relative to fossil fuels is an advantageous physical characteristic ([1]). Some countries have successfully utilized nuclear power for power generation. A typical example is France where nuclear power plants produce almost 80% or $\frac{4}{5}$ of its total electricity! In Korea, about 36% of the share of electricity generation is from nuclear sources, with 12 operating nuclear power plants (10 PWRs and 2 pressurized heavy water reactors). Japan has 53 operating nuclear power units with an installed capacity of 42 400 MW(e), supplying 34% of total electricity.

9.1.1 What Are Nuclear Power and Nuclear Power Plants?

In a nuclear power plant, the nuclear reactor and heat exchanger(s) take the place of conventional boilers, and the steam thus generated is expanded through the conventional turbine and generates electricity via a electrical generator.

The main components of a nuclear power plants and their functions are as follows:

1. Nuclear reactor — to produce heat.

2. Heat exchanger — functions as a boiler or steam generator, in which heat liberated from the reactor core is taken up by the coolant circulated around the core via a heat exchanger to generate steam. The followings components are similar to a steam turbine plant system.

3. Steam turbine — provides expansion of the steam for producing work.

4. Condenser — for heat rejection.

5. Generator — for electricity production.

Apart from electricity generation, depending on the type and operating temperature of the reactor, the associated heat production contributed from a nuclear plant provides a wide range of energy applications, notably:

- High Temperature Gas Cooled Reactor (HTGR). Can operate at about 1000 °C (suitable for water and hydrogen splitting applications)

- Liquid Metal Fast Reactor (LMFR). Operating up to about 550 °C (suitable for petroleum refining applications)

- Pressurized water reactor (PWR). Operating temperature ranging from 270°–310°C PWR has compact reactors, thus fit well nuclear submarines, nuclear ships and steam-powered aircraft application.

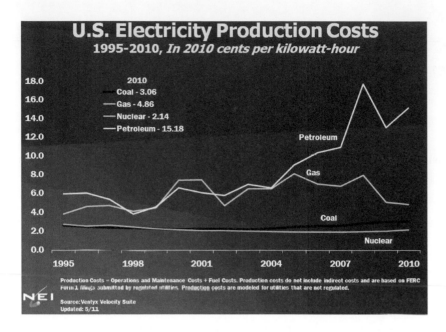

Figure 9.1
Electricity production cost comparison. [Courtesy of Nuclear Energy Institute (NEI).]

- Pressurized heavy water reactor (PHWR). Suitable for Nuclear co-generation plant, in particular for nuclear desalination applications.

Now, although there is wide concern on fossil fuel use — the applications of which causes global adverse environmental effects such as the greenhouse effect and global warming—generally the public at large around the world are more worried about possible nuclear radiation leaks and nuclear catastrophic failures such as those of Chernobyl and Fukushima Nuclear Power Plant.

Despite its controversial reputation, nuclear power is efficient, reliable, and environmentally friendly [24].

9.1.2 What Are the Application Advantages of Nuclear Energy and Nuclear Power Plants?

1. *Nuclear energy has high energy density.* 1 kg of Uranium-235 will generate as much energy as 3,000 tons of coal (and without carbon emissions). As a general comparison, according to IAEA [1], one (1) kg each of coal, oil, and uranium will deliver approximately 3 kWh, 4 kWh, and 50,000 kWh of electricity, respectively.

2. *Security of Supply, i.e., large reserves of uranium around the world.* Known uranium reserves with reactors operating primarily on a once-through cycle (without reprocessing of spent fuel) would be sufficient for fuel supply for at least 50 years at current levels of use. Recycling plutonium from reprocessed spent fuel in thermal reactors as mixed oxide fuel and the introduction of fast breeder reactors would increase the energy potential of today's known uranium reserves by up to 70 times, enough for more than 3,000 years at today's levels of use.

3. *Low fuel production cost.* Nuclear energy is known as having one of the lowest average production costs over recent years (see Figure 9.1).

4. *Mature technology.* The technology of nuclear energy is established.

5. *Environmentally friendly.* Compared with fossil fuel power plants, nuclear power plants contribute no or far less greenhouse gases.

6. *Wide range of nuclear applications.* Depending on the types of nuclear plants and operating temperatures, besides electrical power, nuclear also offers various suitable thermal energy-related applications such as district heating, industrial process, and desalination applications.

However, nuclear energy power plants also have disadvantages, as follows:

1. Generally, nuclear power plants are expensive in construction.

2. Catastrophic failure possibilities (for instance, the incidents at Chernobyl and the recent 2011 Fukushima Nuclear Power Plant failures).

3. Nuclear waste in the form of spent nuclear fuel is extremely hazardous, causing disposal and storage problems (it needs to be stored for thousands of years!).

4. Possibilities of causing radiation and human health problems.

9.1.2.1 Example 9.1

Energy source requirement comparison: nuclear power versus coal energy production.
Consider a nuclear power plant with capacity of 5000 MW of power, with fuel from uranium U_{235}.

(a) Calculate the hourly energy output of the nuclear reactor.

(b) Compute the fuel consumption of the nuclear reactor in one (1) hour. Assume plant efficiency is 25%.

(c) What would be the amount of coal per hour required to obtain the same energy if the efficiency is 75%?

Data:

Uranium: $U_{235} = 8.2 \times 10^{13}$ J/kg; and, average fission energy release of U_{235} nuclide, E_f is 200 MeV.

Coal: Calorific value of coal = 29,300 kJ/kg.

Solution:

(a) Hourly energy output of nuclear reactor:

$$\text{Power output} = 5,000 \text{MW} = 5,000,000 \text{ kW} = 5 \times 10^9 \text{W} = 5 \times 10^9 \text{J/s}$$

$$\text{Hourly energy output} = (5 \times 10^9)\frac{\text{J}}{\text{s}} \times (60 \times 60)\frac{\text{s}}{\text{hour}} = 1.8 \times 10^{13} \text{J}\#$$

(b) Hourly fuel consumption in nuclear reactor:

$$\text{Hourly energy input} = \frac{\text{hourly energy output}}{\text{efficiency}} = \frac{1.8 \times 10^{13}\text{J}}{0.25} = 7.2 \times 10^{13} \text{ J}$$

$$\text{Therefore, hourly fuel consumption} = \frac{7.2 \times 10^{13}\text{J}}{8.2 \times 10^{13}\text{J/kg}} = 0.878 \text{ kg}\#$$

(c) Coal required per hour:

$$\text{Hourly energy input} = \frac{\text{hourly energy output}}{\text{efficiency}} = \frac{1.8 \times 10^{13}\text{J}}{0.75} = 2.4 \times 10^{13} \text{ kJ}$$

$$\text{Therefore, coal required per day} = \frac{2.4 \times 10^{13}\text{kJ}}{29,300 \text{ kJ/kg}} = 819,113 \text{ kg} \#$$

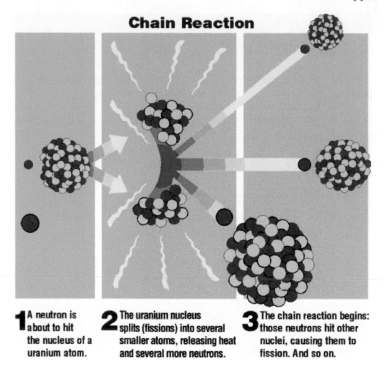

Figure 9.2 (SEE COLOR INSERT)
Nuclear reaction — fission, energy, and heat production. (From United States Nuclear Regulatory Commission, public domain.)

9.1.3   What Is the Working Principle of a Nuclear Power Plant?

9.1.3.1   The Science of Nuclear Fuel: A Quick Overview

Nuclear power is created through a series of nuclear reactions, the process of which is called *fission* or *splliting* (see Figure 9.2) During this process, Uranium-235 isotopes absorb neutrons, causing the U-235 to become unstable and split into two lighter parts called fission fragments. These two new parts are lighter because some of the mass is converted into energy and released as heat. Along with the heat, more neutrons are released. The neutrons continue to disrupt more U-235 atoms in a self-sustaining chain reaction, while the heat from the fission is used to boil water. Boiling water turns turbines and creates electricity, thus generating nuclear power. The leftover fission fragments gather to form nuclear waste, and the safe disposal of this waste is one of the biggest challenges in the nuclear power industry.

9.1.3.2   The Technology of Nuclear Fuel: An Overview of the Nuclear Fuel and Technology Brief

Nuclear fuel fabrication process   According to the World Nuclear Organization [2], nuclear fuel fabrication are generally involves in a three-stages processes (Figure 9.3):

1. Conversion from either UF_6 or UO_3 to UO_2 power.

2. UO_2 powder treatment and pellet production.

3. Fuel rod and fuel assembly production.

9.1.4   The Basic Nuclear Power Plant

Figure 9.4 shows how a typical nuclear power plant is a basic diagram. The plant was designed with two circulation loops with heat transfer at the heat exchanger. In the reactor loop, water leaves the reactor core at high pressure of 16,000 kPa and temperature of around 310 °C, and re-enters the reactor core at about 275 °C. In the steam

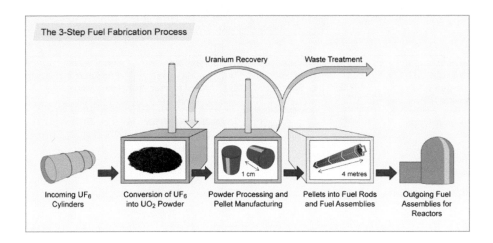

Figure 9.3
The three steps of the nuclear fuel fabrication process. (From World Nuclear Organization [2], public domain.)

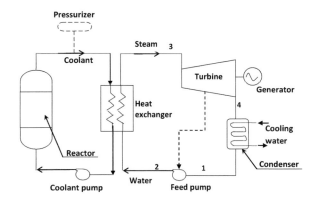

Figure 9.4
Nuclear power plant is a basic design, schematic diagram.

turbine circulation loop, the steam leaves the heat exchanger at 5,000 kPa (dry saturated), and the condensate from the condenser is a saturated liquid at 5 kPa. To increase system efficiency, the feedwater can be preheated to the saturation temperature of the bleed steam at about 500 kPa in an open feed heater (shown in the figure as a dotted line), with suitable bled steam mass flow rate.

9.1.5 Classification of Nuclear Power Plants

Basically, nuclear power plants can be classified based on the coolant type used (Figure 9.5). There are a variety of coolants that can be used for cooling the reactor.

9.1.5.1 Pressurized Water Reactor (PWR)*

A pressurized water reactor (PWR) is the most common type of commercial reactor, that was originally developed in the USA for submarine propulsion applications. Currently, about 60% of the world's commercial reactors are PWRs, and more than BWRs (around 20%). According to British Energy [3], a PWR's thermal efficiency is about 32%.

---

*In 2005, pressurized water reactor (PWR) around 265 in total mostly from the following countries: US, France, Japan, Russia and China.

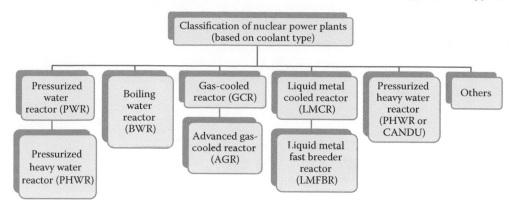

Figure 9.5
Classification of nuclear power plants.

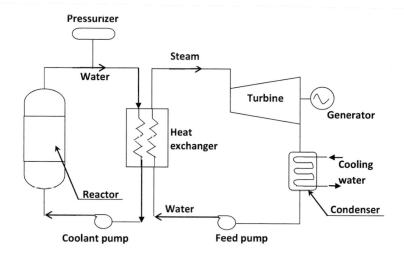

Figure 9.6
Schematic diagram of a Pressurized Water Reactor (PWR).

Pressurized Water Reactor (PWR) is the term used to describe a nuclear reactor that uses conventional water or "light water" as coolant and moderator, and that uses both natural and highly enriched fuel, such as natural uranium (see Figure 9.6.)

9.1.5.1.1 Advantages and Disadvantages of Pressurized Water Reactor (PWR).

Advantages:

1. *Stability.* PWR reactors are very stable at a wide range of temperatures.

2. The PWR turbine cycle loop is separate from the primary loop, so the water in the secondary loop is not contaminated by radioactive materials.

3. PWRs can passively* scram the reactor in the event that offsite power is lost to immediately stop the primary nuclear reaction. The control rods are held by electromagnets and fall by gravity when current is lost; full insertion safely shuts down the primary nuclear reaction (see, e.g., [33]).

---

*Scram refers emergency shutdown (usually) of a BWR.

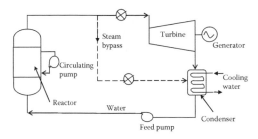

Figure 9.7
Schematic diagram of a Boiling Water Reactor (BWR).

9.1.5.2 Boiling Water Reactor (BWR)[†]

A boiling Water Reactor (BWR) (see, Figure 9.7) refers to a nuclear reactor that uses "heavy water" as coolant and moderator, and that uses enriched uranium fuel. The fuel is kept inside rods, and the rods are put in bundles. Heat is produced by nuclear fission in the reactor core. Water is circulated through the reactor core, picking up heat as the water moves past the fuel assemblies, thus causing the water to heat up and become steam. The steam, passing through the turbines, condenses in the condenser, which is often cooled by huge water bodies, such as sea or river water, for efficient heat rejection.

BWR reactors are generally designed with electrical outputs of 500 to 1500 MW range, with efficiency of $> 30\%$. Examples of BWRs are the General Electric (GE) BWRs and the Japanese BWRs (see Table 9.1) [4].

9.1.5.2.1 Advantages The advantages, among others are:

1. The BWR reactor operates at a substantially lower pressure as compared to a PWR.

2. The BWR has fewer components due to no steam generators and no pressurizer vessel as compared to the PWR.

3. Lower risk of rupture causing loss of coolant compared to a PWR.

9.1.5.2.2 Disadvantages of BWR

1. A BWR requires larger pressure vessels than a PWR of similar power.

Table 9.1 Typical parameters for BWR cores in Japan [4]

No.	Item	Tsuruga Unit-1 (BWR-2)	Fukushima Unit-1 (BWR-3)	Hamaoka Unit-2 (BWR-4)	Tokai Unit-2 (BWR-5)	Kashiwaza ki Unit-6 (ABWR)
(1)	Thermal output (MW)	1064	1380	2436	3293	3926
(2)	Electric output (MW)	357	460	840	1100	1356
(3)	Core equivalent dia. (m)	3.02	3.44	4.07	4.75	5.16
(4)	Core effective height (m)	3.66	3.66	3.71	3.71	3.71
(5)	Fuel assemblies (Number)	308	400	560	764	872
(6)	Control rod (Number)	73	97	137	185	205
(7)	Power density (kw/l)	About 40	About 40	About 50	About 50	About. 50

[†]The total number of Boiling Water Reactors (BWRs) was around 94 in 2005, mostly in the United States, Japan, and Sweden.

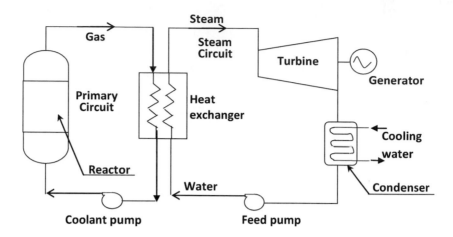

Figure 9.8
Schematic of high temperature, gas-cooled reactor.

2. A BWR requires active cooling following shutdown; if not, the reactor can still overheat to a temperature high enough that zirconium in the fuel cladding may react with water and steam to produce hydrogen, which may result in hydrogen gas explosions.

For further details on BWRs and their applications, readers can refer to, for instance, [4].

9.1.5.2.3   Differences between a PWR and a BWR

1. The arrangement of a BWR power plant is simpler than that of a PWR.

2. A pressurizer and a heat exchanger are required in a PWR.

3. A BWR uses "heavy water" while a PWR uses conventional water or "light water" as coolant and moderator.

4. A PWR requires two water cycles, which are usually called primary circuit and secondary (heat exchanger) circuit, while the heat exchanger circuit is eliminated in a BWR.

9.1.5.3   Gas-Cooled Reactor (GCR) *

In a gas-cooled reactor (see Figure 9.8), the coolant used is generally carbon dioxide (CO_2). Other forms of gas can be air, hydrogen (H_2), and helium. The moderator used is usually graphite [34].

There are two types of gas-cooler reactors, i.e., the gas cooled, graphite moderator reactor (GCGM), and the high temperature gas cooled reactor (HTGC), which operate at pressure and temperature of 7 bar, ~350 °C and 15–30 bar, ~750 °C, respectively. GCR can attain higher thermal efficiency compared with PWRs due to higher operating temperatures.

Advantages of gas-cooled reactor:

1. No or relativly less corrosion problem.

2. The core material, i.e., uranium and graphite, can sustain high temperatures.

3. GCRs can attain higher thermal efficiency compared with PWRs due to higher operating temperatures.

---

*GCRs mostly developed in the United Kingdom. Magnox stations are an older gas-cooled reactor designs; the fuel used is uranium clad in a magnesium alloy called Magnox with a low neutron absorption.

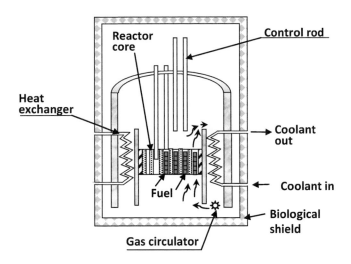

Figure 9.9
Schematic diagram of an AGR reactor core.

Disadvantages:

1. Low power density due to low heat transfer coefficient; thus a large vessel is required.

2. More power is required for gas coolant circulation as compared to water-cooled reactors.

9.1.5.4   Advanced Gas-Cooled Reactor (AGR)

Advanced gas-cooled reactors (AGR) are the second generation of gas cooled reactors (GCRs), using fuel of uranium dioxide enriched to 3% U-235, and clad in stainless steel cans with a graphite moderator and carbon dioxide coolant. In the AGR, the heat exchangers and the carbon dioxide circulators are situated inside the prestressed concrete pressure vessel (Figure 9.9) such that the carbon dioxide is completely contained within the pressure vessel. Operating temperature of an AGR is around 650 °C.

An AGR's thermal efficiency is about 42%, generating a typical electrical output of up to 660MW. This is the type of reactor operated by British Energy and is unique to the United Kingdom (British Energy [3].

Advantages of AGR

1. Unlike in a Magnox design, the quick transit time of feed water into steam makes the AGR far more responsive to demand changes requirements [35].

2. The AGR has a relatively advanced graphite moderator design, with high thermal storage capacity.

3. It has an efficient reactivity control systems, such as coolant flow rates control.

Disdvantages of AGR

1. Subject to thermal stress and reactor instability, due to the use of enriched uranium fuel.

9.1.5.5   Liquid Metal Cooled Reactor (LMCR)

Liquid metal cooled reactors consist of three circuits, i.e., the primary circuit, secondary circuit, and steam circuit (see Figure 9.10).

The primary circuit has liquid metal, e.g., liquid sodium, which circulates through the fuel core in the reactor and gets heated by the fission of the fuel. This liquid sodium gets cooled in the intermediate heat exchanger, transfers the heat to the secondary circuit, and goes back to the reactor vessel.

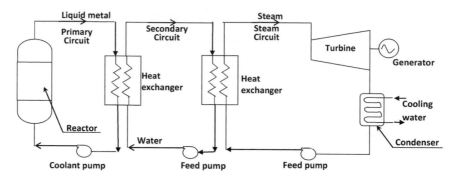

Figure 9.10
Schematic of liquid metal cooled reactor.

The secondary circuit, in a sodium graphite reactor (SGE), for example, has an alloy of sodium and potassium in liquid form. The coolant absorbs heat from the intermediate heat exchanger and gets heat from the liquid sodium of the primary circuit. The heated liquid sodium then passes through a boiler, and the steam generated from the boiler will be superheated.

In the steam circuit, the superheated steam generated will pass through the steam turbine.

In a sodium graphite reactor, sodium works as a coolant and graphite is the moderator.

Advantages:

1. The use of sodium as a coolant in a LMCR reactor does not need to be pressurized

2. High conversion ratio, thermal efficiency, and heat removal

3. Small overall size

Disadvantages:

1. Possibly of thermal stress in related equipment.

2. Heat exchanger must be leak proof, and shielded from radio activity.

3. Complex in cooling. It is necessary to shield the primary and secondary cooling systems with concrete blocks from sodium, which is high radioactive.

9.1.5.6 Liquid Metal Fast Breeder Reactor* (LMFBR)

A fast breeder reactor is basically a reactor in which enriched plutonium is kept without using moderator (see Figure 9.11.) Light water, oil, or graphite is used to provide neutron shielding. The reactor core of LMFBR is usually cooled by liquid metal, in particular, liquid sodium alloy, which is very efficient at removing heat. Sodium has the advantage that it can be heated to 5600 °C without being pressurized, so the reactor does not need a pressure vessel to produce high temperature steam (Nakae et al., 2011 [5]).

---

*Liquid-Metal cooled Fast Breeder Reactors (LMFBRs) were actively developed from the 1970s to the 1980s, in particular, in the United States, France, the United Kingdom, and Germany. In the following years, however, the development of LMFBRs slowed down in almost all countries except in Japan and Russia. There are signs that LMFBR development has been revived. Japan still continues to develop LMFBRs, and the prototype reactor "Monju," which stopped due to a sodium leak accident in the secondary loop from 1995, was successfully restarted in May 2010 (excerpt from Nakae et. al., 2011).

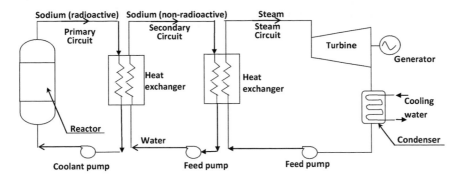

Figure 9.11
Schematic of a liquid metal fast breeder reactor (LMFBR).

Advantages:

1. A moderator is not required.

2. It has a small core size, and is very compact since it has no moderator.

3. *High breeding characteristics.* An LMFBR reactor is called a "breeder" reactor since it can breed its own fuel from the fuel blanket during operation. It is possible to design an LMFBR such the amount of plutonium it makes while running is more than enough to replace the plutonium that is consumed. A breeder reactor can also produce enough plutonium to fuel further reactors, without the need of mining new fuel. Alternatively, it can be designed to use up excess stocks of plutonium.

4. *Safety features.* Like gas cooled AGRs, LMFBRs also have the very attractive safety feature that it can maintain cooling by natural convection should cooling pumps fail.

Disadvantages of LMFBR:

1. Highly enriched fuel (>15%) is required.

2. There is difficult in handling sodium, as it becomes hot and radioactive.

3. It requires extra safety requirements in fuel system design against high neutron flux and against possible melt-down (Nakae et. al., 2011 [5]). Among other things, the following are the important matters in relation to the fuel safety design:

    (i)   Maintaining fuel integrity

    (ii)   Maintaining control rod insertion paths

    (iii)   Securing of geometry requirements to allow proper cooling

9.1.5.7   Pressurized Heavy Water Reactor (PHWR, Better Known as CANDU)

The PHWR is better known as the CANDU reactor. It is so called because it is a Canadian design using deuterium oxide (heavy water) as both coolant and moderator and natural uranium fuel. Heavy water is an oxide of heavy hydrogen or deuterium (D_2O), rather than regular hydrogen (H_2O). Pressurized heavy water is pumped through horizontal fuel tubes and heated by the nuclear reaction before being passed to a steam generator, as in a PWR, but without the need for a pressure vessel as the tubes provide the containment. The fuel is natural uranium oxide contained in Zircaloy tubes. The average power density is about one-tenth that of a PWR or four times that of an AGR. The output of CANDU reactors ranges from 600 MW up to 930MW with a thermal efficiency of about 30% (Raina et al., 2006 [6]; Bhardwaj, 2006 [7]; British Energy, 2011 [3]).

The advantages of PHWR or CANDU:

1. It uses natural uranium; thus enriched fuel is not required.

2. The cost of the reactor is less as it is operating at lower pressure.

3. The moderator fuel is heavy water, which incurs low fuel consumption.

Disadvantages:

1. Although the cost of the reactor is low, the maintenance cost, i.e. heavy water, is very expensive.

2. Large size of reactor, as the power density is low as compared to PWR or BWR.

9.1.5.8 Other Reactor Designs

Other advanced reactor designs are as follows:

- The Pebble Bed Reactor, South Africa, a type of High Temperature Gas-Cooled Reactor (HTGCR); (see, e.g., Walter et al., 2006 [36]).

- Small, Scaled, Transportable, Autonomous Reactor (SSTAR), Lawrence Livermore National Laboratory, USA.

- KAMINI, a unique reactor using Uranium-233 isotope for fuel, Kalpakkam, India.

9.1.6 How to Estimate Nuclear Energy Output

In a nuclear fission power plant, the power density depends on many parameters and can be defined as follows:

$$Power\,density,\ P_{density} = E_f \times A_f \times \phi \tag{9.1}$$

where,

$P_{density}$ = power density, MW_{th}/m^3;
E_f = energy released per fission = 200 MeV/fission = 3.2×10^{-11} J/fission,
    (since 1 eV = 1.602×10^{-19} J or W-s);
A_f = fission cross-section, m^2;
ϕ = average neutron flux.

Then, the power of nuclear, power nuclear, is

$$Power_{nuclear} = P_{density} \times NV \tag{9.2}$$

where,

$Power_{nuclear}$ = nuclear power; MW;
$P_{density}$ = power density, MW_{th}/m^3;
N = Number of fissile nuclei/m^3;
$V = volume\,of$ the nuclear fuel, m^3

Generally, fission power reactors designed with higher power density will have smaller core size and smaller volume. For instance, a high temperature gas-cooled reactor (HTGR) with core volume of $430\,m^3$ will have power density of only 10 MW_{th}/m^3, compared with a pressurized water reactor (PWR) having $40\,m^3$ core volume but with higher power density of 75 MW_{th}/m^3. It is to be noted that higher power densities require rigorous heat transfer systems, and thus structural safety considerations.

The total number of fissile nuclei in the reactor, NV, can be computed by

$$NV = \frac{m \times 6.02 \times 10^{26}}{235} \tag{9.3}$$

where,

m =mass of the U_{235} fuel, kg

Figure 9.12
Nuclear Station Unteweser, Germany. Built in 1972, the plant has one reactor with electricity production capacity of 1410 MW.

Example 9.2   What is the core volume ratio of a boiling water reactor (BWR) to that of a pressurized water reactor (PWR) if the average core power density is 50 MW_{th}/m^3 and 75MW_{th}/m^3 for the BWR and PWR, respectively? Assume both systems have the same capacity of 1000 MW.

Solution   The core volume ratio is

$$\frac{V_{BWR}}{V_{PWR}} = \frac{\left(\frac{Power_{nuclear} \times N}{P_{density}}\right)_{BWR}}{\left(\frac{Power_{nuclear} \times N}{P_{density}}\right)_{PWR}} = \frac{(P_{density})_{PWR}}{(P_{density})_{BWR}} = \frac{75}{50} = 1.5\#$$

In other words, the BWR requires 1.5 times the core volume size of a PWR to accomplish the same power capacity.

Example 9.3   Compute nuclear energy output (in J) in term of energy released per kg fission of U^{235}. Assume 200 MeV is released per fission of uranium nucleus(E_f=200). What is the power produced (in Watts) if a nuclear reactor consumes 5 kg in 1 day?

Solution   Since 1 eV = 1.602×10^{-19}J, and 235 kg of U^{235}(i.e., 1 kg-atom of U^{235}) contains 6.02×10^{26}atoms (Avogadro's number),
energy released per kg of $U^{235}$is,

$$= \left\{200 \times (10^6)\right\} \times \left\{(1.602 \times 10^{-19})\right\} \times \frac{(6.02 \times 10^{26})}{235} J$$

$$= 8.2 \times 10^{13} J \text{ or W.sec (per kg)}\#$$

So, for 5 kg of fuel, power output in 1 day = $5 \times \frac{(8.2 \times 10^{13})J}{24 \times 60 \times 60 \, s} = 4.75 \times 10^9 W\#$

9.1.7   Examples of Nuclear Power Plants around the Globe

Globally, the number of PWRs is more than the number of BWR.
Installed nuclear capacity rose very quickly, from 1 GW (1960s) to 100 GW (late 1970s), to 300 GW (late 1980s), and approaching 400 GW (2010).

- Obninsk Nuclear power plant, Union of Soviet Socialist Republics (USSR) (1954)

- Nuclear Station Unteweser, Germany, (1972), 1 reactor: 1410 MW; Net generation: 284.290 GW h.

Figure 9.13
Three Mile Island Nuclear Power Plant, United States. The plant has one pressurized light-water reactor with net capacity of 810 MW. (From http://www.ohiocitizen.org/campaigns/electric/2004/ph_three_mile_island500.jpg.)

- Magnox gas-cooled reactor, United Kingdom, 1950s

- Susquehanna Nuclear Power Plant, 2 boiling water reactors, Pennsylvania, United States, 2,275 MW (1982)

- Three Mile Island, United States, (see Figure 9.13), 1 pressurized light-water reactor, 810 MW (1974)

- Peach Bottom Power Station, United States, 2 boiling water reactors, 2,224 MW (1966)

- Limerick Generating Station, United States, 2 boiling water reactors; 2,268 MW (1986)

- Chernobyl, 4 High Power channel-type reactors, also named RBMK-100D, Ukraine, Russia, total capacity 4000 MW (reactor number four exploded in 1986).

9.1.8   Recent Advancements and Future of Nuclear Energy

Recent advancements and the future of nuclear power have been observed by many and reported in books, journals, and Web sites, among others: [26]; [25]; [27]; [24], etc.

Recent progress reported is regarding South Africa, which introduced a safer new generation of nuclear power plants called the Pebble Bed Modular Reactors (PBMRs) see Figure 9.14. The reactor encases uranium or plutonium in cermic spheres the size of a tennis ball. The heat that is generated is transferred into a helium gas. The pressure from the safe helium gas is then used to turn a turbine and create electricity. To melt the ceramic spheres containing uranium and release radioactive material, it would take temperatures of 3000 °C, but the inside only ever rises to temperatures of about 900 °C. The PBMR has a passive safety feature in that it requires no operator intervention. Removal of decay heat is achieved by radiation, conduction, and convection. The combination of very low power density of the core and temperature resistance of the fuel provide the safety requirements of the design.

Most recently, according to world nuclear association (www.world-nuclear-news.org), there is a revival of interest in small and simpler units of nuclear plant (for generating electricity and process heat) called small modular reactor (SMR). The primary objectives are to reduce cost, to provide power away from large grid systems, and provide safety.

9.1.8.1   Future of Nuclear Storage and Safety

The recent tsunami and earthquake that struck the coast of Northern Japan earlier in March 2011, crippling the Fukushima Nuclear Power Plant. The devastating 14 m high tsunami knocked out the vital cooling systems causing the reactor to overheat. This event has prompted the International Atomic Energy Agency (IAEA) and all Nations for highest standards of safety and security.

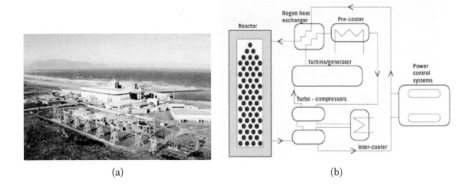

Figure 9.14
Pebble Bed Modular Reactor, Centurion, South Africa. (a) Photo; (b) schematic diagram. (Courtesy of The Institution of Engineering and Technology [29].)

9.1.8.2 Future of Nuclear Energy for Sustainability

Nuclear energy sustainability will depend on the actual capability of reducing the inventory and long-term radiotoxicity of nuclear waste, mainly dominated by the amount of transuranic isotopes remaining on the spent fuel [24].

9.1.8.3 Some Possible or Likely Future Scenarios of Nuclear Energy Applications

1. Wide use of nuclear energy for CHP (cogeneration) plants.

2. *Underwater nuclear cooling.* It was proposed that nuclear plants of the future may be installed under sea water, which provides excellent cooling. The plants may take the shape of pyramids. Undersea facilities have easy access to cooling water while pyramid-shaped plants provide safety against seismic activity [27].

3. *Nuclear policy and standards.* All nations around the world, and the International Atomic Energy Agency (IAEA), will continuously call for the highest standards of safety and security. We have discussed nuclear storage and safety in Chapter 5.8.

4. *Lower cost of construction and plant operations.* According to Joseph Romm's, 2009 Report [28], any nuclear industry would be "likely to pose challenges unless new nuclear plants can be built for under $4,000/kW total cost and provide electricity to the grid at $0.10 per kW or less."

9.2 Energy from Biomass (Bioenergy)

Biomass can be harnessed, similar to the prehistoric plant life that contributed to coal, oil, and gas, for bioenergy production and applications.

9.2.1 What Are Biomass and Bioenergy?

Biomass refers to biological materials derived from living organisms such as woods and residues from agriculture. In terms of applied energy, biomass is used to produce bioenergy — in the form of electricity, heat, or both.

A wide range of biomass sources can be used to produce bioenergy in a variety of forms. For example, food, fiber, and wood process residues from the industrial sector; energy and short-rotation crops and agricultural wastes; and forest and agroforest residues from the forestry sector can all be used to generate electricity, heat, combined heat and power, and other forms of bioenergy (GBEP [8]; Ruane et al., 2010 [10]).

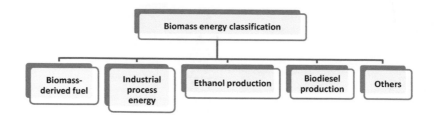

Figure 9.15
Biomass energy classification I.

Traditional biomass materials, including fuelwood, charcoal, and animal dung, continue to be important sources of bioenergy in many parts of the world, and, to date, wood fuels represent by far the most common sources of bioenergy.

Modern bioenergy relies on efficient conversion technologies for applications at the household, small business and industrial scale. Solid or liquid biomass inputs can be processed to be more convenient energy carriers. These include solid biofuels (e.g., firewood, wood chips, pellets, charcoal, and briquettes), gaseous biofuels (biogas, synthesis gas, hydrogen) and liquid biofuels (e.g., bioethanol, biodiesel).

In the United States, corn-based ethanol is currently the largest source of biofuel as a gasoline substitute or additive, and recent energy legislation mandates further growth of both corn-based and advanced biofuels from other sources, see, e.g., [9]. Brazil, the world's largest bioethanol exporter, produced 26 billion litres of sugar cane-based ethanol equivalent to around 30% of the world's total ethanol fuel use in 2010. The success is due mainly to the efficient sugarcane cultivation technology, which uses modern equipment and cheap sugar cane as feedstock, while the residual cane-waste is used to process heat and power. This results in a very competitive price as well as high useful energy balance, that is, energy output per energy input. Now, in regard to soil quality, according to Somerville et al., several hundred years of experience with sugar cane production and recent studies of the effects of sugar cane cultivation on soil carbon indicate that the sugarcane crop can be grown sustainably and can in fact improve terrestrial carbon sequestration [30].*

9.2.2 Biomass Energy Classifications

Biomass energy in the form of biomass fuel includes those of domestic and industrial applications.

Classification 1: Based on fuel application type. They can be essentially divided into a few categories as follows (see Figure 9.15):

• Biomass-derived fuel. The domestic sector is the main user of biomass-derived fuel, particularly for cooking and space heating.

• Industrial process energy. The industries include: agricultural and food processing, metal processing, and mineral-based activities (e.g., brick making, ceramics, and foundry; and forest products and textile process industries (e.g., timber drying, paper making, silk and their textiles).

• Ethanol production

• Biodiesel production

Classification 2: Based on process type. This general classification is based on the processes for biomass energy production (see Figure 9.16). The processes are divided into 2 groups, namely wet bioconversion process and dry bioconversion process. The wet bioconversion process includes extraction, digestion, and fermentation, whilst the dry bioconversion process includes gasification, pyrolysis, combustion and liquifaction.

---

*Because it is inevitable that some mineral nutrients will be removed when biomass is harvested, it will be essential to recycle mineral nutrients, which are not consumed in the production of biofuels, from biomass-processing facilities back onto the land.

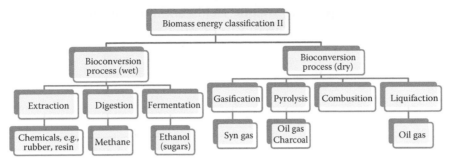

Figure 9.16
Biomass energy classification II.

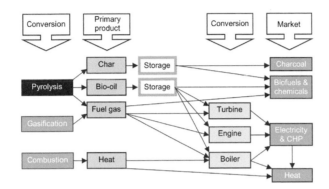

Figure 9.17
Thermal biomass conversion and its products. (From A.V. Bridgwater, Biomass and Bioenergy, 2011, [31] doi:10.1016/j.biombioe.2011.01.048. With permission.)

9.2.2.1 Digestion

Digestion is a biological process that occurs in the absence of O_2 and in the presence of anaerobic organisms at ambient pressure and temperature. In a biogas plant, the container in which the digestion process takes place is called a digester.

9.2.2.2 Fermentation

Fermentation is the biological process of organic matter decomposition by microorganisms, especially by employing bacteria and yeast. It is a well established technology for conversion of sugar from crops into ethanol. Ethanol is usually blended with gasoline for use in internal combustion engines.

9.2.2.3 Gasification

Figure 9.17 summarizes three main thermal processes, namely pyrolysis, gasification, and combustion processes available for converting biomass to a more useful energy form and products.

Gasification is a process that takes place by heating the biomass with limited O_2 to produce *low* heating value gas. Also, it can take place by reacting with steam and O_2 at high pressure (typically greater that 30 Bar) and high temperature (typically reaching 1,500 K) to produce *medium* heating value gas, usually called raw synthesis gas or syngas, a mixture composed primarily of CO and H_2 and some minor by products. Typical syngas produced from coal in an oxygen-blown gasifier is by volume about 30% to 60% CO, 25% to 30% H_2, 0% to 5% methane (CH_4), 5% to 15% carbon dioxide (CO_2), and some water (National Energy Technology Laboratory [6]). The by products are removed to produce a clean syngas that can be used as a fuel to generate electricity or steam, as a basic chemical building block for a large number of uses in the petrochemical and refining industries, and for the production of hydrogen. The medium heating value gas can be used as fuel for direct combustion, or subjected to a liquefaction process by converting it to methanol or ethanol. Or, it can also be converted into high heating value gas [12].

Table 9.2 Typical product weight yields obtained by different modes of pyrolysis of wood

Mode	Conditions	Liquid	Solid	Gas
Fast	~500 °C, short hot vapour residence time ~ 1 s	75%	12% char	13%
Intermediate	~500 °C, hot vapour residence time ~ 10–30 s	50% in 2 phases	25% char	25%
Carbonisation (slow)	~400 °C, long vapour residence hours → days	30%	35% char	35%
Gasification	~750–900 °C	5%	10% char	85%
Torrefaction (slow)	~290 °C, solids residence time ~ 10–60 min	0% unless condensed, then up to 5%	80% solid	20%

(From A.V. Bridgwater, Biomass and Bioenergy, (2011), [31] doi:10.1016/j.biombioe.2011.01.048. With permission.)

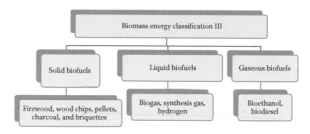

Figure 9.18
Biomass energy classification III.

We will discuss the use of gasification technique for an integrated gasification combined-cycle (IGCC) plant in Chapter 10.

9.2.2.4 Pyrolysis

Pyrolysis is thermal decomposition at high temperatures, i.e., the organic material is converted into low calorific value gas, solids, and liquids, occurring in the absence of oxygen O_2.

Lower process temperatures and longer vapour residence times favour the production of charcoal. High temperatures and longer residence times increase biomass conversion to gas, and moderate temperatures and short vapor residence time are optimum for producing liquids. Three products are always produced, but the proportions can be varied over a wide range by adjustment of the process parameters. Fast pyrolysis for liquids production is currently of particular interest as the liquid can be stored and transported, and used for energy, chemicals, or as an energy carrier. Table 9.2 indicates the product distribution obtained from different modes of pyrolysis, showing the considerable flexibility achievable by changing process conditions [31].

9.2.2.5 Combustion

Combustion is the process of burning in the presence of O_2 to produce heat, light, and by-products. Traditional biomass sources such as fuel wood, charcoal, and animal dung, continue to be important supplies of bioenergy for direct combustion applications in many parts of the world.

Our modern bioenergy is geared towards efficient conversion technologies for applications at the household, small business, SME and as industrial scale.

Incineration is a term used to describe complete combustion to ashes.

The technology of *fluidized bed combustion* can be used for efficient combustion of forestry and agricultural materials such as sawdust, wood chips, rice husks, nutshells, etc. The biomass is fed into the a bed of hot inert particles, such as sand, and kept in fluidized state with air at sufficient flow velocity from below. The operating temperature typically is in the range of 750–950 °C, enabling high combustion efficiency.

Classification **III:** This is based on fuel forms or energy carriers generally called biofuel.

Biofuel can be broadly defined as solid, liquid, or gas fuel consisting of or derived from biomass. As shown in Figure 9.18 these include solid biofuels (e.g., firewood, wood chips, pellets, charcoal, and briquettes), gaseous biofuels (biogas, synthesis gas, hydrogen) and liquid biofuels (e.g., bioethanol, biodiesel).

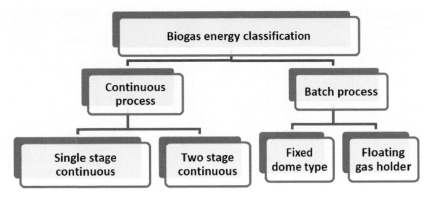

Figure 9.19
Biogas plant technology classification.

9.2.2.5.1 Solid Biofuels Solid biofuels, including firewood, wood chips, pellets, charcoal, and briquettes, are all used normally for direct combustion applications.

9.2.2.5.2 Liquid Biofuels

Biogas. In a biogas plant, also called a bio-digester, biomass like vegetable waste or animal excreta undergoes a decomposition process in the absence of O_2 to form a mixture of gases called *biogas*. Like natural gas, the biogas contains mostly methane. Biogas is very useful for cooking and lighting applications.

The application advantages of biogas are:

1. Biogas can be produced easily.

2. Biogas is generally considered a green gas, because it essentially burns with less smoke and leaves ash as residues.

3. Residential wastes and other biowastes can be disposed of via biogas processes.

Biogas plants or reactors: In order to systematically produce biogas, and to produce it in great quantity, many designs of biogas plants for that purpose are available. Generally, biogas plants can be classified as in Figure 9.19. Here are some simple descriptions of some of the biogas plants:

Figure 9.20 shows schematic diagrams of bioreactor designs of (a) a fixed dome and (b) floating gas holder biogas reactor, respectively.

Fixed dome. Advantages of fixed dome biogas reactors:

1. Low initial and operating costs.

2. Good heat insulation as it is located underground;

3. Less maintenance problems as there are no moving parts;

4. Quantity of gas production is relatively more than for a floating gas holder type of reactor.

Disadvantages of fixed dome biogas reactors:

1. Variable gas pressure

2. Possible problem of scum formations

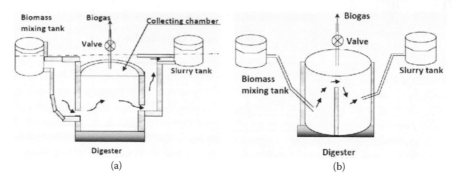

Figure 9.20
Schematic diagram of biogas reactors: (a) fixed dome and (b) floating gas holder.

Floating gas holder.   Advantages of floating gas holder biogas reactors:

1. Constant in gas pressure

2. Less scum problems

3. Compared with fixed dome, the floating gas holder type of biogas reactor is safe from explosions as there is no mixing of biogas and external ambient air.

Disadvantages:

1. High initial construction cost

2. High maintenance cost

Synthesis gases.   Synthesis gas is also called syngas. It is basically a mixture of hydrogen and carbon monoxide. Examples are wood gas, pyrolysis gas, and coke gas.

We shall also note that the production of alcohol fuels (ethanol or butanol) from syngas can be done via thermochemical conversion. Another option is to use microorganisms to ferment the syngas.

Bio-hydrogen   Biohydrogen, bioethanol, and biogas can be produced by using microalgae as feedstock.

9.2.2.5.3   Gaseous Biofuels

Bioethanol   Bioethanol or bio-ethyl alcohol can be produced from biomass that contains appreciable amounts of sugar or materials that can be converted into sugar, such as starch or cellulose. For example, sugar cane and sugar beet are two types of common feedstock that contain sugar, while corn, wheat, and other cereals contain starch that can be converted into sugar.

In producing bioethanol from sugar crops, the sugar from sugar crops is fermented to yield bioethanol. Ethanol fermentation is the biochemical process by which sugars, such as glucose, fructose, and sucrose, are converted into ethanol and carbon dioxide using yeast or other micro-organisms. Glucose and fructose are monosaccharides with six carbon atoms, and are thus termed 6-carbon sugars. Sucrose is a disaccharide made of glucose and fructose joined together. A final step distils (purifies) the ethanol to the desired concentration and usually removes all water to produce "anhydrous ethanol" that can be blended with gasoline. With sugar cane, the "bagasse" (i.e., the crushed stalk of the plant) can be used as a solid fuel and burned for heat and electricity [10].

In producing bioethanol from starchy materials, the process is more difficult compared to that for sugar crops because an additional step, *hydrolysis* of the feedstock, is required. Starch is a polysaccharide consisting of long chains of glucose molecules. Through hydrolysis, where the starch reacts with water, the starch is broken down to fermentable glucose molecules. Hydrolysis, also known as saccharification, can either be enzymatic (using a mixture of enzymes known as amylases) or acid-based [13]. Once the starch is broken down to glucose syrup, the process is similar to that for sugar crops (i.e., the sugars are fermented to ethanol, typically using the yeast called *Saccharomyces cerevisiae*, followed by distillation of the ethanol to the desired concentration and removal of water).

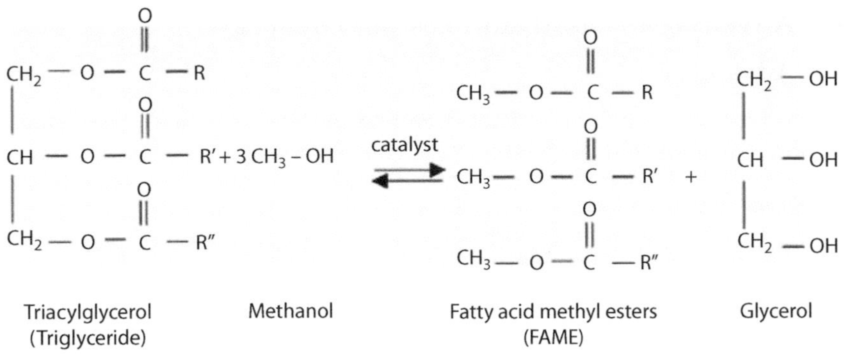

Figure 9.21
Schematic representation of the transesterification of triglycerides (vegetable oil) with methanol to produce fatty acid methyl esters (biodiesel). (From Zhang, Y. et al., *Bioresource Technology* 89, 1–16, 2003. With permission.)

Biodiesel   Biodiesel is made by combining alcohol with oil such as that of vegetable oil or animal fat. Like diesel, biodiesel can be used as an alternative fuel for diesel engines (e.g., transport application for automotive industry) and diesel burners (e.g., for burning oils in SMEs applciations). It can also be used as an additive to reduce vehicle emissions.

For biodiesel production, the feedstocks involved include (e.g., [14]; [16]):

1. *Vegetable oils derived from oilseed crops* (e.g., soybean, sunflower, jatropha, oil palm, rapeseed)

2. *Animal fats.*

3. *Vegetable oils derived from algae.* According to Danielo [36], the possible annual oil yield from algae is about ~ 5000–7000 gallons/acre [4.6–18.4 $\frac{liter}{m^2}$]; about 30 times more oil per hectare than from the terrestrial plants used for the fabrication of biofuels. Algae are excellent bioremediation agents, and have the potential to absorb good amounts of CO_2 (Brown and Zeiler, 1993 [16]). The ability of algae to fix CO_2 has been proposed as a method of removing CO_2 from flue gases from power plants for GHG emission reduction applications.

4. *Used frying oil* (e.g., from restaurants). The use of used or waste frying oil is well justified for bio-diesel applications.

The major components of vegetable oils and animal fats are triacylglycerols (TAGs, also called triglycerides), which consist of three long-chain fatty acids linked to a glycerol backbone. Natural oils are too viscous to be used in modern diesel engines. However, chemical modification of natural oils was introduced in the 1980s. This can help to bring the viscosity of the oils within the range of petroleum diesel, by reacting these TAGs with simple alcohols such as methanol in a chemical process called "transesterification" to form alkyl ester (methyl ester), generically known as biodiesel, the properties of which are very close to those of petroleum diesel (Sheehan et al., 1998 [17]). Figure 9.21 is a diagram representing the transesterification of TAG with methanol to produce biodiesel [18].

Now, the viability of first generation biofuels production has been questioned by many, as it is in conflict with food supply*

The second generation biofuels are derived from non-food feedstock. They are extracted from microalgae and other microbial sources, ligno-cellulosic biomass, rice straw, and bio-ethers, and are a better option for addressing food and energy security and environmental concerns.

Microalgal biofuels could be a viable alternative (Patil et al., 2008 [19]). However, Patil et al., (2008) emphasized *hydrothermal liquefaction* technology for direct conversion of algal biomass to liquid fuel.

---

*The Brazilian bioethanol from sugar cane is a special case, and is accepted by many due to the vast amount of continuous local sugar cane plantations, as well as the efficiency of bioethanol production.

9.2.3 Application Advantages/Disadvantages of Bioenergy

9.2.3.1 Application Advantages

1. "Green" bioenergy production (i.e., via fermentation and pyrolysis processes, etc.) has minimal environmental impact, except when it is used in direct combustion of plant mass.

2. Biomass is available around the globe.

3. Unlike fossil fuel, alcohols, and other fuels, biofuels are efficient, viable, and relatively clean-burning; for instance, for the car fuel application.

4. Biodiesel can be produced from oil products and waste oil products like vegetable oils and animal fats — suitable for waste-to-energy applications.

5. Bio-ethanol conversion can take the waste from wood residues (from wood product plants such as paper mills and sawmills), and municipal solid waste (MSW) and other waste energy sources (e.g., agriculture residue from energy crops).

6. Landfill gases or biogases can be obtained from waste organic materials.

9.2.3.2 Disadvantages

1. Like combustion of fossil fuel, direct burning of biomass could contribute a great deal to global warming and particulate pollution.

2. It is not yet economical for vast applications, both in terms of producing the biomass and converting it to alcohols, ethanol, and other biofuels.

9.2.4 *Case Study 1*: Biodiesel for Oil-Based Mud Drilling

Biofuels are generally known as non-toxic and environmentally safe fuels, and they can be used for many applications to replace the conventional fuels such as diesel and mineral oils. One such application is for petroleum oil rig drilling in oil-based mud drilling applications ([20]; [32]).

The primary functions of drilling fluids include maintaining wellbore stability (by providing hydrostatic pressure to prevent formation fluids from entering into the wellbore), keeping the drill bit cool and clean during drilling, carrying out drill cuttings, and suspending the drill cuttings while drilling is paused and when the drilling assembly is brought in and out of the hole. The drilling fluid used for a particular job is selected to avoid formation damage and to limit corrosion. The three main categories of drilling fluids are water-based mud (which can be dispersed and non-dispersed), non-aqueous mud, usually called oil-based mud, and gaseous drilling fluid.

In the past, the usage of oil especially diesel oil as the continuous phase of oil-based drilling mud was widespread when drilling through sensitive producing formations and troublesome shale zones. However, diesel oil is harmful to the environment—particularly the marine environment in offshore applications, and due to this extensive legislation exists in many countries to regulate this form of oil pollution. Mineral oils have been widely used as an alternatives to diesel oil. Subsequently, various types of low-toxicity oils were introduced to replace the more toxic mineral oils and diesel oil-based mud system, which has proven toxic to the environment.

The use of biodiesel, such as palm oil derivative or oil from other biomass, could be considered as an alternative base fluid that is harmless to the environment.

Tests have been undertaken to evaluate the characteristics of palm oil derivative as the base fluid in the oil based mud and its toxic effect on marine life. The results obtained showed that biodiesel fuel such as palm oil derivative is a suitable alternative to formulate oil based drilling mud with the necessary rheological properties. The main advantage of these vegetable oils is that they are generally regarded as non-toxic have no aromatic content, and are widely used in the food industries. Promising qualities of the methyl esters are that they have a high flash point, are non-toxic, and have good emulsion stability (see Table 9.3). In addition, they are cheaper than any mineral oils (Yassin et al.,1991 [20]).

Table 9.3 Based oil physical properties comparison

Property	Base oil Required Properties	m. e of CPO	m. e of PFAD	Diesel Oil	Mineral Oil	Test Method
Specific Gravity		0.873	0.873	0.8364	0.7906	ASTM D1250
Aniline Point, C	> 65 C	15	12	63	73	ASTM D611
Pour Point, C	< Ambient temp.	14 C	13.5 C	-45 C	-54 C	ASTM D 97
Flash Point, C	> 66 C	172 C	169 C	76 C	79 C	ASTM D93
Fire Point, C	> 80 C	180 C	178 C	108 C	93 C	ASTM D93
Kinematic Viscocity @ 40 deg.C, cST	2.3 - 3.5	4.6	4.7	3.4	1.6	ASTM D445
Aromatic Content, %	4 - 8	-	-	30	0.9	ASTM D1250

(From Yassin, A.M. et al., Society of Petroleum Engineers (SPE) Technical Papers Microfiche No. 23001. pp. 190–206, 1991. With permission.)

9.2.5 *Case Study 2*: Economics of Producing Microalgal Biodiesel (Chisti, 2008 [21])

Yusuf Chisti [21] conducted mathematical estimations to investigate the economics of producing microalgal biodiesel. The approach is to study the competitiveness by estimating the maximum price that could be reasonably paid for algal biomass with a given content of oil if crude petroleum can be purchased at a given price with the same level of energy produced:

The quantity of algal biomass (M, tons), which is the energy equivalent of a barrel of crude petroleum (i.e., has the same energy as a barrel of petroleum), can be estimated as follows:

$$M = \frac{E_{petroleum}}{q(1-w)E_{biogas} + ywE_{biodiesel}} \tag{9.4}$$

where,

M is the quantity of algal biomass (ton);

$E_{petroleum}$ (~6100 MJ) is the energy contained in a barrel of crude petroleum;

q is biogas volume produced by anaerobic digestion of residual algal biomass (m^3/ton);

w is the oil content of the biomass in percent by weight;

E_{biogas} is the energy content of biogas (MJ/m^3);

$E_{biodiesel}$ is the average energy content of biodiesel.

Assuming that converting a barrel of crude oil to various useable transport energy products costs roughly the same as converting M tons of biomass to bioenergy, the maximum acceptable price that could be paid for the biomass would be the same as the price of a barrel of crude petroleum; thus:

$$\text{Acceptable price of biomass, (\$/ton)} = \frac{\text{Price of a barrel of petroleum(\$)}}{M} \tag{9.5}$$

The price of microalgal biomass, estimated using Eq. 9.4 and Eq. 9.5, for biomass that contains various levels of oil (15–55% by weight) is shown in Figure 9.22 for crude petroleum prices of up to \$1000 per barrel. At present the price of crude oil is about \$100 per barrel. At this price, microalgal biomass with an oil content of 55% will need to be produced at less than ~\$340 per ton to be competitive with petroleum diesel. Literature suggests that, in 2007, microalgal biomass could produced for around \$3000/ton [22]. Therefore, the price of producing the biomass needs to decline by a factor of 9, through advances in production technology and algal biology, to make biodiesel from microalgae a feasible option!

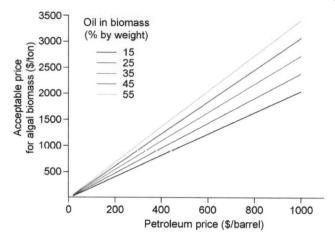

Figure 9.22

Estimation of microalgae price versus petroleum price. Competitiveness of microalgal biomass depends on its oil content and the price of oil [22].

Figure 9.23

Utilization of pyrolysis gas in Chiba, Japan (biomass: 300 ton/day public waste; LHV = 1.5–2 kWh/Nm³; pyrolysis gas: H_2 20–40% CO 35–40% CO_2 25–35% N_2 2–5%). (Courtesy of GE Jenbacher, Jenbach, AustriGasificationa. With permission.)

9.2.6 Examples of Biomass Power Plant Applications around the Globe

1. Utilization of pyrolysis gas in Chiba, Japan, 2001. (See Figure 9.23.)

2. Biomass (Wood) gasification at Güssing, Austria. (See Figure 9.24 and Figure 9.25 for the photo and schematic diagram of the biomass (wood) plant.)

9.2.7 Recent Advancement and Future of Bioenergy

The use of gasifiers to generate power in rural areas or developing countries that have biomass resources but cannot easily access liquid fuels will become popular. The power range envisaged may be between 5 and 300 kW.

Major challenges posed by present fossil and crude-oil-derived fuels are high costs due to insecurity of fuel supplies, and climate change risks from the accumulation of fossil fuel CO_2 and other greenhouse gases in the atmosphere. One option for addressing these challenges simultaneously involves producing ultraclean synthetic fuels from coal and lignocellulosic biomass with CO_2 capture and storage. Systematic comparisons were made to cellulosic ethanol as an alternative low GHG-emitting liquid fuel and to alternative options for decarbonizing stand-alone fossil-fuel

Figure 9.24
Wood gasification at Güssing, Austria. Capacity: wood 8 MW; LHV 10.5 MJ/Nm^3; wood gas: N_2 3 % CH_4 10 % CO_2 23 % H_2 40 % CO 24 %. (Courtesy of GE Jenbacher. With permission.)

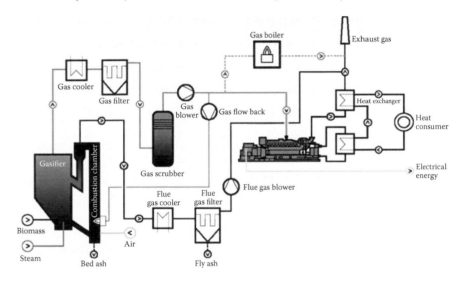

Figure 9.25
Schematic of the wood gasification biomass plant at Güssing, Austria. (Courtesy of GE Jenbacher. With permission.)

power plants. The analysis indicates that FTL fuels are typically less costly to produce when electricity is generated as a major coproduct than when producing mainly liquid fuel (Liu et al., 2001 [23]).

For regions such as sub-Saharan Africa, breakthroughs in the area of liquid biofuels for cooking would be very important as energy for cooking is a priority since 95% of all staple foods must be cooked, and traditional cookstoves, powered by fuelwood and dung, have negative health and social impacts [10].

9.3   *Assignments, Questions, and Problems*

Assignments

1. Individual report: conduct literature review about any *one* of the following topics:

(a) Life cycle study of an energy-efficent biofuel for combined heat and power (CHP/co-generation) application.

(b) Nuclear spent fuels: application and safety.

(c) Pyrolysis from biomass: applications and techno-economical study.

(d) Biodiesel production for future energy application systems.

Questions and Problems

Multiple choice questions

1. "It is the most common type of commercial reactor" and "a nuclear reactor that uses conventional water or light water as coolant and moderator, and that uses both natural and highly enriched fuel such as natural uranium." Which of the following nuclear plants are referred to in the above statements?

    (A) PWR

    (B) BWR

    (C) Gas-cooled reactor (GCR)

    (D) Advanced Gas-cooled reactor (AGCR)

    (E) All of the above

2. The following are the disadvantages of the use of nuclear energy for energy production applications, *except:*

    A) Generally, nuclear plants are expensive in construction.

    B) Nuclear energy has limited applications compared with other energy resources.

    C) Hazardous, causing disposal and storage problems.

    D) Possibilities of causing radiation and human health problems.

    E) Catastrophic failure possibilities.

3. The following are the advantages of the use of nuclear energy for energy production applications, *except:*

    A) Nuclear energy has high energy density.

    B) Nuclear energy is reliable.

    C) Compared with fossil fuel power plants, nuclear power plants contribute far less greenhouse gases.

    D) Nuclear waste is less harmful and can be easily treated and stored.

    E) Compared with fossil fuel power plants, nuclear power plants contribute no or far less greenhouse gases.

4. The following statements correctly describe the advantages of the use of biomass for energy production applications, *except:*

    (A) Biomass is available around the globe.

    (B) Unlike fossil fuel, alcohols, and other fuels, biofuels are efficient, viable, and relatively clean-burning, for instance, for car fuel application.

    (C) Currently, bioenergy is considered tecno-economical for vast applications around the globe, for instance, in terms of converting it to ethanol and other biofuels.

    (D) Solid or liquid biomass can be conveniently transported as "energy carriers."

5. Biomass energy applications include the following:

I. Bio-ethanol production

II Biomass-derived power

III Industrial process energy

IV Biodiesel production

(A) I only

(B) I and IV only

(C) I, II, and III only

(D) III and IV only

(E) I, II, III, and IV

6. One of the most important applications of biomass energy systems is in the fermentation of fuel ethanol from biomass. Which of the following is/are usually considered the biomass resource(s) for ethanol conversion?

I. Wood residues, usually coming from the wood product industry

II Vegetable oil

III Agriculture residue

IV Energy crops

V Municipal solid waste

(A) I only

(B) I and IV only

(C) I, II, and III only

(D) I, III, IV, and V only

(E) All of the above

Theoretical questions

1. List the three steps of nuclear fuel fabrication processes. Also sketch a schematic diagram and label the processes on the diagram.

2. With the aid of a schematic diagram, describe briefly the working principle of a liquid metal cooled nuclear reactor.

3. What are the differences between a boiling water reactor (BWR) and a pressurized water reactor (PWR)?

4. What are the differences between a gas-cooled reactor (GCR) and an advanced gas-cooled reactor AGR?

5. With the aid of a schematic diagram, describe briefly the working principle of a liquid metal cooled nuclear reactor.

6. Combustion is one of the bioconversion processes from dry biomass. Briefly discuss the following:

   (a) Define a combustion process

   (b) What is incineration?

   (c) How can the technology of *fluidized bed combustion* help in the combustion process?

7. What are the differences between fixed dome and floating gas holder biogas reactors?

8. Describe briefly the following processes for bioenergy production applications:

   (a) Fermentation

   (b) Gasification

   (c) Pyrolysis

9. What is a digestion process and what is a digester? With the aid of a schematic diagram, describe either a fixed dome or a floating gas holder gas reactor for batch production of biogas.

10. As biofuels are known as non-toxic and environmentally safe fuels, they can be used for many applications to replace conventional fuels such as diesel and mineral oils. One of such applications is for offshore oil-rig drilling use in oil-based mud drilling applications.

   (a) What are the primary functions of drilling fluids?

   (b) What are the general property requirements of the base fluid and the drilling mud?

   (c) What are the environmental advantages of biodiesel as compared to conventional mineral oil?

Homework problems

1. *Nuclear power for a transport application.* A nuclear submarine is a submarine powered by a nuclear reactor.

   (a) What is its advantages and disadvantages compared to a conventional diesel-driven submarine?

   (b) During 2 days' travel, a nuclear submarine developed an average power of 2,500 kW. If the engine driven by the nuclear plant has efficiency of 30%, how much fuel of U_{235} had been consumed by the submarine. Assume $E_f = 190$ MeV.

   (Answer: Advantages: A nuclear-driven submarine is completely independent of air, thus avoiding the need to surface from the water frequently. Also, high power of the nuclear generator to provide high speed and for long durations. Disadvantages: It is relatively high cost and the passengers may be exposed to nuclear radiation. (b) 18.47 gram)

2. *Nuclear power plant versus coal-fired power plant for electric energy production.* Consider a nuclear power plant capable of producing 300,000 kW of power from uranium U_{235}.

   (a) Calculate the daily energy output of the nuclear reactor.

   (b) Compute the fuel consumption of the nuclear reactor per day if its efficiency is 22%.

   (c) What would be the daily amount of coal required to obtain the same energy if the efficiency is 78%?

   (d) Compare the economics of production, i.e., price per kW if the fuel price is $2000/kg and $0.2/kg for the nuclear fuel and hard coal, respectively.

   Data:

   Uranium: $U_{235} = 8.2 \times 10^{13}$ J/kg; and, average fission energy release of U_{235} nuclide, E_f is 200 MeV.

   Coal: Calorific value of coal $= 29,300$ kJ/kg.

   (Answer: 25.92×10^{11} J; 0.1436 kg; 1.134×10^6 kg)

3. *Nuclear power plant with steam turbine.* For a typical nuclear power plant (Figure 9.26) designed with two circulation loops with heat transfer at the heat exchanger: In the reactor loop, water leaves the reactor core at high pressure of 16,000 kPa. In the steam turbine circulation loop, the steam leaves the heat exchanger at 5,000 kPa (dry saturated); and the condensate is a saturated liquid at 5 kPa. To increase system efficiency, the feedwater is preheated to the saturation temperature of the bleed steam at 500 kPa in the open feed heater, with the bleed steam traveling at a mass flow rate m_2 of 5 kg/s. The operating temperatures are controlled with the enthalpy difference given: $\triangle h_{1-2} = 299$ kJ/kg; $\triangle h_{2-3} = 480$ kJ/kg; $and \triangle h_{1-5} = 2154$ kJ/kg.

   (a) Calculate the heat supply by the nuclear reactor core.

   (b) Determine the steam turbine net work output.

   (c) Determine the overall system efficiency, neglecting pressure losses and pump work.

   Data: The mass flow rate ratio of the working fluids is known to be $\dot{m}_2 = 0.213\,\dot{m}_s$, where s is the steam leaving the heat exchanger.

   (Answer: $50,554\,kJ/kg$; $15,884\,\frac{kJ}{s}$; $31\,\%$)

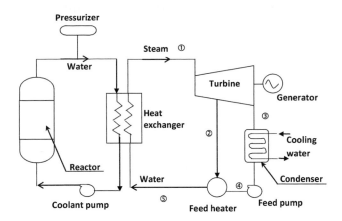

Figure 9.26
Typical PWR reactor (with feed heater) diagram.

Answers to multiple choice questions:

1. (A) PWR

2. (B) Depending on the types of nuclear plants and operating temperature, various applications such as district heating, industrial process, and desalination applications.

3. (D) Nuclear waste in the form of spent nuclear fuel is extremely hazardous, causing disposal and storage problems (need to store for thousands of years!)

4. (C). Compared with fossil fuel, biofuel is not up to the economical stage as yet (except in specific places like Brazil).

5. (E) All of the above

6. (D) I, III, IV, and V only. Note: in (II), vegetable oil is NOT the resource for bio-ethanol but bio-diesel.

Bibliography

[1] International Atomic Energy Agency, 1997. Sustainable Development, and Nuclear Power. IAEA, Vienna.

[2] http://www.world-nuclear.org/info/nuclear_fuel_fabrication-inf127.html, accessed August 29, 2011.

[3] http://www.british-energy.com/documents/Different_types_of_nuclear_power.pdf, accessed Aug. 23, 2011.

[4] Asean Nuclear Safety Network (ANSN), available at http://www.ansn-jp.org/jneslibrary/npp2.pdf accessed August 25, 2011.

[5] Nakae, N., Baba, T., Kamimura, K., Basis of technical standard on fuel for sodium-cooled fast breeder reactor, *Journal of Nuclear Science and Technology*, Volume 48, Issue 4, April 2011, pp. 524–531.

[6] Raina, V.K., Srivenkatesan, R., Khatri, D.C., Lahiri, D.K., 2006. Critical facility for lattice physics experiments for the advanced heavy water reactor and the 500 MWe pressurized heavy water reactors. *Nuclear Engineering and Design*, 236 (78), 758–769.

[7] Bhardwaj, S.A., 2006. The future 700 MWe pressurized heavy water reactor. *Nuclear Engineering and Design*, 236 (78), 861–871.

[8] Global Bioenergy Partnership. A review of the current state of bioenergy development in G8 + 5 countries. Global Bioenergy Partnership; 2007. Available from: http://www.fao.org/docrep/010/a1348e/ a1348e00.htm accessed August 10, 2011.

[9] United State Department of Agriculture (USDA), http://www.ers.usda.gov/features/bioenergy/ accessed July 29, 2011.

[10] Ruane, J., Sonnino A., Agostini, A., Bioenergy and the potential contribution of agricultural biotechnologies in developing countries, biomass and bioenergy 34 (2010) 1427–1439.

[11] NETL, http://www.netl.doe.gov/technologies/coalpower/gasification accessed Aug 4, 2001.

[12] Reed T. B., Das, A., *Handbook of Biomass Downdraft Gasifier Engine System*, The Biomass Energy Foundation Press. 1988.

[13] Balat, M., Balat, H., 2008, Progress in bioethanol processing. *Progress in Energy and Combustion Science*, 34 (5) pp. 551–573.

[14] IEA. Biofuels for transport. International Energy Agency, http://www.iea.org/textbase/nppdf/free/2004/biofuels2004. pdf; 2004, accessed August 23, 2011.

[15] Danielo, O. 2005, An algae-based fuel, *Biofutur*, No. 255, May 2005.

[16] Brown, L.M., Zeiler, B.G., Aquatic Biomass and Carbon Dioxide Trapping. *Energy Convers. Manage.*, 1993, 34, 1005 1013.

[17] Sheehan, J., Dunahay, T., Benemann, J., Roessler, P., A look back at the U.S. Department of Energy's aquatic species program: Biodiesel from algae. National Renewable Energy Laboratory, http://www.nrel.gov/docs/legosti/fy98/24190. pdf; 1998.

[18] Zhang, Y., Dube, M.A., McLean, D.D., Kates, M., Biodiesel production from waste cooking oil: 1. Process design and technological assessment. *Bioresource Technology*, 89 (2003) 1–16.

[19] Patil, V., Tran, K.-Q., Giselrød, H.-R., Towards Sustainable Production of Biofuels from Microalgae, *Int. J. Mol. Sci.* 2008, 9, 1188–1195.

[20] Yassin, A. M., Kamis, A., Abdullah, M. O., (1991) Palm oil diesel as a base fluid in formulating oil-based drilling fluid. Society of Petroleum Engincers (SPE) Technical Papers Microfiche No. 23001. Ann Arbor, Michigan, USA. pp. 190–206.

[21] Yusuf Chisti, Biodiesel from microalgae beats bioethanol, *Trends in Biotechnology.* 2008, Vol.26, No.3, 126–131.

[22] Chisti, Y. (2007) Biodiesel from microalgae. *Biotechnol. Adv.*, 25, 294–306.

[23] Liu, G., Larson, E. D., Williams, R. H., Kreutz, T. G., Guo, X., Making Fischer-Tropsch Fuels and Electricity from Coal and Biomass: Performance and Cost Analysis, *Energy Fuels*, 2011, 25, 415–437.

[24] Abu-Khader, M.M., 2009, Recent advances in nuclear power: A review, *Progress in Nuclear Energy*, 51 (2), pp. 225–235.

[25] Massachusetts Institute of Technology, Update of the 2003 Future of nuclear power, http://web.mit.edu/nuclearpower/pdf/nuclearpower-update2009.pdf, accessed 18 Oct 2011.

[26] Massachusetts Institute of Technology, The Future of Nuclear Power: an Interdisciplinary Study (2003). Available at: http://web.mit.edu/ nuclearpower/

[27] World Nuclear News, Available at: http://www.world-nuclear-news.org/NN_A_look_at_the_future_of_nuclear_power_0311082.html, accessed on Oct 18, 2011.

[28] Romm, J., The Self-Limiting Future of Nuclear Power, Center For American Progress Action Fund. March 16, 2009. Also available at: <http://www.americanprogressaction.org/issues/2008/nuclear_power_report.html> accessed Oct 18, 2011.

[29] The Institute of Engineering and Technology (IET), *Nuclear reactor types*, 2008, United Kingdom.

[30] Somerville, C., Youngs, H., Taylor, C., Davis, S. C., Long, S. P., Feedstocks for Lignocellulosic Biofuels, *Science* August 13, 2010: Vol. 329, no. 5993 pp. 790–792.

[31] Bridgwater, A.V., Review of fast pyrolysis of biomass and product upgrading, *Biomass and Bioenergy*, Available online, March 3, 2011.

[32] Amin, R.A.M., Clapper, D.K., Norfleet, J.E., Otto, M.J., Xiang, T., Hood, C.A., Goodson, J.E., Gerrard, D.P., Joint Development of an Environmentally Acceptable Ester-Based Drilling Fluid, Society of Petroleum Engineers, Trinidad and Tobago Energy Resources Conference, June 27–30, 2010, Port of Spain, Trinidad.

[33] Cilliers, A.C., Nicholls, D., Helberg, A.S.J., Fault detection and characterisation in Pressurised Water Reactors using real-time simulations, *Annals of Nuclear Energy*, Volume 38, Issue 5, May 2011, Pages 1196-1205.

[34] Zhou, X.W., Tang, C.H., Current status and future development of coated fuel particles for high temperature gas-cooled reactors, *Progress in Nuclear Energy*, Volume 53, Issue 2, March 2011, pp. 182–188.

[35] Lewins, J.D., (1978), *Nuclear Reactor Kinetics and Control*, Pergamon Press, Oxford.

[36] Walter, A., Schulz, A., Lohnert, G., Comparison of two models for a pebble bed modular reactor core coupled to a Brayton cycle, *Nuclear Engineering and Design*, Volume 236, Issues 5-6, March 2006, pp. 603–614.

10

Hybrid Energy

In this chapter, we will highlight the importance of hybrid energy applications. As this is an introductory text, we shall attempt to briefly, but comprehensively, discuss the various energy schemes, methodologies, efficiencies, and the usefulness of the selected hybrid energy application systems.

We shall discuss hybrid energy application in the following manner:

1. We shall first define the meaning of *hybrid energy*.

2. We shall then briefly discuss the various selected hybrid energy application schemes in relation to the following five main energy application categories:

   - Energy intensive industries: hybrid geothermal/fossil, hybrid nuclear-fossil fuel, integrated gasification combined-cycle (IGCC), hybrid fuel cell/turbine (FCT) systems.
   - SME industries energy applications (SME small/micro-scale hybrid energy systems).
   - Transportation industries (automotive hybrid electric vehicles or HEVs).
   - Building industries (building integrated PV or BIPV system).
   - Hybrid energy for rural applications (hybrid wind/solar/micro-hydro/fuel cell/diesel energy).

3. Finally, we shall discuss hybrid energy fuel efficiency as well as hybrid energy optimization.

10.1   Hybrid Energy and Hybrid Energy Methods

10.1.1   What Is Hybrid Energy?

A hybrid energy system usually consists of two or more energy sources used together, via suitable energy conversion techniques, to provide enhanced fuel savings or energy recovery, increased overall system efficiency, as well as accomplishing greater balance in energy supply and demand.

Figure 10.1 is a summary of some hybrid energy systems outlined by Energy Efficiency and Renewable Energy for a number of hybrid technologies together with their application range and application priority. Microturbine hybrids, for instance, are suitable and opted for commercial, industrial, and grid-distributed power; reciprocating-engine hybrids can extend the applications for portable power and transportation usages. The renewable hybrids (PV-, wind-, and biomass-hybrids) would be targeted for grid-distributed applications; PV-hybrids would also have the added advantages for residential and building applications, e.g., Building-integrated photovoltaics (BIPV).

10.2   Hybrid Geothermal/Fossil System

Further to our earlier discussion on geothermal power plants and their suitable application in Chapter 8.4, we now proceed to discuss hybrid geothermal systems.

Hybrid geothermal power plant/fossil systems make use of the relatively low temperate (and pressure) of geothermal energy as the low temperature end of the conventional cycle, and combined with the high temperature heat

	Residential	Commercial	Industrial	Grid-distributed	Portable Power	Transportation	Typical Unit Size Range (installation size can be larger)
● Primary Target Market / ○ Secondary Target Market							
Microturbines		●	●	●	○	○	25 – 300 kW
Reciprocating Engines		●	●	●	●	●	5 kW – 50 MW
Low-Temperature Fuel Cells	●	●	○	●	○	●	2 – 250 kW
High-Temperature Fuel Cells		●	●	●	○		100 kW – 3 MW
Fuel Cell/Gas Turbine Hybrids		○	○	●			250 kW – 20 MW
Small Gas Turbines			●	●			500 kW – 5 MW
Photovoltaics	●	○	○	●			1 – 500 kW
Wind Power	○			●			50 kW – 2 MW
Biomass Power			●	●			250 kW – 50 MW

Figure 10.1
Hybrid fuel cell/gas turbine and hybrid fuel cell/microturbine and application range. (Adapted from Energy Efficiency and Renewable Energy [26].)

rejected as the high temperature of the cycle. Undoubtedly, the hybrid approach, by combining two power units with the heat rejected from the first supplying heat to the second, will save energy for the overall hybrid energy systems.

We can classify hybrid geothermal power/fossil energy systems into two types, i.e., either geothermal preheat/fossil or hybrid fossil superheat, as detailed in the following subsections.

10.2.1 Hybrid Geothermal Preheat/Fossil Energy System

As illustrated in Figure 10.2, the low temperature geothermal energy source is used to *preheat* the feed condensate water of a fossil-driven steam power plant by means of a geothermal heat exchanger. The boiler feed pump pumps the preheated fluid from the deaerating heater chamber section to two or three feedwater heaters, the drains of which were designed with cascaded flow backward to provide efficient series heating. The three feedwater heaters receive high temperature heat from the steam bled off from the high pressure turbine. At the economizer section, the high pressure steam bled from the high pressure turbine is again employed to provide further heating or provide reheating to the flow to maximize heat utilization.

10.2.2 Hybrid Geothermal Fossil/Superheat Energy System

In a hybrid geothermal fossil-superheat energy system, the vapor-dominated steam from the geothermal well is superheated in a fossil fuel superheater prior to going into the high-pressure turbine (Figure 10.3). The hybrid system comprises two flash separators, with steam produced from first separator preheated in the regenerator by exhaust steam from the high-pressure turbine. The heated steam from the regenerator is subsequently superheated by the fossil fuel-fired superheater. The condensate from the condenser is pumped and reinjected into the geothermal reservoir.

Advantages of Hybrid Geothermal/Fossil Systems:

1. Energy recovery and energy savings as compared to stand-alone geothermal or stand-alone fossil systems.

2. *High overall hybrid geothermal/fossil power plant efficiency.* Because of their efficient use of the heat energy released from the natural gas, combined-cycle plants are more efficient than steam units or gas turbines alone,

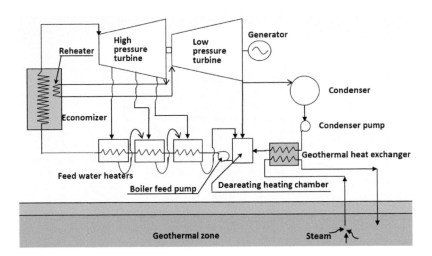

Figure 10.2
Schematic of a typical hybrid geothermal preheat–fossil energy system.

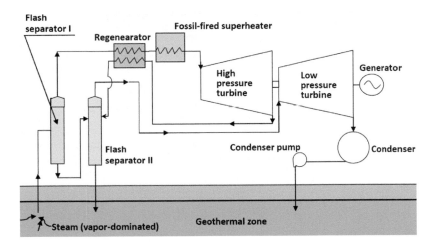

Figure 10.3
Schematic of a typical hybrid geothermal fossil–superheat energy system.

typically with thermal efficiencies in excess of 50–55% [1].

Disadvantages:

1. Complexity and less flexibility of overall plant in both operation and maintenance.

2. High initial cost for the combined systems.

It is to be noted that, apart from the basic single stage, hybrid cycles discussed in the preceding subsections above, for further efficiency improvement, it is also feasible to employ multi-stage compound hybrid systems.

10.2.3 One-Stage Compound Hybrid System

DiPippo and Avalar in 1979 [2] conducted a computer simulation of a one-stage compound hybrid system. Figure 10.4 shows the system component arrangement of the single stage compound hybrid system while the associated T-s diagram is displayed in Figure 10.5.

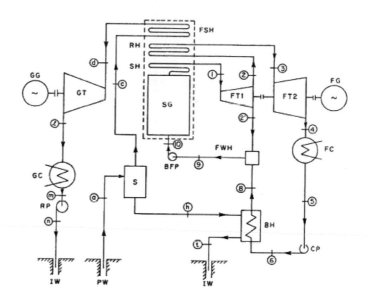

Figure 10.4

One-stage compound hybrid system. (From DiPippo and Avelar, 1979 [2].) Key: PW = Production Well; S = Separator; SG, SH, RH, FSH = Steam Generator and Superheaters; FT1, FT2 = Fossil Turbines; FG = Fossil Generator; FC = Fossil Condenser; CP = Condensate Pump; BH = Brine Heat Exchanger; FIVH = Feedwater Heater; BFP = Boiler Feed Pump; GT = Geothermal Turbine; GG = Geothermal Generator; GC = Geothermal Condenser; RP = Reinjection Pump; I W = Injection Well.

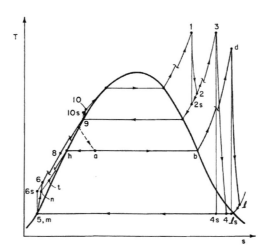

Figure 10.5

One-stage compound hybrid system: the T-s diagram. (From DiPippo and Avelar, 1979.) Pressure losses through the various heat exchangers are denoted by broken isobars in the case of one-stage systems. (After [2].)

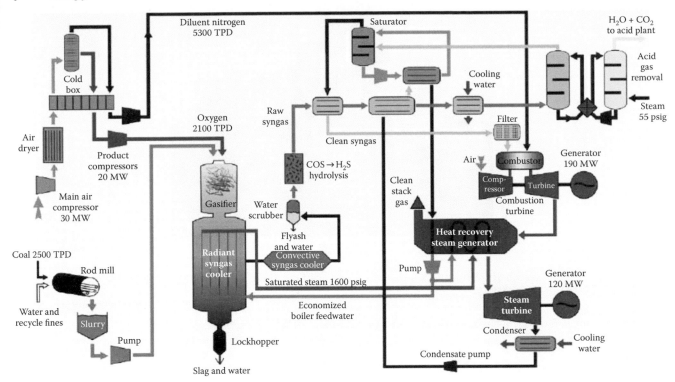

Figure 10.6 (SEE COLOR INSERT)
Schematic of IGCC plant. (From Wikimedia Commons, public domain.)

10.3 Integrated Gasification Combined-Cycle (IGCC)

IGCC is an advanced technology that employs a gasification method to convert coal from a coal power plant into synthesis gas, usually called syngas. It takes in O_2, steam and feed stocks, removes CO_2, and S_2 as well as other impurities well before combustion (rather than attempting to capture CO_2 from the cocktail of exhaust gases), and attempts to turn any pollutants into useful by-products, e.g., hydrogen. This results in lower emissions of sulfur dioxide, mercury, and particulates. Excess high grade heat from the primary combustion and generation is then used to power a steam cycle, similar to a combined cycle gas turbine (see Figure 10.6).

The application of IGCC results in "improved overall efficiency and offers even fewer pollutants compared to conventional pulverized coal, and, potentially, energy for centuries to come." ([3]; Breault, 2010 [4]).

Apart from the above, carbon dioxide can be captured from the coal syngas (carbon monoxide and hydrogen) in an IGCC plant through a water/gas shift process. The CO_2 can be captured in a concentrated stream, making it easier to convert into other products, or to sequester (underground or to depleted oil fields for enhanced oil recovery (EOR) applications). We have already discussed the use of CO_2 for EOR applications in Chapter 5, Section 5.2, Energy Recovery.

10.3.1 The Advantages of IGCC Compared with Conventional Coal-Based Power Generation Systems

1. *Higher efficiencies and lower emissions*. Improvements in efficiency dramatically reduce emissions from coal combustion. Increasing efficiency from 35 to 40%, for example, reduces carbon dioxide emissions by over 10%. With efficiencies currently approaching 50%, IGCC power plants use less coal and produce much lower emissions of carbon dioxide than conventional power plants. With development of new gas turbine concepts and increased process temperatures, efficiencies of more than 60% are being targeted.

2. *Higher output*: Using syngas in a gas turbine increases its output, especially when nitrogen from an oxygen

Figure 10.7
Vřesová IGCC Plant, Czech Republic. (From http://www.gasification.org/Docs/2005_Papers/05CHHO.pdf.)

blown unit is fed to the turbine. Thus a turbine rated at 170MW when fired on natural gas can yield 190MW or more on syngas. Furthermore, output is less dependent on ambient temperature than is the case with natural gas.

3. *Product flexibility* — including carbon capture and hydrogen production: The gasification process in IGCC enables the production of not only electricity, but a range of chemicals, by-products for industrial use, and transport fuels (see graphic).

4. CO_2 *capture*. Carbon dioxide can be captured from the coal syngas (carbon monoxide and hydrogen) through a water/gas shift process and for use, e.g., in EOR recovery of oil/gas fields.

5. H_2 *production*. In addition to electricity generation, hydrogen produced from the process can potentially be used to power fuel cells.

10.3.2   The Disadvantages of IGCC

1. High initial cost, because of plant complexity.

2. Concerns about the need for staff who had experience in the design and operation of chemical processing plants.

3. Concerns about the overall reliability, due to the increased complexity of the integrated plant.

Examples of IGCC stations around the world:

1. Vřesová IGCC Plant, Czech Republic (Figure 10.7), (Feed: dry coal, lignite; net output: 398 MWe)

2. Drym, South Wales, United Kingdom, 450 MW

3. Teesside, United Kingdom - 800 MW

4. Puertollano in Central Spain (Figure 10.8). 2,600 tons/day coal and petcoke; net output: 300 MW

10.4   Hybrid Fuel Cell/Turbine and Fuel Cell/Microturbine System

Hybrid FCT is an emerging advanced energy system making use of clean, high efficiency fuel cell technology.

Figure 10.8
Puertollano IGCC Plant, Spain. (Courtesy of DOE/NETL,seca/doe.gov/technologies/coalpower/gasification/pubs.html.)

As discussed in Chapter 7, Section 7.6, a fuel cell is a special heat engine that uses pure hydrogen or hydrogen-rich fuel and oxygen to create electricity via an electrochemical process that is also known as *inverse electrolysis*, without having undergone any combustion process.

The absence of the combustion process eliminates the formation of pollutants including NOx, SOx, hydrocarbons, and particulates, and significantly improves electrical power generation efficiency. Further efficiency gains are realizable by integration of a turbine, or, a microturbine with the fuel cell. This combined energy conversion technology is called hybrid fuel cell/turbine (FCT) and hybrid fuel cell/microturbine (FCMT) technology.

10.4.1 Hybrid Fuel Cell/Turbine (FCT)

Figure 10.9 shows the FCT hybrid concept in simple form, to provide some understanding of the synergy offered and the basic relationships of components. As shown in Figure 10.9, the high-temperature fuel cell serves as the combustor for the gas turbine. Residual fuel in the already high temperature fuel cell exhaust mixes with the residual oxygen in an exothermic oxidation reaction to further raise the temperature. Both the fuel cell and the gas turbine generate electricity, and the gas turbine provides some balance of plant functions for the fuel cell, such as supplying air under pressure and preheating the fuel and air in a heat exchanger called a recuperator (NETL [6]).

In the *indirect* mode, the recuperator transfers fuel cell exhaust energy to the compressed air supply, which in turn drives the turbine. The expanded air is supplied to the fuel cell. The indirect mode uncouples the turbine compressor pressure and the fuel cell operating pressure, which increases flexibility in turbine selection. Critical issues are the integration of pressure ratios and mass flows and the dynamic control through start-up, shutdown, emergency, and load-following operating scenarios (NETL [6]).

10.4.2 Hybrid Fuel Cell and Microturbine (FCMT)

10.4.2.1 What Are Microturbines and Micropower?

Micropower is a term generally used to describe the power that can be obtained from microturbines, fuel cells, reciprocating engines, and other power systems with capacity less than 1 MW. The turbine that produces less than 1 MW is called a *microturbine*.

10.4.2.2 Introduction to FCMT

Micropower technologies are suitable for residential, commercial, and industrial on site power markets. Fuel cell hybrids can address some of these markets. Distributed power also includes the concept of installing small power generation equipment throughout the distribution grid (e.g., at substations) as an alternative to central station power plants.

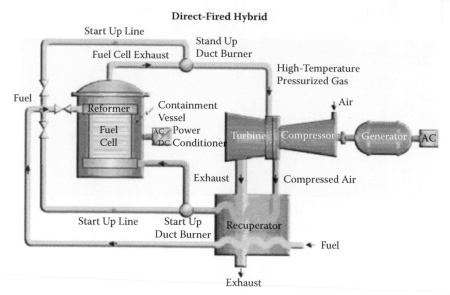

Figure 10.9
Direct FCT hybrid. (From NETL.)

Advantages of Hybrid Fuel Cell/Microturbine (FCMT) System (EERE [26]):

1. *Cost-savings to industry.* The application of micropower and fuel cell hybrid technologies for on-site industrial power generation can be economically attractive relative to current technology (typically grid-power at the size range represented by these technologies). This economic benefit implies a direct cost savings to those industries that implement these advanced technologies, which in turn leads to enhanced competitiveness.

2. *Decreased energy consumption.* Virtually all of the technologies considered herein have, or can achieve, electrical generation efficiencies that are greater than the local conventional electric grid, which was approximately 33% efficient on a national average in 2011. Technologies with lower efficiencies (such as unrecuperated microturbines) may still be attractive in cogeneration applications, where the fuel savings brought about by the recovery of waste heat can more than offset their lower electrical efficiency.

3. *Decreased emissions of criteria pollutants.* Microturbines, fuel cells, and fuel cell hybrids are expected to produce markedly reduced emissions of NOx and SO_2 relative to conventional technologies and the grid average.

4. *Decreased CO_2 emissions.* Micropower technologies have the potential to reduce CO_2 emissions via their increased electrical efficiency, tendency to favor low carbon, natural gas fuels, and by providing opportunities for industries to convert their process wastes into useful energy at scales that have not previously been economic. In the case where micropower technologies are less efficient than the grid, they can still reduce CO_2 emissions when applied in cogeneration.

10.4.3   Hybrid SOFC-PEM/Turbine System

The hybrid system has been described by Dicks et al. (2000) [27] and Mescal (1999) [28]. This hybrid system is making use of two difference types of fuel cells operating under different levels of temperatures, i.e., SOFC operating at a higher temperature and PEMFC at a lower temperature, where the exhaust gas from the SOFC undergoes a shift reaction followed by CO, to produce hydrogen gas that can be reused as a fuel in the PEM stack. This results in energy savings thus higher overall efficiency.

Figure 10.10 shows a typical hybrid SOFC/PEM/turbine system. The overall combined system efficiency is 61% and net electrical output is 489.7 kWe, with a 369.3 kW SOFC stack power and 146.7 kW PEM fuel cell.

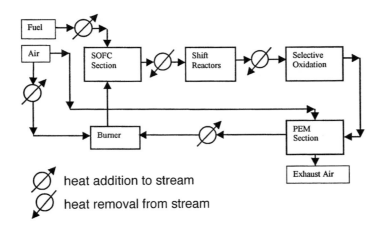

Figure 10.10
Hybrid SOFC-PEM system schematic diagram. (From Dicks, A. L. et al., A study of SOFC–PEM hybrid systems, *Journal of Power Sources*, Volume 86, Issues 1-2, March 2000, pp. 501–506. With permission.)

10.4.4 Hybrid Fuel Cell/Battery System

The hybrid fuel cell/battery systems take advantage of the high efficiencies of both systems, i.e., fuel cell and battery; and that the fuel cell works close to its maximum power at all times. Other feasible combinations are fuel cell, battery, and PV solar panel applications, see, e.g., Gibbard (1999) [29], Li et al. [30].

10.5 Hybrid Electric Vehicles (HEVs)

A hybrid electric vehicle (HEV) is a vehicle that is powered by a conventional fuel (diesel or petrol) driven internal combustion engine (IC) propulsion system combined with an electric-driven propulsion system. Three other variations are the plug-in hybrid electric vehicle (PHEV), battery-electric vehicle (BEV) and fuel-cell vehicle (FCV). Hybrid-electric vehicles (HEVs) are now becoming quite common in usage. Fuel cell vehicles (FCVs), regarded as an advanced electric powertrain, however, are still under development.

10.5.1 Hybrid Electric Vehicles Classification

Classification I: hybrid vehicles based on functionality In recent years, various hybrid electric vehicle propulsion systems for passenger cars and light trucks have been developed and improved in stages by automotive manufacturers for on the road applications. These range from conventional hybrids with minimal regenerative breaking to fully advanced hybrids with added functionality for full power assist ability, i.e., motor-assist and full electric drive ability (see Figure 10.11). Figure 10.11 shows a classification of hybrid electric vehicles based on incremental powertrain functionality. Basically, the functionality evolutions can be well classified into four types, namely micro-hybrid, mild-hybrid, medium-hybrid, and full hybrid [7]. With a full hybrid, storage systems voltage increased from 12 V to > 200 V for added functionality applications.

Classification II: hybrid vehicles based on drivetrain architecture Salmasi classified HEVs into several architectures based on the hybrid drivetrains; however, general series, parallel, and power split, which are shown in Figure 10.12, are the most widely used drivetrain architectures.

Classification III: hybrid vehicles based on control strategies The control strategies for HEVs have been discussed in detail by Salmasi (2007), and these classifications are given in Figure 10.13.

A major challenge for the development of hybrid vehicles is the coordination of multiple energy sources and convertors and, in the case of an HEV, power flow control for both the mechanical and the electrical path. This necessitates the utilization of an appropriate control or energy management strategy.

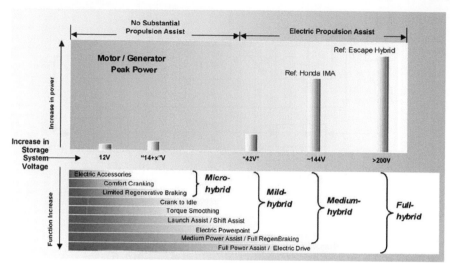

Figure 10.11
Hybrid electric vehicles classification. (After E. Karden et al. 2007 [7].)

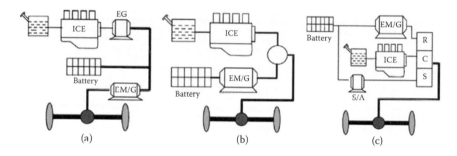

Figure 10.12
General structures for hybrid drive trains. (a) Series, (b) parallel (with torque/speed coupler), and (c) power-split (with planetary gear system). EM = electric motor; G= generator; R = ring; C = carrier; S = sun. (© 2012 IEEE. Reprinted, with permission from Salmasi, F. R., Control Strategies for Hybrid Electric Vehicles: Evolution, Classification, Comparison, and Future Trends, IEEE Transactions on Vehicular Technology, VOL. 56, NO. 5, September 2007, pp. 2393–2404.)

A control strategy, which is usually implemented in the vehicle central controller, is defined as an algorithm, which is a law regulating the operation of the drive train of the vehicle [8]. Generally, it inputs the measurements of the vehicle operating conditions such as speed or acceleration, requested torque by the driver, current roadway type or traffic information, in-advance solutions, and even the information provided by the Global Positioning System (GPS). The outputs of a control strategy are decisions to turn ON or OFF certain components or to modify their operating regions by commanding local component controllers [8].

1 Deterministic rule-based (RB). This is the heuristics method based on analysis of power flow in a hybrid drive train, efficiency/fuel or emission maps of an ICE, and human experiences are utilized to design deterministic rules, generally implemented via lookup tables, to split requested power between power convertors.

2 Fuzzy rule-based (RB). Looking into a hybrid drivetrain as a multidomain, nonlinear, and time-varying plant, fuzzy logic seems to be the most logical approach to the hybrid problem. In fact, instead of using deterministic rules, the decision-making property of fuzzy logic can be adopted to realize a real-time and suboptimal power split. In other words, the fuzzy logic controller is an extension of the conventional rule-based controller. The main advantages of fuzzy rule-based methods are the following: 1) robustness, since they are tolerant of imprecise measurements and component variations, and 2) adaptation, since the fuzzy rules can be easily tuned, if necessary.

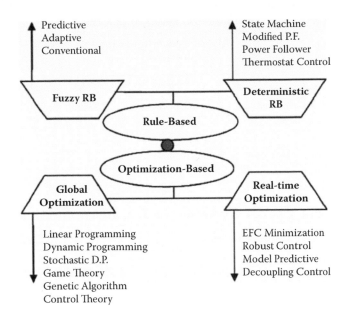

Figure 10.13

Classification of the hybrid power train control strategies. (© 2012 IEEE. Reprinted, with permission from Salmasi, F. R., Control Strategies for Hybrid Electric Vehicles: Evolution, Classification, Comparison, and Future Trends, IEEE Transactions on Vehicular Technology, VOL. 56, NO. 5, September 2007, pp. 2393–2404.)

3. Global optimization-based. In these control strategies, the optimal reference torques for power convertors and optimal gear ratios are calculated by minimization of a cost function generally representing the fuel consumption or emissions. If this optimization is performed over a fixed driving cycle, a global optimum solution can be found. In fact, the global optimal solution is non-casual*in that it finds the minimum fuel consumption using knowledge of future and past power demands.

4 Real-time optimization-based. In order to develop a cost function used in instantaneous optimization, in addition to a measure for fuel consumption, variations of the stored electrical energy should also be taken into account to guarantee electrical self sustainability.

10.5.2   Energy Storage Device Requirements for Hybrid Electric Vehicles

Requirements for the electrical storage of hybrid systems include (Karden, 2011 [7]):

1. Shallow-cycle life.

2. High dynamic charge acceptance. This is essential particularly for regenerative braking.

3. Robust service life in sustained partial-state-of-charge usage (Karden et al., 2007 [7]).

10.6   Hybrid Energy Systems for Rural Application

Currently, the common application areas for rural electrification purposes include rural villages, islands, highlands, wildlife sanctuaries, and national parks, as depicted in Figure 10.14.

---

*Non-casual optimization methods generally yield better control prediction results compared with the causal non-optimized methodology.

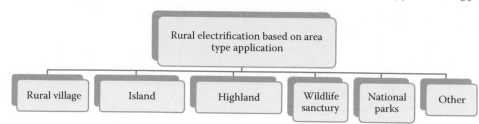

Figure 10.14
Rural electrification classification.

10.6.1   Overview of Hybrid Energy Systems for Rural Applications

In this subsection, we will provide a quick overview of various hybrid energy systems used as discussed in the literature.

An overview of PV decentralized rural electrification has been discussed by Chaurey and Kandpal (2010 [10]).

A wind/diesel/battery hybrid energy system was proposed by Nayar et al. in 1993 [16] for remote area, in Australia.

A 2.24 kW hybrid photovoltaic (PV)- proton exchanged membrane (PEM) fuel cell generation system employing an electrolyzer for hydrogen generation was designed and simulated by El-Shatter et al. 2002 [11]

Lee et al., 2004 [12] studied and analyzed a 40W direct methanol fuel cell (DMFC) combined with battery hybrid power for portable application. The DMFC was used for the main power source at average load while the battery was applied for auxiliary power at overload. The experiments were done on hybrid DMFCs using a lead-acid battery, Ni-Cd battery, and Ni-MH battery. Typical results show a stable transient characteristic for some seconds in the beginning of the load growth. Abdullah et al., 2011 [13] constructed a mini fuel cell/battery/solar portable device, which provided a laboratory study for small, multipurpose energy kit applications. To meet pulse-power demand, in which the power demand is impulsive rather than constant, a design and experimental tests of control strategies for an active hybrid fuel cell/battery power source had been investigated by Li et al., 2004 [14]. A hybrid energy system consisting of a 5 kW wind turbine and a fuel cell system is presented by Iqbal, 2003 [15]. In summary, hybrid energy systems for rural applications have been attempted by many due to the many advantages a hybrid system could offer, such as sustainability, better resources utilization, and so forth, compared with a standalone system, which is dependent only on one single resource.

10.6.2   Energy Sizing and Control of Hybrid Energy Systems for Electricity Production

The design process of hybrid energy systems requires the selection and sizing of the most suitable combination of energy sources, power conditioning devices, and energy storage systems together with the implementation of an efficient energy dispatch strategy. "In order to achieve a larger penetration of renewable resources, without deterioration of the quality of service offered to customers, the development of control tools is necessary" [17]. For that purpose, System simulation software such as Hybrid 2 and Homer are essential tools to analyze and compare possible system combinations. The results to be obtained from hybrid energy simulation software are limited by [17]:

1. Environmental conditions at the location (irradiance, temperature, wind speed, humidity)

2. Customer requirements and preferences

3. Financial resources of customers and availability of government support schemes

4. Availability of reliable system technology and technical support

Two main aspects of control strategies need to be considered for hybrid energy systems, as pointed out by Wichert (1997):

- *Energy dispatch strategy*, which is concerned with the allocation of resources and the direction of power flow in the system on a particular timescale (of minutes to days);

- *Power quality control*, which is concerned with the stability of the AC voltage and its total harmonic distortion (THD) on a timescale below the time period defined by the supply frequency.

10.6.2.1 PV–Diesel Hybrid Energy System Configurations

PV–diesel hybrid energy systems generate AC electricity by combining a photovoltaic array with an inverter, which can operate alternately or in parallel with a conventional engine-driven generator. They can be classified according to their configuration as ([16]; [19]):

- Series hybrid energy systems

- Switched hybrid energy systems

- Parallel hybrid energy systems

Series configuration   All the energy is passed through the battery bank and the AC power delivered to the load is converted from DC to regulated AC by an inverter or a motor generator unit. The system can be operated in manual or automatic mode, with the addition of an appropriate battery voltage sensing and start/stop control of the engine-driven generator.

The summary of the three most common system topologies is presented below.

Switched configuration   Despite its limitations, the switched configuration remains one of the most common installations today. It allows operation with either the engine-driven alternator or the inverter as the AC source, yet no parallel operation of the main generation sources is possible. The battery bank can be charged by the diesel generator and the renewable energy source. The load can be supplied directly by the engine-driven generator, which results in reduced cycling of the battery bank. It can be operated in manual mode, although the increased complexity of the system makes it highly desirable to include an automatic controller, which can be implemented with the addition of appropriate battery voltage sensing and start/stop control of the engine-driven alternator.

Parallel configuration   The parallel configuration allows all energy sources to supply the load separately at low or medium load demand, as well as supplying peak loads from combined sources by synchronizing the inverter with the alternator output waveform. The bi-directional inverter can charge the battery bank (rectifier) when excess energy is available from the engine-driven generator, as well as act as a DC-AC convertor (inverter) under normal operation. The bi-directional inverter may provide peak shaving as part of the control strategy when the engine-driven alternator is overloaded.

10.6.2.1.1 Advantages and Disadvantages of the Configuration of the Systems

For series configuration:
Advantages:

- No switching of AC power between the different energy sources is required, which simplifies the output interface.

- The power supplied to the load is not interrupted when the diesel generator is started.

- The inverter can be sine, quasi-sine or square-wave, depending on the application4.

- The engine-driven generator can be sized to be optimally loaded while charging the battery bank, until a battery state-of-charge of 75–85% is reached.

Disadvantages:

- The inverter cannot operate in parallel with the engine-driven generator, therefore, the inverter must be sized to supply the peak load of the system.

- The battery bank is cycled frequently, which shortens its lifetime.

- The cycling profile requires a large battery bank to reduce the depth of discharge.

- Reduced overall efficiency, since all energy flows through the battery and the inverter.

- Inverter failure results in complete loss of power to the load.

Switched configuration:
Advantages

- The inverter can be sine-, quasi-sine or square-wave, depending on the application.

- Both energy sources can power the load directly.

Disadvantages:

- Power to the load is interrupted momentarily when the AC power sources are transferred.

- The engine-driven alternator and inverter have to be designed to cope with the peak load.

- No optimized allocation of fuel-based and renewable resources is possible.

Parallel configuration:
Advantages

- The parallel configuration offers a number of advantages over other system topologies.

- The system load can be met in an optimal way.

- Diesel efficiency can be maximized.

- Diesel maintenance can be minimized.

- Reduction in the rated capacities of the diesel generator, battery bank, inverter, and renewable resources is feasible, while also meeting the peak loads.

10.6.3 Hybrid Energy Systems and System Optimization

Here, we shall briefly discuss and provide an overview on the hybrid energy system optimization aspects and the various associated computer techniques for solving the optimization problems.

The various common computer software/control techniques are:

1. Artificial intelligent (AI) techniques, i.e., fuzzy-logic controller, neural network, and genetic algorithms.

2. Power management optimization.

Now, a popular issue arising in hybrid energy research is how these hybrid technologies can be combined for an optimized performance, either on design phase (e.g., Muselli et al., 1999 [5]; Ashok, 2007 [20]) or operating phase (e.g., [21]; [23] and [24]). This line of study is sometimes referred to as energy management (Muselli et al., 1999 [25]). In [25], a method for calculating the correct size of a photovoltaic (PV)-hybrid system and for optimizing its management is developed. In [20], several energy technologies are discussed and a general model to find an optimal combination of energy components for a typical rural community minimizing the life cycle cost is presented. The model helps in sizing hybrid energy system hardware and in selecting the operating options.

The importance of optimizing the performance of a hybrid energy system in operating phase has been highlighted in hybrid vehicles research. In [21], a fuzzy-logic-based energy management and power control strategy for parallel hybrid vehicles is proposed. It optimizes the fuel economy and the operational efficiency of all its components. The proposed approach optimizes the power output of the electric motor/generator and the internal combustion engine by using vehicle speed, driver commands from accelerator and braking pedals, state of charge (SOC) of the battery, and the electric motor/generator speed. Separate controllers optimize braking and gear shifting.

In [22] and [23], an intelligent energy management agent (IEMA) for parallel hybrid vehicles is proposed. It incorporates a driving situation identification component whose role is to assess the driving environment, the driving style of the driver, and the operating mode of the vehicle, using long and short term statistical features of the drive cycle.

In [24], a procedure for the design of a near-optimal power management strategy is proposed. The design procedure starts by defining a cost function, such as minimizing a combination of fuel consumption and selected emission species over a driving cycle. Dynamic programming (DP) is then utilized to find the optimal control actions including the gear-shifting sequence and the power split between the engine and motor while subject to a battery SOC-sustaining constraint.

10.6.4 Application Advantages of Renewable Hybrids for Rural Village Electrification

Power range of interest is from 10 W to 100 KW.

1. Higher availability. Hybrid systems reduce daily and seasonal resource variations by making use of more energy resources available.

2. Lower overall energy and fuel cost. Resource diversity minimizes battery size and fuel usage.

3. Higher power motor loads. Hybrid systems are suitable as more energy loads are increasing gradually for most village electrification requirements, due to refrigeration, grain grinding, carpentry, etc.

4. Improving energy sustainability.

5. Higher Quality of service and possible extension to 24-hour applications.

6. Single resource utilities such as diesel or hydro stand-alone have higher supply risks.

Some difficulties or complexities of rural electrifications, and some suggestions for remedies

1. Hybrid systems are a potentially significant solution to rural AC electricity needs, but further technology development, systems integration, simplification, and industry expansion will be required.

2. A maintenance support infrastructure for hybrid energy applications in the rural areas must be established and nurtured from the very conception of a project. Repairing equipment in remote locations is difficult and expensive. Multiple systems in a region are required to develop and sustain a cost-effective support infrastructure.We shall further notice that nothing is maintenance-free.

3. Retrofitting expensive hybrid power systems in a village without first addressing end-use appliances, metering, and switches is a mistake (NREL).

10.6.4.1 Future Hybrid Energy Systems

Hybrid electric vehicle (HEVs) applications:

In term of control aspects, according to Salmasi, 2007 [8], here are four primary issues that should be addressed in the future:

- *Durability of the energy sources*: Imagine that someone has purchased an HEV. While saving 70% per gallon, he has to pay much more than that to repair or change the battery cells, fuel cell, or ultracapacitor in the car. Therefore, it would be wise to design a control strategy that weighs heavily on the durability extensions of the energy sources in the overall cost function.

- *More components in the drivetrain*: There can be more components in a hybrid drivetrain besides the conventional ones such as ultracapacitor, ISA, continuous variable transmission, and multiple side loads in the car, which necessitates the development of new power split systems.

- *Complex structures*: The main literature in this field is devoted to parallel structures. Development of control strategies for more complex structures can be cumbersome, specifically for analytical approaches.

- *More control objectives*: In addition to traditional goals such as fuel/efficiency optimization, or charge sustaining, some other objectives can be taken into account, such as vibration control of the vehicle.

Rural hybrid-power electrifications applications:

According to Wichert, 1997 [18], future hybrid energy systems may replace the engine-driven generator with *modern fuel cells*, which have a number of advantages, including:

1. High efficiency of energy conversion (40–60%), particularly at part load operation

2. Little or no noxious emissions, depending on the fuel cell type

3. Silent operation

4. Modular construction

5. Ability to utilize most fuels

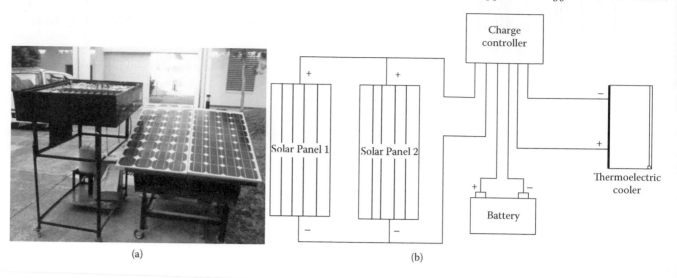

(a)   (b)

Figure 10.15 (SEE COLOR INSERT)
Hybrid solar thermoelectric-adsorption cooling system. (a) Photo of the hybrid system. (b) Schematic diagram of the hybrid system. (From Abdullah et al. 2009 [32].)

10.7   Other Hybrid and Integrated Energy Application Systems

1. Hybrid hydro/steam power plants. The hybrid hydro/steam power plants take advantage of hydro plants (such as hydro pump storage) which can take up load quickly and follow the peak variations much better than power plants. Depending on the water availability, two possible operation modes are as follows: During availability of high water flow rate, the steam plant serves as peak load and the hydro plant serves as base (constant) load. During low water flow rate availability, the steam plant can be used as base load and hydro plant as peak load.

2. Building integrated PV or BIPV system. BIPV is a hybrid system in which photo votaic (PV) materials are used in parts of the building envelope such as roofs, facades, and skylights. Some of the noted advantages of BIPV are: Reduction of fossil fuel in buildings thus greenhouse gas reduction. The use of PV provides energy savings potential. BIPV provides new aesthetics effects for modern building construction.

10.8   Fuel Efficiency of Hybrid Energy Systems: Case Study

In this subsection, we will consider a simple hybrid system, the combined thermoelectric-adsorption system.

Case Study: SME and Rural Electrification Application

Combined Thermoelectric-Adsorption System   In a typical combined thermoelectric-adsorption system (see Figure 10.15), cooling is produced via the Peltier effect during the day, by means of thermoelectric elements, and through an adsorption process at night.

The COP for adsorption cycle [32] is represented by

$$COP_{ad} = \frac{Q_e}{Q_g} = \frac{m\,c_p \triangle T_e}{m\,c_p \triangle T_g} = \frac{\triangle T_e}{\triangle T_g} \tag{10.1}$$

where,

Q_e = cooling output of evaporator (J)

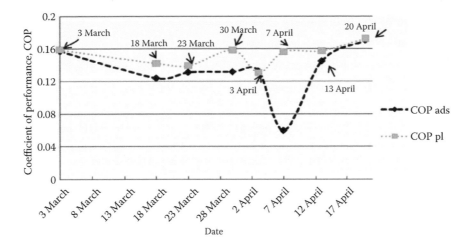

Figure 10.16
Variations of the COP on various days. (After Abdullah et al., 2009 [32].)

Q_g = heat supplied to the adsorber (J)
m = mass;
C_p = specific heat capacity (J/(g K))
$\triangle T_e$ = temperature change in the evaporator °C;
$\triangle T_g$ = temperature change in the adsorber °C.

For the thermoelectric cooling system, its COP can be estimated by Eq. 10.2:

$$COP_{th} = \frac{Q_h}{W_{in}} = \frac{m\, c_p \triangle T_{load}}{I\, V\, t} \tag{10.2}$$

where,
Q_h = heat removed from the hot side (J);
W_{in} = the work supplied to the thermoelectric cooler (J);
I = amperage (A);
V = voltage (V);
Cp = specific heat capacity (J/(g °K));
t = the time needed for the load to reach its lowest cooling state (s).

COPs of a solar-powered combined thermoelectric-adsorption system had been evaluated and a typical result is as shown in Figure 10.16. It was found that the COP of this type of combined thermoelectric-adsorption conversion, powered by solar PV, is rather low, i.e., around 0.16 maximum. This is partly due to the the low solar irradiation that is caused by cloudy skies.

10.9 *Application Projects, Questions, and Problems*

Application Project

Linking knowledge with applied energy design and practice

1. *Industrial visit (energy intensive industry) group project.* We are going to visit one of the plants that deals with hybrid energy applications. You will be entertained by the Technical Manager/Senior Engineer of the company. Each group (five persons per group) is supposed to write a technical report and do a presentation (marks for presentation: 10%) after the industrial trip. Contents of the report include:

(a)   Description of the hybrid energy systems [30%]

(b)   Comparison of the various energy resources use [10%]

(c)   Techno-economical advantages of the system, including system efficiency [30%]

(d)   Your suggestions for further improvements of the industry, including environmental aspect [20%]

2. *Industrial visit (SME energy industry) group project* . We are going to visit one of the SME plants that deals with one of the following: (I) industrial fryer; (II) boiler; or a nearby (III) small scale smoked-fish plant. Each group (five persons per group) is supposed to write a technical report and do a presentation (marks for presentation: 10%) after the industrial visit. Contents of the report include:

(a)   Description of the hybrid energy systems used by the company [30%]

(b)   Comparison of the various energy resources used (consider a Sankey diagram as well) [10%]

(c)   Techno-economical advantages of the system, including system efficiency [30%]

(d)   Your suggestions for further improvements of the SME company [20%]

Questions and Problems

Multiple choice questions

1. In a hybrid geothermal power plant/fossil system, the hybrid plant is making use of the _____ as the low temperature end of the conventional cycle, and combined with the _____ as the high temperature of the cycle.

   (A) relatively low temperature (and pressure) of geothermal energy; high temperature heat rejected

   (B) relatively high temperature (and pressure) of geothermal energy; low temperature heat rejected

   (C) relatively low temperature (and pressure) of fossil-fuel; high temperature heat rejected

   (D) relatively high temperature (and pressure) of fossil-fuel energy; low temperature heat rejected

2. Which of the following statements *correctly* discusses the hybrid geothermal fossil/superheat energy system?

   I The vapor-dominated steam from a geothermal well is superheated in a fossil fuel superheater prior to going into the high-pressure turbine.

   II The hybrid system can be comprisesd of two or more flash separators, with steam produced from the first separator preheated in the regenerator by exhaust steam from the high-pressure turbine.

   III The heated steam from the regenerator is subsequently superheated by the fossil fuel-fired superheater.

   IV The condensate from the condenser is pumped and reinjected into the geothermal reservoir.

   (A) I and II only

   (B) II and V only

   (C) I, II, and III

   (D) I and III only

   (E) I, II, III, and IV

3. IGCC is an advanced technology that employs a gasification method to convert coal from a coal power plant into synthesis gas, usually called syngas. The following is the list of impurities that IGCC possibly can remove, *except,*

   (a) CO_2

   (b) S_2

(c) O_2

(d) Particulate

4. All of the following are the most important control features of future electric car (HEV) applications, *except*,

(A) Durability of the energy sources

(B) More control components in the drivetrain

(C) Complex structures

(D) More control objectives abilities, such as vibration control of the vehicle.

(E) Speed control ability

5. Which of the following are the applications advantages of renewable hybrids for rural village electrification?

I Higher availability

II Lower overall energy and fuel cost

III Single resource utilities such as diesel or hydro stand-alone have higher supply risks

IV Improving energy sustainability

(A) I and II only

(B) II and V only

(C) I, II, and III

(D) I and III only

(E) All the above

6. Which of the following are important common features or components of the IGCC plants for power applications?

I Sulfur/CO_2 removal

II Gasification plant

III Gas turbine

IV Steam turbine

V Heat recovery steam generator

(A) I and II only

(B) II and V only

(C) I, II, and III

(D) I, III, and IV only

(E) All the above

7. In a hybrid direct fuel cell/turbine (FCT hybrid) system,

I The high-temperature fuel cell serves as the combustor for the gas turbine.

II Both the fuel cell and the gas turbine generate electricity.

III The gas turbine provides some balance-of-plant functions for the fuel cell, such as supplying air under pressure and preheating the fuel and air in a heat exchanger called a recuperator.

IV The recuperator (heat exchanger) transfers fuel cell exhaust energy to the compressed air supply.

(A) I and II only

(B) II and V only

(C) I, II, and III only

(D) I, III, and IV only

(E) All the above

Theoretical questions

1. What is a *hybrid* energy system?

2. What are the advantages of hybrid geothermal/fossil systems? With the aid of a schematic diagram, discuss *any one* of the following methods for hybrid geothermal/fossil system applications.

    (a) Hybrid geothermal preheat/fossil energy system
    (b) Hybrid geothermal fossil/superheat energy system

3. Why are IGCC systems becoming popular for industrial applications, in particular, in coal power plants?

4. Name and briefly discuss the three energy storage device requirements for hybrid electric vehicles.

5. Sketch schematic diagrams of *any two* of the following drive-trains architectures for HEV applications.

    (a) Series
    (b) Parallel (with torque/speed coupler)
    (c) Power-split (with planetary gear system)

6. Briefly discuss *any two* of the following control strategies for HEVs:

    (a) Deterministic rule-based (RB)
    (b) Fuzzy rule-based (RB)
    (c) Global optimization-based
    (d) Real-time optimization-based

7. Briefly discuss *any two* of the following PV–diesel hybrid energy system configurations for rural hybrid electrification applications.

    (a) Series hybrid energy systems
    (b) Switched hybrid energy systems
    (c) Parallel hybrid energy systems

8. What are the application advantages of renewable hybrids for rural village electrification, compared with stand-alone renewable systems?

9. With the aid of a schematic diagram, discuss the working principle of a hybrid geothermal preheat–fossil energy system.

10. Why is an IGCC plant more environmental-friendly than the conventional plans? List and briefly explain the advantages of IGCC systems compared with conventional coal-based power generation systems.

Homework problems

1. *Combined thermoelectric-adsorption system.* A typical combined thermoelectric-adsorption system has adsorption cycle efficiency of 0.4.

    (a) What is its cooling output if heat supply to the system is 500 J?
    (b) Estimate the thermoelectric efficiency when the system has undergone cooling of 3 kg of fruit juice from 25 °C to -3 °C within a period of 2 hours, with electrical input of 5 A and 12 V. Assume C_p=4.184 $kJ/kg°C$.
    (Answer: 2 kJ; 0.81)

2. *Hybrid SOFC/PEM/gas turbine system.* A solid oxide-PEM fuel cell-gas turbine (SOFC/PEM/GT) hybrid system is designed for combined power and heat applications. The SOFC stack power, the PEM stack power, and the turbine power are recorded to be 553.95 kW, 220.05 kW, and 150.45 kW, respectively.

    (a) If the net power output from the hybrid system is 773.25 kW, what is the compressor power?
    (b) Now, if the electrical output is 734.63 kW, with overall system efficiency of 65%, what is the heat output produced?
    (Answer: −151.2 kW; −133.74 kW)

Answers to multiple choice questions:

1. (A)

2. (E) all of the above

3. (c) O_2 is not an impurity but a necessity for an IGCC.

4. (E) Speed control ability is not the main concern.

5. (E) all of the above — I,II, III, and IV

6. (E) all of the above — I,II, III, and IV

7. (C) I, II, and III only

Bibliography

[1] MIT, 2006, *The Future of Geothermal Energy*, Cambridge, MA.

[2] DiPippo, R., Avelar, E. M., *Compound hybrid geothermal-fossil power plants*, June 1979, C00-4051-44. U.S. Department of Energy.

[3] Professional Engineer Publishing, Coal can come clean, May 11, 2005, *Professional Engineering*, Volume 18, Number 9, pp. 26–27.

[4] Breault, R. W., Gasification Processes Old and New: A Basic Review of the Major Technologies, *Energies*, 2010, 3, 216–240.

[5] Muselli, M., Notton, G., Louche, A., Design of Hybrid-Photovoltaic power generator, with optimization of energy management, *Solar Energy*, 1999; 65 (3): 143–157.

[6] National Energy Technology Laboratory (NETL), U.S. Department of Energy, http://www.netl.doe.gov/technologies/coalpower/fuelcells/hybrids.html accessed August 12, 2011.

[7] Karden, Eckhard, Ploumen, S., Fricke, B., Miller, T., Snyder, K., Energy storage devices for future hybrid electric vehicles, *Journal of Power Sources*, 168 (2007) 2–11.

[8] Salmasi, F. R., Control Strategies for Hybrid Electric Vehicles: Evolution, Classification, Comparison, and Future Trends, IEEE Transactions on Vehicular Technology, VOL. 56, NO. 5, September 2007, pp. 2393–2404.

[9] Salmasi, F. R., Designing control strategies for hybrid electric vehicles, in Proc. Tutorial Presentation EuroPes, Benalmadena, Spain, Jun. 15–17, 2005.

[10] Chaurey, A., Kandpal, T.C., Assessment and evaluation of PV based decentralized rural electrification: an overview. *Renewable and Sustainable Energy Reviews*, 2010; 14:2266-78.

[11] El-Shatter, T.F., Eskandar, M.N., El-Hagry, M.T., Hybrid PV/fuel cell system design and simulation. *Renew Energ*, 2002; 27:479-485.

[12] Lee, B.D., Jung, D.H., Ko, Y.H., Analysis of DMFC/battery hybrid power system for portable application. *J. Power Sources*, 2004; 131:207-212.

[13] Abdullah, M.O., Meng, T. K., Gan, Y. K., A multi-purpose mini hybrid fuel cell- solar portable device for rural application: laboratory testing, *International Journal of Research and Reviews in Applied Sciences* 7(4) June, 2011.

[14] Jiang, Z., Gao, L., Blackwelder, M. J., Dougal, R.A., Design and experimental tests of control strategies for active hybrid fuel cell/battery power sources, *J. Power Sources* 2004; 130:163–171.

[15] Iqbal, M.T., Modeling and control of a wind fuel cell hybrid energy system, *Renewable Energy*, 2003; 28(2): 223-237.

[16] Nayar, C.V., Phillips, S.J., James, W.L., Pryor T.L., Remmer, D., Novel wind/diesel/battery hybrid energy system, *Solar Energy*, 1993; 51(1): 65–78.

[17] Nogaret, E., Stavrakakis, J. C., A new expert system based control tool for power systems with large integration of PVs & wind power plants. Paper presented at 1st WCPEC, Hawaii, 1994.

[18] Wichert, B., PV-Diesel Hybrid Energy Systems for Remote area power generation - a review of current practice and future developments, *Renewable and sustainable energy reviews*, Vol. 1, No. 3, pp. 209–228, 1997.

[19] Bower, W., Merging photovoltaic hardware development with hybrid applications in the U.S.A. Paper presented at Solar '93 ANZSES, Fremantle, Western Australia. 1993.

[20] Ashok, S., Optimised model for community-based hybrid energy system, *Renewable Energy* 2007; 32(7): 1155–1164.

[21] Schouten, N. J. Salman, M. A., Kheir, N. A., Energy management strategies for parallel hybrid vehicles using fuzzy logic, *Control Engineering Practice*, 2003; 11(2): 171-177.

[22] R. Langari, R. and J.-S. Won, Intelligent energy management agent for a parallel hybrid vehicle-part I: system architecture and design of the driving situation identification process, *IEEE Transactions on Vehicular Technology*, 2005; 54(3): 925–934.

[23] Won, J.-S. Langari, R., Intelligent energy management agent for a parallel hybrid vehicle-part II: torque distribution, charge sustenance strategies, and performance results, *IEEE Transactions on Vehicular Technology*, 2005; 54(3): 935–953.

[24] Lin, C.-C., Peng, H., Grizzle, J.W., Kang, J.-M., Power management strategy for a parallel hybrid electric truck, *IEEE Transactions on Control Systems Technology*, 2003; 11(6): 839–849.

[25] Muselli, M., Notton, G., Louche, A., Design of Hybrid-Photovoltaic power generator, with optimization of energy management, *Solar Energy*, 1999; 65 (3): 143–157.

[26] Energy Efficiency and Renewable Energy, www1.eere.energy.gov/industry/distributedenergy/pdfs/micropower_opp_vol1_text.pdf, accessed on Oct. 15, 2011.

[27] Dicks, A. L., Mescal, M., Seymour, C., A study of SOFC–PEM hybrid systems, *Journal of Power Sources*, Volume 86, Issues 1-2, March 2000, pp. 501–506.

[28] Mescal, C.M., (1999) SOFC-PEM hybrid fuel cell systems, *ETSU Report No. F/03/00177/REP*, AEA Technology, U.K.

[29] Gibbard, H.K., 1999, Highly reliable 50 Watt fuel cell system for variable message signs, Proceedings of the European fuel cell forum. Portable Fuel Cell Conference, Lucerne, pp. 107–112.

[30] Li, C.-H., Zhu, X.-J., Cao, G.-Y., Sui, S., Hu, M.-R., Dynamic modeling and sizing optimization of stand-alone photovoltaic power systems using hybrid energy storage technology, *Renewable Energy*, Volume 34, Issue 3, March 2009, pp. 815–826.

[31] Energy Efficiency and Renewable Energy, available at: www1.eere.energy.gov/industry/distributedenergy/pdfs/micropower_opp_vol1_text.pdf. Accessed on August 30th, 2011

[32] Abdullah, M. O., Ngui, J. L., Abd-Hamid, K., Leo, S. L., Tie, S. H., Cooling performance of a combined solar thermoelectric - adsorption cooling system: an experimental study, *Energy and Fuels*, 23 (2009), pp. 5677–5683.

11

Other Energy Conversion Methods

In this chapter, we shall examine other energy conversion methods as follows:

- Absorption and adsorption energy conversion
- Thermoelectric and thermionic energy conversion
- Magnetohydrodynamic (MHD) power conversion

11.1   Absorption Energy Conversion

Both absorption and adsorption processes are well established energy conversion technologies, usually for cooling applications. Unlike the desiccant process, which is operating on an open cycle, both absorption and adsorption cycles energy conversion utilize closed-cycle systems (see Figure 11.1).

Typically, an adsorption system has some features in common with the conventional vapor-compression system but differs in a few aspects. The main difference is that the mechanical compressor in vapor-compression systems is replaced by a thermally driven absorption/adsorption compressor, also called a chemical or thermal compressor. Hence, the ability to be driven by low grade thermal heat energy makes an absorption/adsorption system an attractive option for electric energy savers. Another difference between the systems is that a heat source must be introduced in absorption/adsorption systems to retrieve the refrigerant vapor from the adsorbent before the refrigerant enters the condenser. This process thus eliminates necessary shaft work of the engine to drive a mechanical compressor as in the case of a conventional compression cycle conversion system, which increases the engine load and fuel consumption.

11.1.1   Absorption Energy Conversion

11.1.2   What Is an Absorption Thermal Energy Convertor?

An absorption thermal energy convertor is a heat engine that uses low grade thermal energy to produce desired cooling via a process of absorption cooling in a chemical absorber, as compared to the common refrigeration technique, which uses an electrical compressor to accomplish the same cooling effect.

11.1.2.1   Working Principle of an Absorption Energy System

The vapor compression systems use a mechanical compressor to compress the vaporized refrigerant in order to elevate its temperature approximately $10^{\circ}C$ higher than the ambient temperature surrounding the condenser for efficient heat rejection. To accomplish the same function, the vapor absorption system utilizes a secondary fluid, usually called the absorbent, to absorb the refrigerant at a lower temperature level and then delivers the liquid to the higher temperature level, usually by means of a liquid pump. At higher temperatures, the water evaporates into steam, which flows in the refrigeration part of the cycle (see Figure 11.2).

The efficiency of the absorption system usually is evaluated by coefficient of performance (COP), a term defined by the cooling output rate per energy input at the generator, Eq. 11.1.

$$\text{COP} = \frac{\text{Cooling output rate}}{\text{Heat rate input at generator}} \tag{11.1}$$

The COP of a single stage absorption cycle is usually around 0.6–0.8.

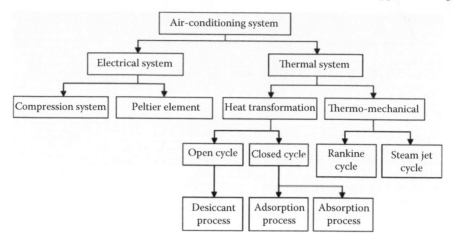

Figure 11.1
Absorption and adsorption technologies for thermal energy conversion. (After Abdullah et al. 2011.)

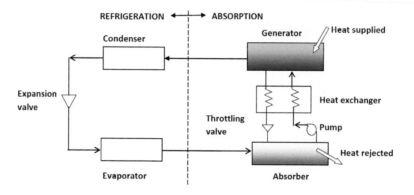

Figure 11.2
Vapor absorption energy cycle.

The maximum theoretical COP is the reversible Carnot refrigerator efficiency

$$COP_{max,\,Carnot} = \left(1 - \frac{T_{ambient}}{T_s}\right)\left(\frac{T_L}{T_{ambient} - T_L}\right)$$

where,
T_L = temperatures of the refrigerated space, °R
$T_{ambient}$ = temperatures of the environment (ambient temperature), °R
T_s = temperatures of the heat source, °R.

11.1.2.2 Materials Selection and Working Pair

The two most common working pairs used are:

1. Lithium-bromide and water

2. Ammonia-water

11.1.2.3 Application Advantages of Absorption Energy Systems

1. An absorption refrigerator could be conveniently driven by various energy sources and via different conversion methods [34], e.g.:

   • It can be powered by conventional electric energy, by means of a heating coil, to heat the solution mixture in the generator.

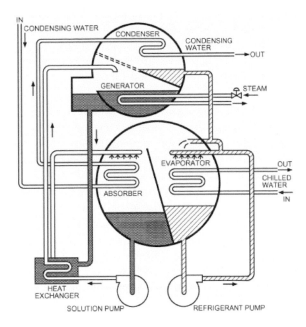

Figure 11.3
Two-shell, single-effect lithium bromide cycle – water chiller. (Courtesy of ASHRAE, 2010 [3].)

- It can be driven by low grade heat source energy, to heat the solution mixture in the generator.
- It can be driven by external heating of fossil fuel, by means of a gas burner, such as LPG.

2. The absorption cycle can be powered by waste heat. The ability to use unwanted heat has caused such systems to gain popular in commercial cooling for energy recycleing applications.

3. Absolute low noise is also one of the added advantages of the absorption refrigerator, as pointed out by Horuz and Callander [1].

4. It had been reported to cause zero or minimum ozone depletion by Misra et al., 2005 [2], etc., because it does not use any CFC or HFCs refrigerant as the working fluids.

11.1.2.4 Industrial Design and Commercial Absorption Systems

Since the generator and condenser operate at similar pressure, and the evaporator and absorber operate at the same lower pressure level, the commercial adsorption cooling units combine the generator and condenser in one vessel, and the evaporator and absorber in another vessel (see Figure 11.3 for a typical two-shell absorption cycle employing lithium bromide–water as the working fluid pair).

11.1.2.5 Some Advanced Absorption Cycles

Advanced absorption cycles include double-effect cycles, triple-effect cycles, GAX (generator-absorber-heat exchanger) cycles, etc. Theoretically, the advanced cycles would deliver higher COPs for cooling applications. Some brief discussions of the systems are given as follows:

11.1.2.5.1 Double-Effect Cycle

Figure 11.4 is a schematic of a double-effect indirect-fired liquid absorption chiller. All the major components are similar to the single-effect chiller except for an added generator (first-stage or primary generator), condenser, heat exchanger, and optional condensate subcooling heat exchanger. Operation of the double-effect absorption machine is similar to that for the single-effect machine. The primary generator receives heat from the external heat source, which boils dilute absorbent solution. Pressure in the primary generator's vapor space is about 100 kPa. This vapor flows to the inside of the tubes in the second-effect generator. At this pressure, the refrigerant vapor has a condensing temperature high enough to boil and concentrate absorbent solution on the outside of these tubes, thus creating additional refrigerant vapor with no additional primary heat input. These machines are typically fired with medium-pressure steam of 550 to 990 kPa (gage) or hot liquids of 150 to 200 °C. Typical operating COPs are 1.1 to 1.2. These machines are available commercially from several manufacturers and have capacities ranging from 350 to 6000 kW of refrigeration [3].

11.1.2.5.2   Triple-Effect Cycles

Triple-effect absorption cooling can be classified as single-loop or dual-loop cycles. Single-loop triple-effect cycles are basically double-effect cycles with an additional generator and condenser. The resulting system with three generators and three condensers operates similarly to the double-effect system. Primary heat (from a natural gas or fuel oil burner) concentrates absorbent solution in a first-stage generator at about 200 to 230 °C. A fluid pair other than water–lithium bromide must be used for the high-temperature cycle. The refrigerant vapor produced is then used to concentrate additional absorbent solution in a second-stage generator at about 150 °C. Finally, the refrigerant vapor produced in the second-stage generator concentrates additional absorbent solution in a third-stage generator at about 93 °C. The usual internal heat recovery devices (solution heat exchangers) can be used to improve cycle efficiency. As with double-effect cycles, several variations of solution flow paths through the generators are possible. Theoretically, these triple-effect cycles can obtain COPs of about 1.7 (not taking into account burner efficiency). Difficulties with these cycles include the following:

- High solution temperatures pose problems to solution stability, performance additive stability, and material corrosion.
- High pressure in the first-stage generator vapor space requires costly pressure vessel design and high-pressure solution pump(s).

A double-loop triple-effect cycle consists of two cascaded single-effect cycles. One cycle operates at normal single-effect operating temperatures and the other at higher temperatures. The smaller high-temperature topping cycle is direct-fired with natural gas or fuel oil and has a generator temperature of about 200 to 230 °C. A fluid pair other than water/lithium bromide must be used for the high-temperature cycle. Heat is rejected from the high temperature cycle at 93 °C and is used as the energy input for the conventional single-effect bottoming cycle. Both the high and low temperature cycles remove heat from the cooling load at about 7 °C. Theoretically, this triple-effect cycle can obtain an overall COP of about 1.8 (not taking into account burner efficiency). As with the single-loop triple-effect cycle, high temperatures create problems with solution and additive stability and material corrosion. Also, using a second loop requires additional heat exchange vessels and additional pumps. However, both loops operate below atmospheric pressure and, therefore, do not require costly pressure vessel designs [3].

11.1.2.5.3   GAX (Generator-Absorber Heat Exchange) Cycles

The GAX cycle is a heat-recovering cycle in which absorber heat is used to heat the lower-temperature section of the generator as well as the rich ammonia solution being pumped to the generator. This cycle has potential gas-fired COPs of 0.7 in cooling mode and 1.5 in heating mode, making it capable of significant annual energy savings. ([3])

GAX systems can be classified into two types: simple GAX and branched GAX cycle. Their brief descriptions are given in the following paragraphs:

Simple GAX Cycle   A simple GAX cycle is shown in Figure 11.5. The solid and dotted lines represent a GAX cycle and single effect cycle, respectively. In the absorber and generator, the pressure and concentration are maintained in such a way as to cause a temperature overlap between the absorber and the generator. This provides the possibility that some of the heat of absorption may be rejected to the generator within the cycle, leading to a higher COP. This overlapping of heat is an attractive characteristic of the GAX cycle using ammonia–water, which cannot be realized in the water–lithium bromide absorption cooling systems.

Branched GAX Cycle   Figure 11.6 shows the schematic of the branched GAX cycle. The amount of heat supplied by the absorber to the generator is less than the requirement of the generator in a simple GAX cycle. This can be increased by increasing the mass flow rate in the absorber. This is accomplished in the branched GAX cycle by an additional solution pump. When the solution flow rate in the high temperature section of both the absorber and generator is increased, the high temperature section of the absorber supplies more heat to the low temperature section of the generator. At the same time, the heat requirement of the high temperature generator is increased, but the amount of heat that has to be supplied by the external source is decreased (Jawahar and Saravanan, 2010 [5]). In other words, the branched GAX cycle eliminates the mismatch between the heat available in the absorber and

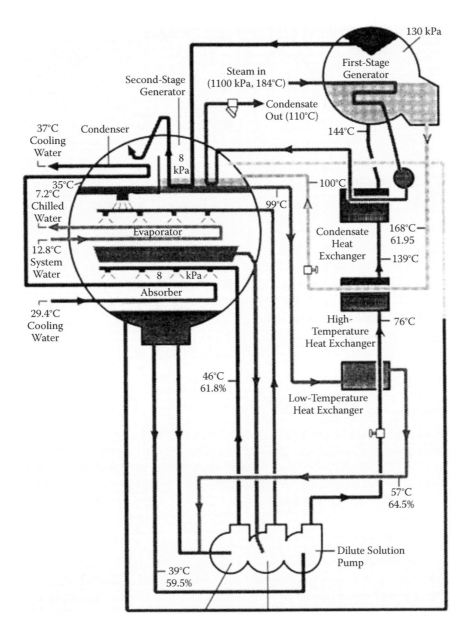

Figure 11.4
Double-effect indirect-fired absorption chiller. (From *ASHRAE Handbook*, 2010 [3]. With permission.)

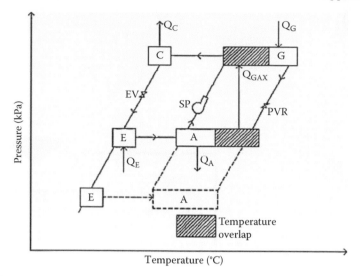

Figure 11.5
Schematic of simple GAX cycle–temperature overlap. (From Kang and Kashiwagi [4].)

the heat required in the generator. Due to this, some amount of refrigerated vapor is generated in the GAX section of the generator, which reduces the external heat input requirement of the generator further.

In terms of performance, Jawahar and Saravanan (2010) recently revealed that an improvement in the COP of about 10–20%, 20–30%, and 30–40% in absorber heat recovery cycle, simple GAX, and branched GAX cycle can be achieved compared to a single stage absorption cycle [5].

11.2   Adsorption Energy Conversion

The efficient utilization of energy at low grade heat source temperature is a key issue of energy wastage, contributed especially from our industries and residential applications. The heat-activated adsorption energy conversions are suitable for energy recycling or regeneration, i.e., in converting low-temperature waste heat into useful effects — such as cooling, and refrigeration, as well as heating. Such "thermal" or "chemical-operated" cycles have direct environmental benefits in terms of lower emissions, etc.

11.2.1   Basics of Adsorption

Generally, adsorption is a process where molecules of a gas or liquid contact and adhere to a solid surface. This process is always exothermic, where heat is liberated. The substrate on which adsorption take place is called an adsorbent, whereas the material concentrated on the surface of the adsorbent is called an adsorbate. Adsorbents are the materials that contain a lot of miniscule internal pores as small as nanometers. Adsorption mechanisms can be categorized into two types, namely physical adsorption and chemical adsorption. Physical adsorption is the type of adsorption in which the forces involved are intermolecular forces or Van der Waals forces. Chemical adsorption, on the other hand, is the type of adsorption in which the forces involved are covalence or ionic forces between the adsorbing molecules and the adsorbent. As covalence or ionic bonding is normally greater than Van der Waals bonding, more heat is liberated when chemical adsorption occurrs. Besides, the process of physical adsorption is reversible, while the chemical adsorption process is irreversible.

11.2.1.1   The Adsorption System

Typically, an adsorption cycle applied in air-conditioning or refrigeration does not use any mechanical energy, but uses only heat energy from waste heat, solar, or any means of low grade thermal heat. Adsorption units usually

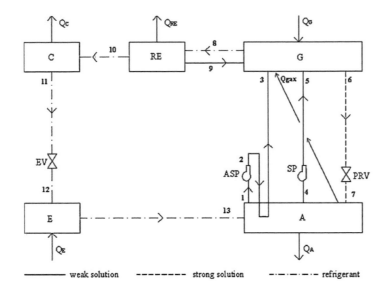

Figure 11.6
Schematic of branched GAX cycle. (After Herold et al.,1996 [6].)

consist of one or several adsorbers, a condenser, and an evaporator, as shown in Figure 11.7(a).

This cycle is basically an intermittent one because cold production proceeds only during part of the cycle if only one adsorber is utilized. Nevertheless, when more than one adsorber is used, this cycle can be operated out of phase and generate a quasi-continuous cooling effect. For situations where all the energy required for heating the adsorber is provided by the heat source, the cycle is called a single effect cycle. Likewise, a double effect cycle can be processed by using two or more adsorbers. In double effect cycles some heat is internally recovered between the adsorbers, which enhances the cycle performance. Various types of adsorption cycles have been studied extensively by Wang in 2001 [7]. Some of the common adsorption cycles are basic cycle, mass recovery cycle, continuous heat recovery cycle, thermal wave cycle, cascade multi effect cycle, and hybrid heating and cooling cycle.

11.2.1.2 Adsorbent–Adsorbate Pairs Selection

Figure 11.8 shows a typical adsorption system called "chemical" compressor, specially designed for vehicle compressor applications.

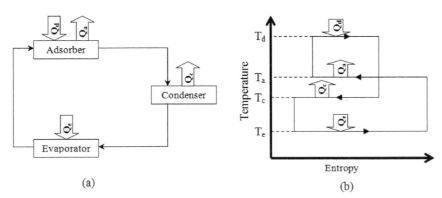

Figure 11.7
The adsorption system. (a) Block diagram of a simple adsorption cycle system; (b)an ideal T-S diagram. (After Leo and Abdullah, 2010 [11].)

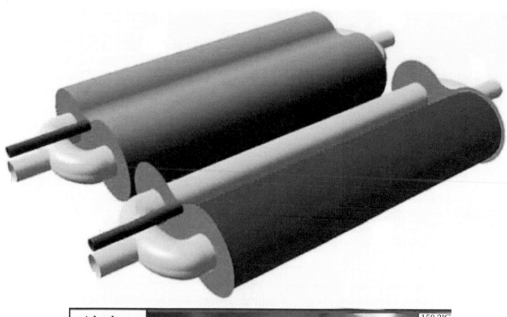

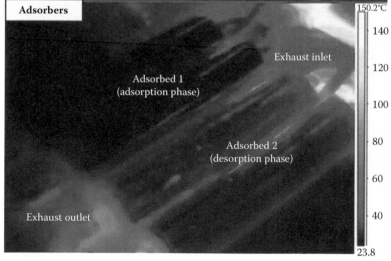

Figure 11.8 (SEE COLOR INSERT)

A typical adsorption system. (*Top*): the "chemical" compressor, 3-D diagram view. (*Bottom*): the thermograph photo of the chemical adsorber system showing temperature variation of the adsorbers during operation. The system works intermittently where a pair of the adsorbers is cold while the other pair is heated, and vice versa. (From Leo, S.L., and Abdullah, M.O., HVAC&R Research 16 (2) (2010), pp. 221–231. With permission.)

Figure 11.9 (SEE COLOR INSERT)
SEM image of a typical activated carbon used in an adsorption system. Activated carbon properties: type = granular; average size = around 3.0 mm; density = 0.431 g/cm^3; heat of adsorption = 1800 kJ/kg; surface area = 1000-1100 m^2/g. (From Leo and Abdullah, HVAC&R Research 16 (2) (2010). With permission.)

In general, selection of an appropriate working medium is a crucial factor for the successful operation of an adsorption air-conditioning system, as the performance of this system varies over a wide range of operating temperature by using different working pairs at different temperatures. In order to choose the best adsorbent, the following factors need to be considered:

1. Good thermal conductivity and low specific heat capacity to decrease the cycle time

2. High adsorption and desorption capacity to achieve higher cooling effect

3. No chemical reaction with the adsorbate used

4. Widely available and at and affordable cost

The preferable adsorbate should have the following desirable thermodynamics and heat transfer properties:

1. High latent heat per unit volume to increase the cooling effect

2. High thermal conductivity to decrease the cycle time

3. Chemically stable within the working temperature range

4. Non-toxic and non-corrosive

Examples of commonly used adsorbate are activated carbon and silica gel. The scanning electron microscope (SEM) image of a typical granular palm-derived activated carbon is as shown in Figure 11.9.

11.2.1.3 Dubinin–Astakhov (D–A) Equation and Heat of Adsorption

Adsorption properties of activated carbon generally follow the Dubinin–Astakhov (D-A) equation.

$$x = x_0 \exp\left[-k\left(\frac{T}{T_s} - 1\right)^n\right] \tag{11.2}$$

where x_0, k, and n are coefficients that are specific for different activated carbons and refrigerants, T is the adsorption temperature (in Kelvin), T_s is the saturated temperature of methanol (also in Kelvin), and x is the adsorption quantity of refrigerant in the activated carbon (expressed in units of kg/ kg).

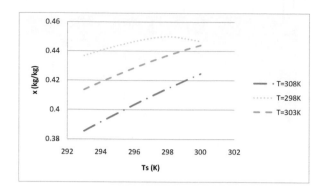

Figure 11.10
Estimated adsorption quantity of refrigerant in the activated carbon versus saturated refrigerant temperature, at three different adsorption temperatures of activated carbon. (Adopted from Abdullah et al. 2009 [10].)

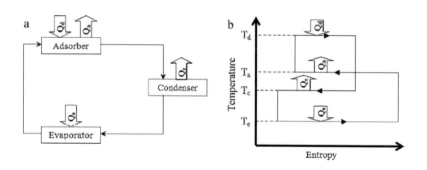

Figure 11.11
Thermodynamic analysis of adsorption cycle. (a) A simple adsorption cycle; (b) T-S diagram. (From Abdullah et al. 2011[8].)

Example 11.1   For a typical activated carbon-methanol adsorption system, estimate the adsorption quantity of refrigerant required in the activated carbon versus saturated refrigerant temperature, at three different adsorption temperatures of activated carbon at 298K, 303K, and 308K, respectively. Given that $x_0 = 0.45$, $k = 13.38$, and $n = 1.5$, plot your result on the graph x as a function of saturation temperature (Ts).

Solution: Employing the D-A Equation, Eq. 11.2, we can estimate and plot the result as in Figure 11.10.

11.2.1.4   Thermodynamic Analysis of a Typical Adsorption Cycle

In general, the operation of an adsorption cycle involves the processes of isosteric heating, desorption, isosteric cooling, and adsorption. A simple adsorption cycle indicating all the heat transfer for a complete cycle and its T-S diagram (Leo and Abdullah, 2010 [11]) are as illustrated in Figure 11.11(a) and (b), respectively. Heat (Q_d) is supplied to release adsorbates from the adsorbent at high temperature Td during the desorption process. The desorbed adsorbate vapors then travel to the condenser, where they are condensed by releasing heat (Q_c) to the surroundings at temperature T_c. When the liquid adsorbates reach the evaporator, it evaporates and heat (Q_e) is adsorbed from the surroundings to produce cooling effects at lowest temperature T_e. During the adsorption process, these vapors are adsorbed back by the adsorbent by released heat (Q_a) at temperature Ta.

11.2.2 Energy Performance of Adsorption Energy System

The performance of the adsorption system is usually assessed by using two performance factors, namely coefficient of performance (COP) and specific cooling power (SCP). COP is defined as the amount of cooling produced by the adsorption cooling system per unit of heat supplied ([12]; [13]) as given below:

$$COP = \frac{Q_e}{Q_d} \tag{11.3}$$

where,

Q_e is the quantity of heat transferred through the evaporator, and

Q_d is the quantity of heat adsorbed by the adsorber during desorption phase.

SCP is defined as the ratio between the cooling production and the cycle time per unit of adsorbent weight, given as follows ([13]; [12]; [11]):

$$SCP = \frac{Q_e}{t_c m_a} \tag{11.4}$$

where,

t_c is the cycle time, and

m_a is mass of the adsorbent.

Since SCP is correlated to both the mass of adsorbent and the cooling power, it determines the size of the energy system. For a small cooling load, higher SCP values indicate the compactness of the system.

11.2.3 Performance of Adsorption Systems for Different Applications

Table 11.1 summarizes the performance of some adsorption systems for different applications namely icc making [35], chilled water [35], space air-conditioning [36], and vehicle cooling from exhaust heat ([11], [8]). As the results were obtained under different working conditions, they should not be compared with one another, but only serve as a general reference for feasibilities study on wide applications of adsorption systems, in particular, those driven by waste heat or other low thermal grade energy such as solar energy.

11.2.4 Application Advantages of Adsorption Energy System

The employment of adsorption air-conditioning technology could be one of the best options to replace conventional compression cooling system. The reasons follow:

Table 11.1 Performance of adsorption refrigeration systems for different applications

Application	Heat source temperature	Working pair	COP	SCP	Year
Ice making	105 °C	$AC-NH_3$	0.10	35 W/kg	1997
	< 120 °C	AC–Methanol	0.18	27 W/kg	2005
	115 °C	$(AC+ CaCl_2-NH_3)$	0.39	770 W/kg	2006
Chilled water	55 °C	Silica gel$-H_2O$	0.36	3.2 kW/unit	2001
	75 – 90 °C	Silica gel$-H_2O$	0.35–0.60	15kW/m^3	2004
	80 °C	Silica gel$-H_2O$	0.33 to 0.50	92 to 172 W/kg	2005
Space air-conditioning	204 °C	Zeolite$-H_2O$	0.60–1.60	36 to 144 W/kg	1988
	230 °C	Zeolite$-H_2O$	0.41	97 W/kg	1999
	310 °C	Zeolite$-H_2O$	0.38	25.7 W/kg	2000
	100 °C	$AC-NH_3$	0.20–0.21	21–30W/kg	2004
Vehicle cooling from exhaust heat	250–400 °C	$AC-NH_3$	0.19	396.6 W/kg	2011
	Exhaust heat	Zeolite 13X-H_2O	0.38	25.7	2000
	Exhaust heat	$AC-NH_3$	0.16	20.8	2001

1. The potential use of low-grade heat to decrease the quantity of carbon dioxide emission from combustion of the fossil fuels as the engine load was decreased, thusing reduce the overall operational cost.

2. Clean refrigerants (instead of using CFCs and HCFCs) that have zero ozone depletion and global warming potentials can be used as a working fluid.

3. Low regeneration temperature (less than 150 °C) could be used to operate the system.

4. Less moving parts, low maintenance cost, and simple system structure make it attractive for many cooling applications.

11.3 Thermoelectric and Thermionic Energy Conversion

Both thermoelectric and thermionic convert heat directly into electricity via an interface-involving no moving parts, no expensive catalyst (as in th of fuel cell), and no working fluids (as in the adsorption/adsorptive conversions).

11.3.1 Thermoelectric Energy Conversion

Thermoelectric energy conversion is applied in many industries including aerospace, medician, telecom and other industries that needs lower a temperature, and for waste-heat recovery applications.

11.3.1.1 What Is a Thermoelectric Energy Convertor?

We can define a thermoelectric energy generator as a heat engine that takes up heat at the upper temperature level (called hot junction) converts part of the heat into electricity via the Seebeck effect, and discharges the rest at the lower temperature (called cold junction).

11.3.1.1.1 What Is the Seebeck (Thermoelectric) Effect? We shall notice that when two different materials are joined to form a loop and the two junctions are maintained at different temperatures, it will result in an electromotive force (e.m.f.) around the loop (see Figure 11.12). We usually call this kind of thermo-electric effect a Seebeck effect, named after the German scientist seebeck who discovered such phenomena in 1822. The magnitude of e.m.f. E is,

$$E = \alpha \triangle T \tag{11.5}$$

where
α = the Seebeck effect coefficient;
$\triangle T$ = the temperature difference between the two junctions.

11.3.1.2 The Working Principle of a Thermoelectric Generator

In a thermoelectric generator, the thermoelectric generator module is made up of many thermoelectric cells or "thermocouple modules" connected in series to increase overall voltage and power. The heat input is supplied externally from, for example, a gas burner (as in the case of a thermoelectric refrigerator), direct solar radiation, etc.

Unfortunately, until recently, the efficiency of the thermoelectric generator has low been around 5–7%. A 1-kW thermoelectric generator would require about 5000 thermocouples in series with temperature difference $\triangle T$ of 400 °C.

11.3.1.3 What is the Estimated Energy Output of a Thermoelectric Module?

When a temperature difference is established across the module, the maximum power output of the module can be calculated using [14],

$$P_{max} = \frac{V_1 2}{4R_L(\frac{V_1}{V_2} - 1)} \tag{11.6}$$

where R_L is the load resistance, which includes contributions from all the wires and connections in the circuit. V_1 and V_2 are voltages measured at the terminal ends 1 and 2 across the load resistance, respectively.

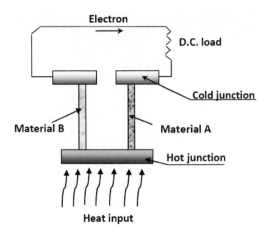

Figure 11.12

Thermoelectric generator: schematic diagram of a thermoelectric cell. A thermoelectric generator module is made up of many thermoelectric cells or "thermocouple modules" connected in series to increase overall voltage.

From Figure 11.13(a), it can be seen that the maximum power output increases parabolically with an increase in temperature difference. For a given temperature difference, there is a significant variation in maximum power output for different modules due to variation in thermoelectric materials, module geometry, and contact properties. However, as shown in Figure 11.13(b), the maximum power output follows a clear trend and increases with an decrease in thermoelement length for a given cross-sectional area.

When we consider the effects of thermal and electrical contact resistances, the power output per unit area of the thermoelectric module $(p = \frac{P}{NA})$ and the conversion efficiency of a module are given by (Min et al., 1995 [15]) Eq. 11.7 and Eq. 11.8, respectively:

$$P = \frac{\alpha^2}{2\rho} \frac{NA\triangle T^2}{(1+n)(1+2r\frac{l_c}{l})} \tag{11.7}$$

$$\phi = (\frac{T_h - T_c}{T_h}) \left\{ (1 + 2r\frac{l_c}{l})^2 [2 - \frac{1}{2}(\frac{T_h - T_c}{T_h}) + \frac{4}{zT_h}(\frac{1+n}{1+2rl_c})] \right\}^{-1} \tag{11.8}$$

where, α, ρ, and z are the Seebeck coefficient, electrical resistivity, and figure-of-merit, respectively, of the thermocouple material; N the number of thermocouples in a module; A the cross-sectional area of thermoelements; T_h and T_c are the respective temperatures at the hot and cold sides of the module; l_c the thickness of the insulating ceramic layers; $n = \frac{\rho_c}{\rho}$ and $r = \frac{\lambda}{\lambda_c}$, where ρ_c and λ_c are the electrical and thermal contact resistivities.

In Figure 11.14 the conversion efficiency is shown, together with the corresponding maximum power output, as a function of thermoelement length for different temperature differences at $n = 0.1$ mm and $r = 0.2$ for a typical commercially available module [14].

11.3.1.4 Materials Selection for Thermoelectric Generator

The material selection for a thermoelectric generator depends on many factors. Some of the primary properties are given as follows:

1. *Figure of merit, Z* should be as high as possible. The Z value for bismute telluride (Bi_2Te_3) is 0.004 per °K, while lead telluride (PbTe), Germanium telluride (GeTe), and zink antimonide (ZnSb) have Z values of $\sim$ 0.0015 per °K, and cesium sulphide has Z value of only 0.001 per °K.

2. *Thermal conductivity.* Should be as low as possible.

3. *Mobility of current carries.* Effectiveness and current carries should be as high as possible.

Apart from the above, the thermoelectric material should have good mechanical strength and good chemical resistance, for instance, from oxidation, etc.

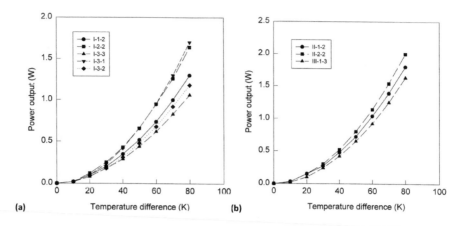

Figure 11.13
Maximum power output as a function of temperature differences. (a) Modules with 127 thermocouples and a cross-sectional area of $1.4 = 1.4$ mm^2; (b) Modules denoted II possess 31 thermocouples and a cross-sectional area of $4.5 = 4.5$ mm^2. Module denoted III possesses 49 thermocouples and a cross-sectional area $5.0 = 5.0$ mm^2. (After Rowe and Gao Min, 1998 [14].)

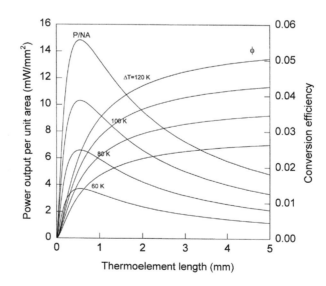

Figure 11.14
Power output per unit area and conversion efficiency as a function of thermoelement length. (After Row and Min, 1998 [14].)

Figure 11.15
A typical thermoelectric cooler. This cooler is operating at 12 V power supply. (Courtesy of Koolatron.)

11.3.1.5 Application Advantages of a Thermoelectric Generator

The application advantages of a thermoelectric generator are as follows:

1. No moving parts, thus quiet in operation.

2. No carbon emission, thus environmentally friendly.

3. Can be useful for waste heat recovery.

4. Similar to a battery, a thermoelectric generator can be connected in series to increase voltage and meet the total energy requirement.

11.3.1.6 What Is the Potential Application Area of a Thermoelectric Energy Convertor?

Thermoelectric modules, which were originally developed for cooling applications, also exhibit a promising performance for electrical power generation using waste heat in the temperature range 300–400 K. Rowe et al. reported that a cost-per-watt of about £4SD/W can readily be obtained using commercially available modules with an appropriate thermoelement length.

The relatively low conversion efficiency of thermoelectric modules (around 5%) has been a major factor in limiting their applications in electrical power generation, and has restricted their use to only specialized situations [16]. However, one exception is the thermoelectric recovery of waste heat when it is unnecessary to consider the cost of the thermal input ([14]; [17]). Such waste-to-energy applications are an attractive area for future thermoelectric energy conversion.

11.3.1.7 Application Examples

Thermoelectric usually is used for cooling purposes, in particular, at hotels and sound-sensitive areas due to their silent operation advantage.

• Refrigerators

• Coolers (see example, Figure 11.15)

11.3.2 Thermionic Energy Conversion

11.3.2.1 What Is a Thermionic Energy Convertor?

Thermionic energy convertor is a type of heat engine that uses an electron gas as a working fluid. It is quite similar to a diode that transforms heat to electricity via a process usually known as *thermionic emission* — emission of electrons from a metal when it is subjected to heat.

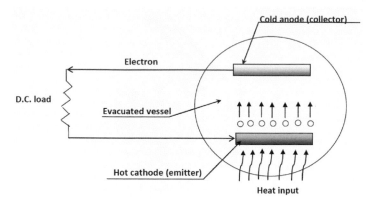

Figure 11.16
Thermionic energy generator — schematic diagram.

11.3.2.1.1 What Is Work Function?
Work function refers to the energy required to extract an electron from the metal in the thermionic emission process. Work function values are a function of the nature of the metal and its surface conditions. It is usually measured in electron volts.

11.3.2.2 Working Principle of a Thermionic Generator

A thermionic generator (see Figure 11.16) essentially consists of two metals (electrodes) with different work functions sealed into an evacuated vessel. The electrode with a large work function is maintained at a higher temperature, while the electrode with a smaller work function is at a lower temperature. Both electrodes are separated by a vacuum. When we supply heat to the cathode, this raises the energy of the electrons enabling them to "escape" from the cathode surface, and flow to the cold anode via the vacuum medium. The electrons subsequently transfer from the anode to the electric circuit.

In order to achieve significant current output with efficiency of around 10%, a temperature of 1000 °C usually is needed to produce a substantial electron emission rate. Higher efficiency (as high as 40%) can be obtained at higher temperatures. This is because higher temperatures increase the current per unit area of the emitter, resulting in an increase in power.

$$Power = Voltage \times current \tag{11.9}$$

11.3.2.3 Materials Selection for Electrodes of a Thermionic Generator

Anode materials are usually made from barium and strontium oxides, which have low work function; cathodes are normally made from tungsten impregnated with a barium compound.

11.3.2.4 Applications of a Thermionic Generator

We can use any energy sources (fossil fuels or renewables) for powering a thermionic generator. And, we could use it at many locations, even in remote areas and in space.

11.3.2.5 Thermoelectric Generator and Thermionic Generator: Application Comparisons

1. A thermionic generator based on the ballistic current flow is highly efficient, and its theoretical efficiency is close to the Carnot efficiency [18]. A thermoelectric generator, however, has poor efficiency due to the diffusive current flow ([19]; [20]).

2. A thermionic generator usually requires a high-temperature heat source (e.g., 1500 K) to generate a practically useful current [1,3]. A thermoelectric generator, however, can produce electrical power from low quality heat energy sources [21]; [20]

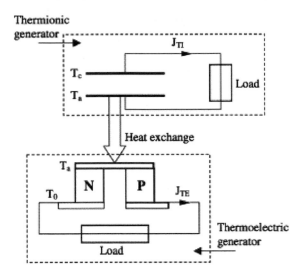

Figure 11.17
Combined thermionic–thermoelectric generator. (From Xuan and Li, 2003 [20].)

11.3.3 Combined Thermionic–Thermoelectric Generator

In a combined thermionic–thermoelectric generator system, the thermoelectric generator makes use of the rejected heat from the anode of the thermionic generator, and produces additional electrical power.

Figure 11.17 shows a schematic of a combined thermionic–thermoelectric generator.

Figure 11.18 and Figure 11.19 show the performances with respect to Ta in terms of the maximum power output and efficiency, respectively. As expected, there exists an optimum value of Ta at which P or Z is maximized. In comparison with the performance of a vacuum thermionic generator, the performance improvement of the combined thermionic–thermoelectric generator system is obvious, especially the output power density.

Xuan and Lun (2003) concluded that "the combined thermionic–thermoelectric generator system can provide not only more electrical power but also higher efficiency than a vacuum thermionic convertor or a thermoelectric generator. Since this combined generator system requires separate thermionic and thermoelectric generating circuits, this may complicate the system and its matching with external loads. Better thermoelectric materials over the

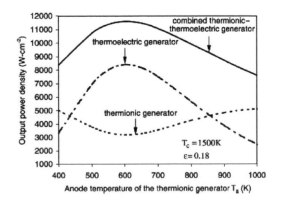

Figure 11.18
Output power density of the combined thermionic–thermoelectric generator against the anode temperature T_a of the thermionic generator. The other two curves for the thermionic and thermoelectric devices are calculated from the parameters corresponding to the optimal combined generator. (From Xuan and Li, 2003 [20].)

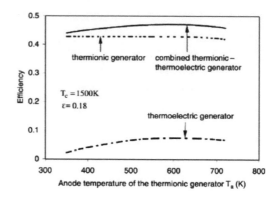

Figure 11.19
Efficiency of the combined generator. (After Xuan and Li, 2003 [20].)

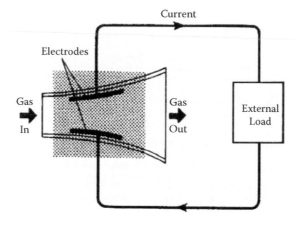

Figure 11.20
Schematic diagram of MHD generator. Shaded area represents magnetic field into paper. (After R.A. Coombe, A62 [22].)

temperature range of 300–600 K or higher can further improve the performance of the combined generator system."

11.4   Magnetohydrodynamic (MHD) Power Conversion

Magnetohydrodynamic (MHD) power generator is a heat engine that uses a gaseous conductor to convert heat energy of a fuel *directly* into electrical energy without the need of a conventional electrical generator. It offers much greater efficiency than the conventional turbine generator by eliminating the intermediate steps of heat energy conversions to electricity energy. Also, it can operate at high temperatures without moving parts, making it suitable for use as a topping cycle.

11.4.1   Working Principle of an MHD

An MHD generator works based on the Faraday's Law, which states that when a conductor and a magnetic field move with respect to each other, an electric voltage is induced in the conductor, the process of which is usually called induction. Unlike conventional induction in most generators, an MHD (see schematic diagram by Coombe, 1962, Figure 11.20) uses a *moving* conductive fluid (either *gas* conductor or *liquid metal*), usually by forcing a high pressure, high temperature combustion gas through a strong magnetic field. The induction creates forces on the fluid and also changes the magnetic field itself.

Governing equations of MHD systems. Governing equations are used to describe the complex MHD systems and processes, which basically consist of a combination of the standard Navier–Stokes equations of fluid dynamics, and the well known Maxwell's equations of electromagnetism applied in plasma physics and fluid dynamics. The resulting differential equations can then be solved simultaneously by numerical methods. For Detailed discussions, readers can refer to the numerous publications available, e.g., Dedner et al., 2002 [23] and Mignone, 2007 [24].

Nevertheless, for completeness and introductory purposes, the basic MHD equations are given below [25]:

The continuity equation is

$$\frac{\partial \rho}{\partial t} + \nabla \cdot (\rho V) = 0 \tag{11.10}$$

where,
V = is the bulk plasma velocity.
ρ = is the mass density

Navier–Stokes equations (momentum equation) with the Lorentz force on the right-hand side:

$$\rho \left(\frac{\partial \mathbf{v}}{\partial t} + (\mathbf{v} \cdot \nabla) \mathbf{v} \right) = -\nabla p + \mathbf{J} \times \mathbf{B} + \rho \mathbf{g} \tag{11.11}$$

Where,
$\mathbf{B}$ = is the magnetic field
J = is the current density
p = is the plasma pressure.

The Lorentz force term $\mathbf{J} \times \mathbf{B}$ can be expanded to give

$$\mathbf{J} \times \mathbf{B} = \frac{(\mathbf{B} \cdot \nabla) \mathbf{B}}{\mu_0} - \nabla \left(\frac{B^2}{2\mu_0} \right) \tag{11.12}$$

where the first term on the right-hand side is the magnetic tension force and the second term is the magnetic pressure force.

The ideal Ohm's law for a plasma is given by

$$\mathbf{E} + \mathbf{V} \times \mathbf{B} = 0. \tag{11.13}$$

Faraday's Law is

$$\frac{\partial \mathbf{B}}{\partial t} = -\nabla \times \mathbf{E}. \tag{11.14}$$

The low-frequency Ampere's law neglects displacement current and is given by

$$\mu_0 \mathbf{J} = \nabla \times \mathbf{B}. \tag{11.15}$$

The magnetic divergence constraint is

$$\nabla \cdot \mathbf{B} = 0. \tag{11.16}$$

The energy equation is given by

$$\frac{\mathrm{d}}{\mathrm{d}t} \left(\frac{p}{\rho^\gamma} \right) = 0, \tag{11.17}$$

where γ is the ratio of specific heats for an adiabatic equation of state.

11.4.2 Classification of MHD Power Systems

MHD power systems can be classified into two main categories, i.e., open cycle and closed cycle (Figure 11.21).

A typical open cycle scheme for a coal-fired MHD system and a typical closed cycle scheme for a nuclear MHD power plant system are given in Figure 11.22 and Figure 11.23, respectively.

11.4.3 Application Advantages of MHD

MHD is one of the most promising direct energy convertors.

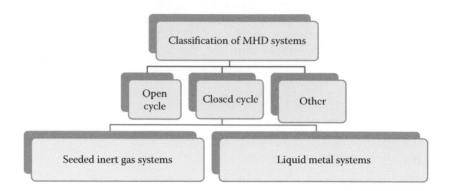

Figure 11.21
Classification of MHD power systems.

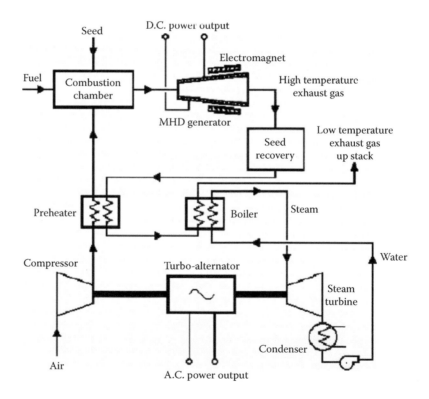

Figure 11.22
A typical open cycle scheme for a coal-fired MHD system. (From Roberto Pintus, Development of an inductive magnetohydrodynamics generator, Università degli Studi di Cagliari, PhD Thesis, with permission.)

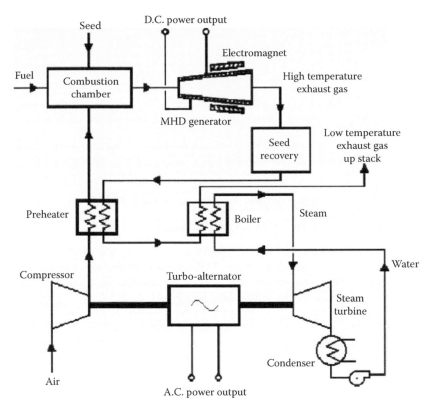

Figure 11.23
A typical closed cycle scheme for a nuclear MHD power plant system. (From Roberto Pintus, Development of an inductive magnetohydrodynamics generator, Università degli Studi di Cagliari, PhD Thesis, with permission.)

The advantages of MHD are:

1. No moving parts.

2. Good start up and suitable for peak power generation.

3. *Compact in size.* The size of an MHD generator is smaller than that of conventional fossil fuel generators.

4. *High efficiency and good fuel utilization when combined with steam plant.* MHD offers much greater efficiency than the conventional turbine generator, by eliminating the intermediate steps of heat energy conversions to electricity energy. Utilizing MHD hot exhaust gases in a boiler to make steam, for instance, results in a combined MHD–steam Rankine cycle plant that can convert fossil fuels into electricity with 60% energy efficiency.

5. *High power and potential for high-power propulsion applications.* MHD and superconducting magnets have an important role in the development of high power plasma rockets for fast and highly efficient nuclear or solar powered missions through space [26]. In these rockets, superheated plasma made up of high energy charged particles, instead of the traditional chemical fuel, is expelled to provide useful high-power thrust.*

6. *Good energy storage.* Superconducting magnets carry a large current; therefore, in a superconducting magnet system, a large amount of energy can be stored indefinitely.† Such energy storage systems have been developed

---

*Superconducting magnets are used to (a) accelerate and control the plasma; and (b) to insulate the surrounding structures from its high temperatures. The strong field required for a Variable Specific Impulse Magnetoplasma Rocket (VASIMR, developed by NASA) as well as other advanced space propulsion systems, is provided by superconducting magnets, and the power is provided by MHD generators.
†A typical experimental magnet at Conseil Européen pour la Recherche Nucléaire (CERN) has enough stored energy to provide 1-kW of continuous power for one month. On the surface of the Moon or Mars, such a superconducting energy storage system will provide life sustaining energy for long periods of time (after AMS and power generation).

and used on the ground to ensure continuous power supply without interruption; an example is the Chubu Electric Power Co. in Japan.

The disadvantages of MHD are:

1. High cost and technical demand due to high temperature. The use of highly conductive and heat-resistant materials such as zirconium dioxide to retard oxidation may impose high cost to the plant.

2. Low efficiency of around 20% if operated by itself (without combination with an other power generator, in particular steam generator), thus not suitable for large scale utility power generation.

11.4.4   Some Current Usage and Future Applications of MHD

11.4.4.1   *CHP (Co-Generation) with MHD Applications*

CHP (Cogeneration) of electricity and process heat appears to be one of the most attractive applications of MHD. This is because the combined systems can deliver few MWs power ranges compatible with CHP applications, in particular from the incorporation of a steam turbine operating at the low pressure part of the plant.

Some of the application advantages cited (e.g. [32]) are as follows:

1. *A relatively low temperature cycle.* Featuring nearly isothermal expansion, the MHD Rankine cycle enables expanding the steam to a given pressure starting with a significantly lower steam temperature than required for the corresponding turbine cycle. Moreover, the MHD cycle does not require a steam superheater, i.e., it enables using simple boilers.

2. *A relatively high temperature heat source.* For the same initial steam conditions and back pressure, the steam coming out of the expansion in the MHD system is at a significantly higher temperature than the steam coming out of the conventional turbine. This unique feature of the MHD technology can be attractive to customers in need of a high temperature heat source or of a high temperature, low pressure steam.

3. *Extended range of electrical-to-thermal energy supply (E/T) ratio.* The E/T energy supply ratio required by different industries spans a wide range. The MHD technology can significantly extend the range of E/T accessible with steam turbines. This arrangement could be attractive as a retrofit to existing industrial plants that are equipped with a moderate pressure ST. The E/T ratio of the combined cycle is about 1.4 times that of the ST only system. The cycles featuring the MHD systems offer significantly more electricity generation per unit flow rate of steam than the ST only cycles.

4. *Heat recovery from liquid metals.* A number of industrial processes require, or could benefit from heat recovery from processes involving liquid metals. Such processes might be directly integrated with the MHD system and thus possibly become more efficient and economical.

5. Good match with heat sources that deliver their energy at a nearly constant temperature. Examples for such heat sources include fluidized bed combustion systems. The good match with such heat sources, due to the near isothermal expansion process of the MHD technology, is a prerequisite for high thermodynamic efficiency.

6. *Versatility in type of fuel used.* Unlike gas turbines and reciprocating engines, but like steam turbines, MHD technology can be used with any kind of fuel, including waste, as well as with renewable energy sources such as solar and biomass.

11.4.4.2   Nuclear Powered MHD Energy Systems

According to Smith and Anghaie [28] and S. Anghaie et al. [29], both the United States and Russia have invested substantially in the development of many devices to convert nuclear power directly to electricity without the necessity of traditional heavy turbines using MHD techniques. Since an MHD generator can directly convert nuclear energy to electricity without degrading its high quality, as is common in conventional turbine-generators, the power conversion efficiency is found to be significantly higher than conventional nuclear reactor power plants. Other notable advantages cited are:

1. Nuclear powered MHD systems operate at much lower nuclear fuel inventory, which results in significant enhancement of safety and reduction of long-lived radioisotope actinides in the waste stream.

2. Higher conversion efficiency in nuclear powered MHD proportionally reduces the initial capital cost, the cost of power generation, and the generated nuclear waste as compared with state-of-the-art conventional nuclear power plant systems.

3. Using the MHD and superconducting magnet technology, nuclear power plants could optimally reach nearly 70% energy efficiency.

11.4.4.3 Coal Fired MHD Technology (Kulkarni and Gong, 2003a [30]; Kulkarni and Gong, 2003b [31])

By passing the combustion gases through a magnetic field, a magneto generator converts heat directly to electricity power without moving parts. The inherent characteristic of high efficiency and low emission makes CHP coal fired electric power generation technology attractive as a potential alternative power generation method in the future. However, there are still some challenges on the way to the possibility of wide commercial use.

11.4.4.4 MHD Enhanced Solar Energy Applications

The use of MHD for enhanced solar energy conversion applications is promising due to the increasing concern about environmental pollution and increase of fossil fuel prices.

The advantages of using liquid metal from MHD in solar energy receivers can be due to the following [32]:

1. The liquid metal of MHD has promising properties such as low vapor pressure, good heat transfer properties, and high heat capacity.

2. The liquid metal in the MHD system provides good thermal energy storage capability, which could compensate for fluctuations in the solar energy density reaching the receiver. Several of the experimental solar tower facilities are using liquid metal for the working fluid of their receiver.

3. The compatibility of the MHD technology with battery energy storage is encouraging.

11.4.4.5 Thermo-Acoustic–MHD Electrical Generator

Recently, Alemany et al. (2011) proposed a thermo-acoustic – MHD electrical generator, with an induction generator of MHD for obtaining electric current with adjustable voltage and current strength, delivered at the thermo-acoustic wave frequency, via a solar parabolic collector (Figure 11.24). The MHD generator can be placed between two thermo-acoustic tubes. The return circuit corresponding to the use of traveling waves can be removed. In that case, the MHD generator would be sandwiched between two standing wave tubes. The pressure fluctuations generated on both sides of the MHD generator produce oscillations of liquid metal which, by the interaction with the applied magnetic field, imposed by a permanent magnet, produce electric current (Figure 11.25). The simulation studies conducted by Alemany et al. show that there is a solution for coupling the thermo-acoustic and MHD engines at approximately 80 Hz frequency, corresponding to the temperature level of about 1300 K. Total power produced on the load is around 501.2 W and at 31.32% in efficiency.

11.4.4.6 MHD Plasma Devices for Vehicle Applications

The use of plasma devices for vehicle applications is being explored and discussed by, e.g., Van Wie [33] for applications ranging from drag reduction and steering control to boundary layer modification to ignition and combustion system enhancement. Advanced plasma technologies with application to hypersonic flight vehicles were investigated and a system-level study was conducted to demonstrate the potential performance benefits of these technologies.

After evaluating about 18 technologies in relation to vehicle application and control, several general trends emerged, and are summarized as follows:

- The mass and volume requirements of the magnet needed for large-scale inlet flow control technique are currently too large for practical vehicle applications. A two-order-of-magnitude reduction in conventional superconducting magnet mass will be required to achieve performance benefits.

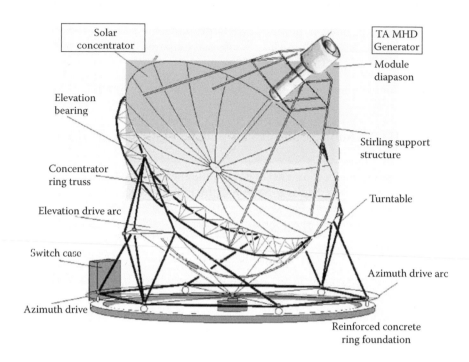

Figure 11.24
The proposed TA–MHD generator using concentric solar energy. [From Alemany et al., Energy Procedia 6 (2011) 92–100. With permission.]

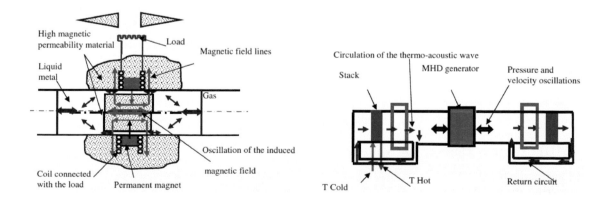

Figure 11.25
The conceptual design of the MHD induction generator. (a) Schematic diagram; (b) the working concept of the electrical generator, a combination of thermo-acoustic and MHD generators. [From Alemany A. et al., Energy Procedia 6 (2011) 92–100. With permission.]

- The mass and volume requirements for the power generation system has a first-order impact on the viability of the drag reduction schemes, so efficient generation of on-board power will be required.

- Both experimental and numerical simulations have demonstrated promising ranges of drag reduction energetic efficiencies, which begin to have positive system impacts with a lightweight power generation subsystem.

- A plasma assisted ignition and flame holding system offer increased system operatibility for small mass, volume, and power penalties.

11.5 *Questions and Problems*

Questions and Problems

Multiple choice questions

1. Which of the following statements does *not* correctly describe the absorption energy system in terms of energy sources application?

   (A) It cannot be powered by conventional electric energy.

   (B) It can be driven by low grade heat source energy.

   (C) It can be powered by external heating of fossil fuel.

   (D) It can be driven by waste heat.

2. Which one of the following statements *correctly* describes the application advantages of an MHD generator as compared to a conventional turbine generator?

   I No moving parts.

   II High efficiency and good fuel utilization.

   III The size of the MHD generator is smaller than that of conventional fossil fuel generators.

   IV Good start up and suitable for peak power generation.

   V One of the most promising direct energy convertors.

   (A) I only

   (B) I and II only

   (C) I, II, and III only

   (D) I, II, III, and IV

   (E) All of the above

3. Which one of the following statements does *not* correctly describe the application advantages of a thermoelectric generator?

   (A) High efficiency

   (B) Can be useful for waste heat recovery

   (C) No moving parts, thus quiet in operation

   (D) Environmentally friendly

4. MHD is one of the most promising direct energy convertors. The following are the main advantages of an MHD, *except*:

A) No moving parts

B) Good start up and suitable for peak power generation

C) The size of the MHD generator is smaller than that of conventional fossil fuel generators

D) Low cost

E) High power and potential for high-power propulsion application

F) Good energy storage

5. In recent years, many efforts have been initiated around the world to develop Generator-Absorber Heat Exchange (GAX) cycle systems. Which of the following are the characteristics of a GAX system?

I The GAX cycle is a heat-recovering cycle in which absorber heat is used to heat the lower-temperature section of the generator as well as the rich ammonia solution being pumped to the generator.

II The GAX cycle has potential gas-fired COPs of 0.7 in cooling mode, and 1.5 in heating mode, which makes it capable of significant annual energy savings.

III The amount of external heat required to power a GAX system is less than the amount of heat required to drive a conventional absorption system, as a result of which a GAX system can have higher COP than a conventional absorption system.

IV A GAX cycle system is a type of adsorption energy convertor that requires the use of suitable adsorbents.

(A) I only

(B) I and II only

(C) I, II, and III only

(D) I, II, III, and IV

(E) All of the above

6. The triple-effect absorption cooling cycle is one of the advanced absorption cycles. Single-loop triple-effect cycles are basically double-effect cycles with an additional generator and condenser. Which of the following statements *correctly* describes the characteristics of single-loop triple-effect cycles?

I Primary heat from a natural gas or fuel oil burner concentrates absorbent solution in a first-stage generator at about 200 to 230°C.

II High pressure in the first-stage generator vapor space usually requires costly pressure vessel design and high-pressure solution pumps.

III The refrigerant vapor produced from the first-stage generator is used to concentrate additional absorbent solution in a second-stage generator at about 150 °C.

IV The refrigerant vapor produced in the second-stage generator concentrates additional absorbent solution in a third-stage generator at about 93 °C.

V Theoretically, these triple-effect cycles can obtain COPs of about 1.7.

(A) I only

(B) I and II only

(C) I, II, and III only

(D) I, II, III, and IV

(E) All of the above

7. The material selection for a thermoelectric generator depends on many factors. All of the following are the primary properties and their associated requirements, correctly described *except,*

(A) Figure of merit, Z. Should be as high as possible.

(B) Thermal conductivity. Should be as high as possible.

(C) Mobility of current carries. Effectiveness and current carries should be as high as possible.

(D) Chemical resistance. Must have good chemical resistance.

8. What is the SCP of a 3 kW adsorption system if cycle time is 30 sec and mass of adsorbent used is 1 kg?

(A) 0.1 kW/(sec.kg)

(B) 10 kW/(sec.kg)

(C) 90 kW/(sec.kg)

(D) 0.9 kW/(sec.kg)

(E) 9 kW/(sec.kg)

Theoretical questions

1. Why do the commercial adsorption cooling units combine the generator and condenser in one vessel?

2. What are the main differences between the vapor adsorption system and the conventional vapor-compression system?

3. Discuss the *Seebeck Effect*, and how a thermoelectric generator or "Seebeck" generator works.

4. With the aid of a schematic diagram, discuss the working principle of a thermionic generator.

5. Both thermoelectric generators and thermionic generators employ electron gas as the working fluid, and convert thermal energy directly into electrical energy without mechanical moving parts. Compare and discuss the application differences between a thermoelectric generator and a thermionic generator.

6. Sketch a typical combined thermionic–thermoelectric generator. Is the combined thermionic-thermoelectric generator superior in efficiency than a stand-alone thermionic generator, and why?

7. Discuss the application advantages of adsorption energy systems as compared to compression systems for cooling usage.

8. Sketch a schematic diagram of an MHD generator, and describe how power is generated by an MHD.

9. Why can an MHD power plant have efficiency much higher than a conventional turbine power plant?

10. Name three advantages of MHD systems, and briefly discuss the following applications.

    (a) CHP (co-generation) with MHD applications

    (b) MHD enhanced solar energy applications

Homework problems

1. *Absorption*: A solar-driven absorption refrigeration system receives heat from a solar flate-plate collector at a temperature of 95 °C. The energy thus obtained is used to provide steam to maintain a typical refrigerated space at -18 °C. Assume the ambient temperature is at a constant temperature of 25 °C.

    (a) What is the suitable absorption type for the above application (single effect or double effect) and what are the suitable working pair (fluids) to use?

    (b) What is the maximum theoretical COP this absorption system can have?

    (Answer: *single effect, ammonia-water absorption chiller; 1.128)*

2. *Absorption*: In a typical flash steam geothermal power plant, steam with heat of 190 °C is supplied constantly to an absorption refrigeration system from the production well at a production rate of 9×10^4 kJ/h. If a refrigerated space is to be cooled and maintained at -20 °C, what is the maximum rate at which this absorption system can remove heat from the refrigerated space? The ambient temperature is at 27 °C. *(Answer: 1.151×10^{-5} kJ/kg)*

3. *Absorption*: Heat is available in the form of steam from a waste heat source of a factory with pressure at 1.5 bar and saturated temperature of 135 °C. If a 750 kW chiller with chilled water flow temperature of 7 °C is required,

   (a) What is the suitable absorption type for the above application (single effect or double effect) and what are the suitable working pair (fluids) to use?

   (b) What is the heat input required?

   (c) Determine the flow rate of the steam.

   Assumption: Assume COP of the chiller = 0.68, and that condensate from the steam is subcooled by 20 °C at the condenser.

   (Answer: $1,103\,kW$; $0.49\,kg/s)$

4. *Thermoelectric*. A thermoelectric refrigerator powered by a 12-V 3A lead acid battery of a car cools 2 kg of canned drinks, from hot ambient temperature of 35 °C to 3 °C, in 10 hours. Knowing that for water $C_p = 4.184\,kJ/kg°C$, determine:

   (a) The cooling capacity of the thermoelectric refrigerator (in kJ)

   (b) The average cooling rate of the thermoelectric refrigerator (in Watts)

   (c) The electric power consumed by the refrigerator

   (d) The average COP or efficiency of the refrigerator

   (Answer: $267.78\,kJ$; $7.4\,W$; $36W$; $21\%)$

Answers to multiple choice questions:

1. (A) It can be powered by conventional electric energy, by means of a heating coil, to heat the solution mixture in the generator.

2. (E) I, II, III, IV, and V

3. (A) The efficiency of thermoelectric generator is less than 10%

4. (D) Currently, MHD power generator is still considered a high cost energy convertor mainly due to high temperature requirements.

5. (C) I, II, and III only

6. (E) All of the above: I, II, III, IV, and V

7. (B) Thermal conductivity. Should be as *low* as possible.

8. (A) From $SCP = \frac{Q_e}{t_c m_a} = \frac{3\,kW}{30\,sec \times 1\,kg} = 0.1\,kW/(sec.kg)\#$

Bibliography

[1] Horuz, I., Callander, T.M.S., The investigation of a vapor absorption refrigeration system, *Int J Refrig*, 2004;27:10–6.

[2] Misra, R.D., Sahoo, P.K., Gupta, A., Thermoeconomic evaluation and optimization of an aqua-ammonia vapour-absorption refrigeration system, *Int J Refrig*, 2005;29:47–59.

[3] *ASHRAE Handbook 2010*, Refrigeration, SI Edition, page 18.2.

[4] Kang, Y.T., Kashiwagi, T., An environmentally friendly GAX cycle for panel heating, PGAX cycle, *International Journal of Refrigeration*, 23 (5) (2000), pp. 378–387.

[5] Jawahar, C.P., Saravanan, R., Generator absorber heat exchange based absorption cycle—A review, *Renewable and Sustainable Energy Reviews* 14 (2010) 2372–2382.

[6] Herold, K.E., Radermacher, R., Klein, S., *Absorption chillers and heat pumps*, New York: CRC Press; 1996.

[7] Wang, R.Z., Adsorption refrigeration research in Shanghai Jiao Tong University, *Renew Sust Energy Rev*, 5 (1) (2001), pp. 1–37.

[8] Abdullah, M. O., Tana, I. A. W., Lim, L. S., *Renewable and Sustainable Energy Reviews*, Volume 15, Issue 4, May 2011, pp. 2061–2072.

[9] Sumathy, K., Yeung, K.H., Yong, L., Technology development in solar adsorption refrigeration systems. *Prog Energy Combust Sci*, 2003; 29(4):301–27.

[10] Abdullah, M. O., Ngui, J. L., Abd. Hamid, K., Leo S. L., Tie, S. H., 2009. Cooling Performance of a Combined Solar Thermoelectric–Adsorption Cooling System: An Experimental Study, *Energy & Fuels*, Vol. 23, Issue 11, pp. 5677–5683. ISSN 0887-0624 (Print) 1520-5029 (Online). American Chemical Society (ACS).

[11] S.L. Leo and M.O. Abdullah, Experimental study of an automobile exhaust heat-driven adsorption air-conditioning laboratory prototype by using palm activated carbon–methanol, *HVAC&R Res* 16 (2) (2010), pp. 221–231.

[12] Anyanwu, E.E., Review of solid adsorption solar refrigeration. II. An overview of the principles and theory, *Energy Convers Manage* 45 (7) (2004), pp. 1279–1295.

[13] Suzuki, M., Application of adsorption cooling systems to automobiles, *Heat Recov Syst CHP*, 13 (4) (1993), pp. 335–340.

[14] Rowe, D.M. Min, G., Evaluation of thermoelectric modules for power generation, *Journal of Power Sources*, 73 1998. 193–198.

[15] Min, G., Rowe, D.M., in Rowe D.M., Ed., *CRC Handbook of Thermoelectrics*, Chap. 38, CRC Press, London, 1995.

[16] Rowe, D.M., Bhandari, C.M., *Modern Thermoelectrics*, Holt, Rinehart and Winston, London, 1983.

[17] Shakouri, A., Thermoelectric and Thermionic energy conversion, 2005, The 24th International Conference on Thermoelectrics (ICT) (IEEE Cat. No.05TH8854C), June 19-23, 2005. Clemson University.

[18] Houston, J.M., *J. Appl. Phys.*, 30 (1959) 481.

[19] Goldsmid, H.J., *Electronic Refrigeration*, Pion Limited, London, 1986.

[20] Xuan, X. C., D. Li, Optimization of a combined thermionic–thermoelectric generator, *Journal of Power Sources* Volume 115, Issue 1, March 27, 2003, pp. 167–170.

[21] Mahan, G.D., Thermionic Refrigeration, *J. Appl. Phys.*, 76 (1994) 4362–4366.

[22] Coombe, R.A., Electric power by Magnetohydrodynamic. *New Scientist*, No. 272, 1 February 1962, pp 254-256.

[23] Dedner, A., Kemmy, F. Křoner, D., Munzy, C.-D., Schnitzer, T., Wesenberg, M., Hyperbolic Divergence Cleaning for the MHD Equations, *Journal of Computational Physics*, 175, 645–673 (2002).

[24] Mignone, A., A simple and accurate Riemann solver for isothermal MHD, *Journal of Computational Physics*, Volume 225, Issue 2, August 10, 2007, pp. 1427–1441.

[25] Čertík, O., Theoretical Physics Reference, http://theoretical-physics.net/dev/src/fluid-dynamics/mhd.html#finite-element-formulation, accessed March 31, 2011.

[26] Smith, B., Anghaie, S., Gas Core Reactor with Magnetohydrodynamic Power System and Cascading Power Cycle, *Nuclear Technology*, Volume 145, Number 3, March 2004, pp. 311–318.

[27] Alemany, A., Krauze, A., Al Radi, M., Thermo acoustic – MHD electrical Generator, *Energy Procedia*, 6 (2011), 92–100.

[28] Smith, B., Anghaie, S., Gas Core Reactor with Magnetohydrodynamic Power System and Cascading Power Cycle, Nuclear Technology, Volume 145, Number 3, March 2004, pp. 311–318.

[29] Anghaie, S., et al., Optimum Utilization of Nuclear Fuel with Gas and Vapor Core Reactors, *Progress in Nuclear Energy*, Volume 47, No. 1-4, pp. 74–90, 2005.

[30] Kulkarni, M., Gong, X., Coal fired MHD technology: A prospective power generation method, Proceedings of the 2003 International Joint Power Generation Conference 2003, Pages 847–856.

[31] Kulkarni, M., Gong, X., Cogeneration via magneto-hydro-dynamic (MHD) power generator, *Cogeneration and Distributed Generation Journal*, Volume 18, Issue 4, September 2003, pp. 58–80.

[32] Branover, H., El-Boher, A., Greenspan, E., Barak, A., Promising applications of the liquid metal MHD energy conversion technology, Energy Conversion Engineering Conference, 1989. IECEC-89., Proceedings of the 24th Intersociety, pp. 1051–1058, vol. 2.

[33] Van Wie, D. M., (2005), Future Technologies – *Application of Plasma Devices for Vehicle Systems*, The Johns Hopkins University, Applied Physics Laboratory – Laurel, Maryland – NATO Document.

[34] Abdullah, M. O., Hieng, T. C., Comparative analysis of performance and techno-economics for a $H_2O - NH_3 - H_2$ absorption refrigerator driven by different energy sources, *Applied Energy*, 87 (2010) 1535–1545.

[35] Wang, K., Vineyard, K., Adsorption refrigeration - new opportunities for solar, *ASHRAE Journal*, September 2011, pp. 14–22.

[36] Wang, R.Z., Oliveira, R.G., 2009, Adsorption refrigeration - an efficient way to make good use of waste heat and solar energy, *Progress in Energy and Combustion Science*, 32(4) pp. 424–458.

12

Some Applied Energy Related Issues

Similar to our daily life and the associated activities that are always full of ease and difficulty, health and sickness; for applied energy we deal with strength and weakness, advantages and disadvantages, and issues. This final chapter will bring us to the various energy application issues.

12.1 "Energy Storage" Issues

Figure 12.1 gives a simple classification of energy storage.

Note that we could also divide energy storage by either grid connected (electricity supply network) or small-scale stand-alone (distributed generation of energy).

12.1.1 Fossil Fuel Storage Issues

As far as the conventional fossil fuels, that is, oil and gas, are concerned energy storage technologies are relatively established, such that most issues have been substantially solved over the years. The main issues are the integrity and safety of the storage vessels and energy systems in relation to prolonged storage applications. One example is ammonia storage, given in the following subsections.

12.1.1.1 Ammonia Storage and Ammonia Storage Tanks Issues

We had discussed briefly an ammonia storage tank in Chapter 5, Section 5.8.2, and pointed out some important factors that many cause tank failures. The primary issue with ammonia storage is the integrity of the ammonia tank, especially when the storage tanks are subject to prolonged use, and without a proper periodical tank commissioning, there will be the possibility of stress corrosion cracking (SCC) induced which is the main concern. Although it happen extremely rarely, severe SCC may cause tank rupture (see, e.g., Lunde and Nyborg (1989) [1]; Abdullah et al. (2011a) [33]). The details of two historical SCC-related failures are described in the following paragraphs.

A tanker ruptured killing five people in France about midday on Wednesday, August 21, 1968. The tanker filled with ammonia suddenly ruptured in the yard of a factory at Lievin (Pas de Calais), France. The almost instantaneous escape of ammonia that resulted, caused burns to the respiratory organs of twenty persons, 5 of whom died. The root cause was stress corrosion, which caused numerous cracks, especially in the plane where the rupture took place,

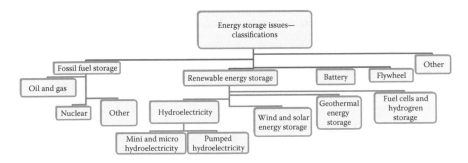

Figure 12.1
Energy storage issues—classifications.

and it can be seen that SCC had been considerable due to additional stresses in the region of the welds since there had been no post-weld heat treatment of the tank. The rupture was due to corrosion cracking aggravated by fatigue of the tank in the plane of the rupture, situated just above the rear wheels, in an area subjected to abrupt variations in thickness. The construction of road tankers needs to improve the method of attaching fixtures such as internal baffles and external stiffening beams. The position of the rear wheels is equally important.

On July 13, 1973, at a Potchefstroom, South Africa, fertilizer plant, one of four 50-ton (metric) pressure-storage tanks (horizontal bullet type) failed. The failure resulted from brittle fracture of a dished end. No specific source of cracking, or a "triggering incident" was identified. The incident occurred at 16:15 on July 13, when the storage tank failed while being filled from a railroad tank car. One employee, 45 m from the tank, was killed outright by the blast; 8 were killed by gas while attempting to escape from points within 100 m of the tank; and 3 others died within a few days as a direct result of having been gassed. Outside the plant fence, 4 people died immediately, and 2 others died several days later. In addition to the 18 deaths, approximately 65 people required medical treatment in the hospital and an unknown number were treated by private doctors.

12.1.1.2 Nuclear Energy Storage Issues

For nuclear energy storage issues, further to the discussions already given in Section 5.8, there are two parts that allow us to briefly discuss, the storage of nuclear energy itself, and the controversial nuclear waste storage especially in relation to long term storage.

The storage issues of nuclear energy and nuclear waste

1. *Permanent storage issues.* Until today, some have argued that there is no "perfect" storage of radioactive materials to prevent them from finding their way into the soil, water, plants, animals, fish, and humans.

2. *Corrosion issues.* There is always eventual corrosion of the metal barrels containing the nuclear waste.

3. *Temporary storage and safety issues.* Another concern is the reactor site if located in an earthquake region that contains many interconnected faults and fractures (such as in the case of Japan). These could move groundwater and any escaping radioactive material through the repository to the aquifer below and then to the outside environment.

In relation to temporary storage above, the *temporary storage and safety issue*, also known as *spent nuclear fuel (SNF)* issue, indeed belongs to the most difficult and controversial fields of nuclear power policy. The debate—both within the nuclear community and in society at large—has been and continues to be characterized by a striking diversity of ideas, proposals, and arguments concerning SNF approaches [34].

12.1.2 Renewable Energy Storage Issues

As far as renewable energy applications are concerned, we noticed that a number of technologies for energy storage already exist, including some that have been around for decades. The main challenge, however, is to make them robust, reliable, and techno-economically competitive— while matching the most suitable technology to each energy source or to the particular location applications.

In the following sub-sections, we will discuss the issues pertaining to various storage methods such as battery storage, compressed air, pumped hydroelectric, flywheel storage, geothermal storage, wind storage, solar storage, fuel cell, hydrogen storage, and so forth.

12.1.2.1 Battery Storage Issues

12.1.2.1.1 Lead–Acid Batteries Issues Until today, large-scale battery storage has been an unsolved problem even though lead–acid technology that has been the basis of car batteries had been established for nearly a century. Lead–acid batteries still have some problems and issues, such as:

1. Lead–acid batteries have a low energy density such that they are bulky and heavy for the amount of energy they store [3].

2. Lead–acid batteries do not stand up well to repeated charge–discharge cycles [3].

3. Lead–acid batteries have been chosen for wind- or solar-powered installations on account of their wide availability in a range of sizes and their acceptable cost. For the storage of renewable energy, the chief disadvantages of these batteries are [2]:

- the need for periodic water maintenance, i.e., water "top-up" (valve-regulated cells being the exception)
- relatively poor performance at low and high ambient temperatures
- a variable but limited charge–discharge cycle-life (typically, 500 deep-discharge cycles)

The importance of the different limitations of lead–acid batteries will depend upon the application, and it is necessary to weigh them carefully against the performance and cost of competitive batteries.

12.1.2.1.2 Sodium–Sulphur (NaS) Battery Issues The sodium–sulphur (NaS) battery can store energy by chemically dissociating sodium polysulphide into sodium and sulphur. The energy can then be released by allowing the two elements to react again. NaS batteries have a high energy density and can last through thousands of charge–discharge cycles compared with lead acid batteries. However, their drawbacks and notable storage issues are:

1. The sodium and sulphur have to be kept in separate reservoirs in the molten state, at about 300 °C.

2. The batteries suffer irreparable damage if they discharge completely and grow cold. The resulting need for a robust container, along with other technical requirements, means that NaS batteries cost about US$3,000 per kilowatt (kW) of available power. That compares unfavorably with standard gas-powered plants, which cost about $1,000/kW. Nonetheless, NaS batteries have been developed commercially by NGK Insulators in Nagoya, Japan. Japan now has an installed capacity able to supply its grid with about 300 MW when extra power is needed, for up to six hours at a stretch. Other countries are also picking up the pace. The United States, for example, has about 10 MW of NaS capacity in place and a similar amount on the way, led by companies such as American Electric Power and Xcel Energy in Minneapolis, Minnesota [3].

3. In the future, large-scale NaS storage could face a challenge from lithium-ion technology.

12.1.2.1.3 Lithium-ion Batteries and Associated Issues Already in widespread use for mobile phones and laptops, and under development for electric cars, lithium-ion batteries have a high energy density and efficiencies of > 90%. Their currents issues, however, are:

1. *High cost and safety* issues. Their big drawback is cost, which is in part driven by safety considerations: the batteries use a lithium salt in an organic solution, which is flammable, necessitating robust construction to minimize fire hazards. Lithium-ion batteries made for consumer electronics currently cost a few hundred dollars per stored kW hour. Safety issues are more easily and cheaply met for batteries in secure and fixed installations than in hand-held devices [3]. The earlier version of the lithium cell contains metallic lithium and is therefore prone to *thermal runaway* and *fire* due to reaction of the lithium with the electrolyte.

2. *Electric-vehicle power and grid application issues.* A few large lithium-ion battery modules have been made as prototypes for electric-vehicle power sources, but as yet these have technical problems and are expensive in cost [2]. The key barrier may be the high-energy, high-power application requirements. According to David Lindley, in order for widespread vehicle applications, the cost must come down closer to $100 per kW hour, and for grid applications it needs to be even lower [3]. According to Cairs and Albertus, system costs may be brought down by using cathode materials less expensive than those presently employed (e.g., sulfur or air), but reversibility will remain a key challenge. Nevertheless, continued improvements in the ability to synthesize and characterize materials at desired length scales, as well as to use computations to predict new structures and their properties, are facilitating the development of a better understanding and improved systems [41].

12.1.2.1.4 "Alkaline" Batteries and Their Related Issues Alkaline batteries have three general types, that is, nickel–iron battery, nickel–cadmium battery, and nickel–metal-hydride system.

Nickel–iron battery. The "nickel–iron battery" suffers from major defects. The iron negative electrode is subject to appreciable corrosion and self-discharge on standing, and its low overpotential for hydrogen evolution gives rise to excessive gassing during recharge. These effects result in a low overall electrical efficiency (Wh_{out}/Wh_{in}) and a high water-maintenance requirement, neither of which is acceptable for the storage of electricity from renewable sources. Despite considerable research on this battery, the problems have not been resolved [2].

Nickel–cadmium battery. The "nickel–cadmium batteries" are widely used to a moderate extent in both mobile and stationary applications. The high-rate and low-temperature performances of the battery are better than those of lead–acid. Other beneficial features are a flat discharge voltage, long life (2000 cycles), continuous overcharge capability, low water maintenance, and high reliability. However, the battery has a high cost (up to 10 times that of lead–acid) and a low voltage (1.2 V), and there are environmental concerns associated with the disposal of toxic cadmium in spent batteries. Nevertheless, the higher cost may be acceptable for sites that are remote, unmanned, and difficult to access. Also, the long life of the battery and its freedom from maintenance will then present cost savings to be weighed against the higher capital outlay [2].

Nickel-metal-hydride battery. This battery has recently been commercialized and now finds widespread application in portable telephones and other electronic devices. Large versions of the battery have been produced and used to power prototype electric and HEVs. Nickel–metal-hydride is, however, even more expensive than nickel–cadmium. Thus, unless and until the cost falls substantially, the technology is unlikely to be a strong candidate for the storage of renewable electricity [2].

12.1.2.1.5 High-Temperature Battery Issues

High-temperature batteries are based upon molten sodium as the negative-electrode reactant and make use of a solid electrolyte, beta-alumna, in the form of a ceramic tube. Beta-alumna is an electronic insulator, but has a high conductivity for sodium ions at elevated temperatures. The positive electrode reactant is either molten sulfur or solid nickel chloride. Large batteries, made up of several hundred cells, are contained within a double-walled, steel, vacuum enclosure so as to minimize the heat loss. The "sodium–sulfur battery" operates at 300–400 °C. The discharge reaction, which produces various sodium sulfides, takes place in two stages. During the first stage, the open circuit voltage remains steady at 2.076 V, but then declines progressively to 1.78 V during the second stage. This battery was the subject of intense research and development for almost 30 years in various countries, e.g., Canada, Germany, Japan, the United Kingdom, and the United States of America. Except in Japan,* all attention was directed towards electric-vehicle applications, and all the programs were abandoned in the mid-1990s for a combination of technical, commercial, and safety reasons, even though the battery had been demonstrated in numerous electric vehicles. A major concern of sodium–sulfur batteries is the safety hazard of having molten sodium in close proximity to molten sulfur and separated only by a brittle ceramic tube. Should a tube crack or fracture, there will be a runaway thermal reaction and associated fire and this can extend beyond the cell and propagate throughout the battery. To avoid such a situation, numerous safety features are incorporated in the cell and these serve to limit the uncontrolled reaction. Nevertheless, a concern remains. Other problems with the sodium–sulfur battery are materials' compatibility and corrosion, and the inability of cells to pass current when fully charged. Since these problems all stem from the sulfur electrode, it was concluded that liquid sulfur was not a satisfactory material for the positive electrode and needed to be replaced by a non-volatile solid. In summary, whether or not the sodium–sulfur battery becomes commercially successful is a question of reliability, safety, and cost [2].

The sodium–nickel-chloride ("ZEBRA") battery grew out of the sodium–sulfur program, and works on a relatively lower operating temperature range of 200–400 °C. The open-circuit voltage is 2.59 V. The cell was assembled in the discharged state using a mixture of nickel metal and sodium chloride. Liquid sodium chloraluminate, $NaAlCl_4$, is added to the mix as a second electrolyte so as to make good electrical contact between the surface of the beta-alumna tube and the positive reactant mix. The assembly of the cell in the discharged state has the added advantage that the sodium, when formed, is ultra-pure as only sodium ions diffuse through the betaalumna lattice. The battery has many advantages over sodium–sulfur, namely, no corrosion problems, no volatile constituents (thus relatively safe), and tolerance of overcharge and overdischarge. Thousands of cells have been assembled on a pilot production line and many vehicle-sized traction batteries have been built. The performance of both the batteries and the electric vehicles they power has been reported to be excellent. As well as vehicle traction batteries, the ZEBRA system is suitable for building stationary storage batteries, although stringent cost targets need to be addressed. Again, according to Della and Rans, the battery is unlikely to meet the cost targets for bulk electricity storage, but may be suitable for some RAPS applications [2].

---

*In Japan, the focus was on stationary energy-storage applications.

12.1.2.2 Compressed-Air Storage Issues

The main problem with compressed-air energy storage facilities is due to the fact that gas heats up when it is compressed, limiting the amount of air that can be pumped underground before it becomes too hot to be stored safely. Moreover, the longer that hot air is left in place, the more of its heat will be lost into the walls of the surrounding cavern. And then when it is released again, the expanding air cools down. In the Huntorf and McIntosh facilities, the released air is fed into a standard natural gas turbine, boosting its efficiency.

12.1.2.3 Pumped Hydroelectricity Storage Issues

As pointed out by David Lindley, pumped hydroelectricity has a storage efficiency of 70–85%, and it is one of the most mature and widespread renewable energy technology being used for large-scale electricity storage. China, Japan, and the United States, for example, have numerous installations with generating capacities ranging from tens of megawatts (MW) to several gigawatts (GW). Pumped hydro is particularly appropriate for a network that has a large nuclear component, since, for both technical and economic reasons, nuclear reactors are best operated on baseload. As the response time of pumped-hydro is *rapid*, it may contribute to all of the network applications. A typical pumped-hydro plant, as operated by a utility, is capable of generating many megawatts of electricity. To date, this is the most practical and economic means of storing electricity on the megawatt–hour scale (Della and Rand, 2001) [2]. Pumped storage hydroelectricity is a particularly good match for wind power because water pumped into an upper reservoir will stay there for a long time, making up for potentially large gaps in wind generation.

But in its conventional form, pumped storage hydroelectricity has some limitations and the related issues are enumerated as follows [3]:

1. *Geography limitation.* Pumped hydroelectricity requires mountains, so opportunities are limited by geography.

2. *High cost.* Building such storage also tends to be expensive.

3. *Environmental and social issues.* Building pumped hydroelectricity storage onto the mountains is believed by many to be environmentally destructive, and installing high-voltage transmission lines to connect remote storage sites to grids often triggers opposition on environmental grounds.

4. *Not for portable power use.* Like many renewables, pumped hydroelectricity is only suited in specific sites, and could not be used for portable applications, for example, to power a car.

12.1.2.4 Flywheel and Kinetic Energy Storage Issues

Flywheels have one of the most straightforward ways to store energy: electrical energy gets converted into the kinetic energy of rotation by running it through a motor, which accelerates the flywheel. The kinetic energy is extracted when it is needed by coupling the flywheel to a generator, which slows the wheel down and produces electricity. It has a long list of advantages compared with other storage methods as outlined in Chapter 5, Section 5.8. However, its actual application is more complex and some of the issues or challenges are enumerated as follows:

1. The flywheel has to spin very fast yet be strong enough to keep from flying apart [3]. The maximum energy which can be stored is dependent upon the tensile strength of the material from which the flywheel is constructed. The highest tensile flywheels are made from fiber-reinforced composites.

2. Flywheel storage systems are commercially available as uninterruptable power supplies that can deliver modest amounts of power for seconds or minutes, but they are not competitive for the longer storage times needed by the electric utility companies [3].

3. Beacon Power of Tyngsboro, Massachusetts, has innovated a high-tech huge flywheel that is optimized for frequency regulation with an efficiency of 85%, and that can spin up and down for perhaps millions of cycles during it entire life, being far more durable than batteries. The challenge now, however, is to bring the cost down. [3].

4. The most significant limitation of flywheels lies in their relatively modest capability for energy storage. They are essentially surge-power devices rather than energy-storage devices, and are best suited to applications that involve the frequent charge and discharge of modest quantities of energy at high-power ratings [2].

12.1.2.5 Geothermal Energy Storage Issues

Geothermal energy can be considered one of the energy storage systems that are in the form of thermal reservoir energy, and it is a very huge means of energy storage. The National Renewable Energy Laboratory (NREL), Colorado, United States of America, estimates that geothermal could provide 4–20 % of current U.S. electricity needs by 2025. Also, it is already price competitive with coal-fired power plants in many places. A 2006 study by Tester et al. from MIT concluded that geothermal energy has the potential to generate about 100 gigawatts or more in the United States by 2050, and calculated that the world's total EGS resources are sufficient to provide all the world's energy needs for thousands of years (Tester, J. et al., 2006 [4]). Several projects using the so-called *enhanced geothermal systems* (EGS) or *heat mining* are under development in France, Australia, Japan, Switzerland, and the United States. The race is on to establish the world's first commercial hot dry rock power plant.

Some issues in relation to geothermal applications cited in literature are summarized below:

1. *Possible geology and seismic effects.* Many geologists are concerned about possible seismic consequences connected to drilling through the earth's surface to access hot dry rock resources. In 2007, a geothermal power plant in Switzerland triggered a historical 3.4 Richter-scale earthquake.

2. *High geothermal exploration cost.* Another major barrier is the cost involved in finding viable geothermal resources. Like oil well exploration and subsequent drilling, this is expensive and increases investment risk.

3. *Geothermal system design issue.* The issues are the requirement for high injection pressures, water losses potential, unwanted scaling or deposition, as well as geochemical impacts including seismicity.

4. *Social issue.* In recent years, when planning initiatives of great interest, or when discussing the opportunity of starting a development project in a given area, the issue of "social acceptability" of the project has come to the fore for policy-makers, investors, local administrators, and people living in the area concerned. The debate on these issues is usually significant, in particular, in the case of large geothermal developments due to their impact on relatively small areas ([32]; [5]).

To tackle the above issues, however, some steps that have been taken by geothermal operators or suggested for geothermal applications are:

1. System design issues usually can be either fully resolved or made manageable with proper monitoring and operational changes. Likewise, scaling or deposition issues are solved by thorough understanding of the rock-fluid geochemistry and by sound prediction of scaling and proper actions of remedy.

2. Advanced exploration techniques and monitoring making geothermal testings more economical and sound, reducing the risk of investing money "dry holes."

3. The binary geothermal systems had been suggested as an efficient alternative to tap cooler hot springs and other geothermal resources where direct and flash steam plants are not possible.

4. There is a need to provide more efficient geothermal power grid transmission networks and the associated capacity to open up remote or "stranded" geothermal sites.

12.1.2.6 Solar Energy Storage Issues

12.1.2.6.1 Solar Photovoltaic PV Storage Issues

The science and technology of solar photovoltaic (PV) electricity generation are advancing, so much so that the efficiency of PV solar cells has improved from only 10% to almost 20% recently; and the cost of PV modules has fallen as the solar PV manufacturing industries have developed and evolved gradually around the globe.

The principal applications for solar-generated electricity to date have been in situations where mains electricity is not available—such a facility is known as a "remote-area power supply" (RAPS). One such area is in marine operations, where solar modules are employed to power navigation buoys, on drilling platforms, for the cathodic protection of structures, and as independent electricity supplies for small boats. In terrestrial applications, RAPS systems now operate microwave relay stations, telecommunications networks, railway signaling, street lighting, irrigation equipment, pipeline monitoring, and cathodic protection; and provide power to remote communities, homesteads, and holiday caravans. In the space field, solar electricity is used to power all satellites—a minor use by market volume, but vital for modern telecommunications. Most of the RAPS applications are small, i.e., in the 0.1–10 kW power range, but require a sizable back-up battery with a storage capacity of 1–100 kWh. Collectively, however, RAPS systems represent a huge potential market, world-wide, for solar installations and battery manufacturers [2].

By definition, RAPS systems do not act as substitutes for mains electricity and so do not directly reduce the consumption of fossil fuels in power stations. Nevertheless, in the absence of a photovoltaic facility, if the alternative is to use a small petrol or diesel generator to charge batteries, the reduction in the consumption of fossil fuel and the concomitant lower emissions of CO2 start to become significant. Obviously, these benefits will increase as the price of solar modules falls further and the modules are ever more widely used in RAPS applications. In some locations, it may be desirable to employ a diesel generator and/or a wind turbine in conjunction with a photovoltaic array, either because the insolation is inadequate or because an array of the required size is too expensive. To a degree, solar power and wind power are complementary in high latitudes, since winds tend to be stronger in winter when sunlight is at a premium [2].

Optimizing the output from a PV array is quite a sophisticated procedure. Much is dependent upon the latitude, which determines the angle of the array to the sun, as well as upon whether the loads are seasonal or constant throughout the year and whether the peak demand is during the week (as with a rural school or workplace) or at the weekend (holiday cottages, caravans, etc.). Another critical factor is whether the array is fixed in position, or steerable to follow the sun during the day. A steerable array is more expensive to construct but, for a given electrical output, can be smaller than a fixed array. [2].

The back-up battery to a solar installation performs three roles: (i) to store electricity from daytime, when it is generated, to evening or night when it is required; (ii) to meet power surges during the day ("peak shaving"); (iii) to smooth fluctuations in the current and voltage output from the array (for instance, as the day progresses or as the sun disappears behind clouds). Thus, the system is one in which both the electrical output of the array and the electrical demand for the application are quite variable, and in which the battery has to be sized to smooth these fluctuations over a period of, typically, up to a week. Optimization of the size and the cost of both the solar array and the battery, working together, is quite a complex exercise and much effort has gone into the development of system designs and associated control algorithms [2].

For communication purposes, storage batteries powered by solar energy are used in many telecommunication industries. Telecommunication systems need incessant power that assures continuous operation of the system even during cold or cloudy weather and hazy days when there is no sunlight. Hence the need for energy storage with sufficient capacity seems to be necessary, which poses some special operational demands in addition to the requirements of batteries operating in conventional ways (Gutzeit, 1986 [6]).

12.1.2.6.2 Solar Thermal Energy Storage Issues

The application areas are suited for process industry and SME industries applications. The solar thermal energy storage issues are very much subject to the sun's availability. The location, type of collector, working fluid to determine required storage volume, size of the system, and storage volume to determine the heat exchanger size and the load are the factors that need to be considered for the specific applications (Kalogirou, 2003 [7])

A few aspects worth considering are: method of storage, material of storage, and operating cost. To account for low efficiency, or insufficiency of storage, some means have to be in place, such as:

1. *There is a need for an auxiliary (backup) heater* (Constantinos et al., 2007 [8]). For big systems about 10% -15% of natural gas (or propane) back-up necessary to maintain the temperature of the storage tank.

2. *Supplemented energy from integration schemes.* Solar water heaters (SWH) usually also use passive *integrated* collector storage to supplement the solar thermal energy (Schnitzer et al., 2007) [25].

3. *High cost of storage at high pressure.* Industrial heat applications usually use hot water (low pressure steam) or pressurized steam corresponding to the heat required for the systems operation. Water is usually the running fluid in thermal applications depending on its availability, thermal capacity, storage convenience, and low cost. Nevertheless, cost of the storage system increases remarkably when higher pressure is required. For temperatures above 100 °C, the system needs to be pressurized. For medium temperature > 100 °C applications, mineral oils are used. However, higher costs, tendency of cracking, and oxidation are a few issues involved in such high temperature systems [26].

12.1.2.7 Fuel Cells and Hydrogen Storage Issues

Fuel cells are highly efficient heat engines with efficiency of 50% up to around 80%. Clean hydrogen gas can also be used to fuel vehicles and CHPs. Research is ongoing to fuel convert the sun's energy directly into a affordably by means of fuel cell technology, that is, to produce pure H_2 directly from sunlight and water. Some have suggested bio-energy conversion through MFCs technology for future potential conversion of vast available marine or ocean sources into energy.

There are various issues in relation to fuel cells and hydrogen storage as discussed in various literature. At present, the first issue is regarding cost, although the overall cost can be < 50\$ per kW if the fuel cells are mass produced.

The variety of fuel cell types have significant differences in operating conditions, the result of which their materials of construction, fabrication techniques, and system requirements differ from each other. These distinctions result in individual characteristics and associated potential of the various cells to be used for different applications. Therefore the issues of fuel cells are numerous, the main issues of which follows.

For the future, electrode materials other than precious metals (platinum) will most probably be used due to cost issues and energy density delivery capability.

PEMFC: The PEMFC, like the SOFC, has a solid electrolyte which enables it to exhibit excellent resistance to gas crossover. In contrast to the SOFC, the cell can operates at a low temperature as low as 80 °C. This results in excellent startup — a capability to bring the cell to its operating temperature quickly; however, the rejected heat is of low grade thermal energy and thus cannot be used for cogeneration or additional power applications. We can see from various literature and related industrial reviews that the PEMFC cell can operate at very high current densities compared to the other cells; however, the main concern now is heat and water management issues that may limit the operating power density for practical energy applications. Apart from that, the PEMFC tolerance for CO is in the low ppm level (50ppm) and sensitive to the impurities of hydrogen, limiting it to be able to function soundly at high purity of H_2 fuel input.

Micro-solid oxide fuel cells (μ − SOFC) are reported as promising power sources for portable electronic device applications; high μ − SOFC power performances of $677\,\text{mW/cm}_2$ were reported at lower temperatures of 400 oC [28]. Apart from that, SOFCs have potential to achieve three to four times the energy density of lithium-ion or nickel-metalhydride batteries, rendering them an attractive alternative clean power supply source [29]. Nevertheless, there are a few issues regarding the suitable substrate material use, such that the substrate must be physically stable at a wide range of application temperatures. When it comes to the substrate material, many different materials have been suggested by the different groups, ranging from silicon and glass ceramics to metals. So far there is no consensus on which is the best substrate material. The electrodes for -SOFC application are also far from optimum—the cathode, in particular, limits the -SOFC performance when approaching lower operating temperatures. Coarsening by Ostwald ripening of nanostructured materials is a major concern whenever -SOFC membranes are operated at elevated temperatures. The substrate must be chemically and electrically inert. Foturan® glass-ceramic fulfills these criteria, but wet etching with HF-acid may remain a problem, as it may attack the deposited thin films. Silicon wafers are semi-conducting and therefore require an insulating layer (Si3N4 or SiO2) to avoid electrical short-circuiting; this seems to be more difficult than expected [28].

MFC: The performance of microbial fuel cells (MFCs) is affected by issues such as mass transport, reaction kinetics and ohmic resistance. According to Wanga et al., 2011, these factors are manipulated in micro-sized MFCs using specially allocated electrodes constructed with specified materials having physically or chemically modified surfaces. Both two-chamber and air-breathing cathodes are promising configurations for mL-scale MFCs. However, most of the existing lL-scale MFCs generate significantly lower volumetric power density compared with their mL-counterparts because of the high internal resistance. The lL-scale MFCs have yet to provide sufficient power for operating conventional equipment; however, they show great potential in rapid screening of electrochemical microbes and electrode performance [30].

In general, one of the problems with hydrogen, being a gas, is that it is not readily transported in bulk, except by dedicated pipeline networks. The conveyance of hydrogen in cylinders is both inconvenient and cumbersome, while liquid hydrogen is very expensive to manufacture and transport.

If an interest is to develop in the local generation of electricity using fuel cells, in particular for vehicle transport applications, then it will be necessary to have a source of hydrogen for each generation site. For that purpose, Della & Rand [2] have outlined three solutions for the supply of hydrogen to stationary fuel cells, namely:

- Employing a gas grid

- Construction of a dedicated water–electrolysis or solar–thermal plant for mass production of H_2

- The central manufacture of liquid fuel (probably methanol), conveyance of this fuel by road tanker to the fuel cell facility, and conversion back to hydrogen in a reformer when needed

12.2 "Energy Sustainability" Issues

The famous definition of "energy sustainability" outlined by the Bruntland Commission in 1987 defined "sustainability" as "meeting the needs of the present generation without compromising the needs of future generations" [27]. In other words, sustainability effort is geared towards continuous economic growth while providing energy security and environmental protection.

As already pointed out in Chapter 3, due to our world's limited resources, it is globally accepted that the present production and use of energy pose a serious threat both to the future energy supply as well as the global environment, particularly in relation to emissions of CO_2 and other unwanted greenhouse gases, which are discussed in Chapter 4. Therefore, the consequences of *in-sustainability* are tremendous, and include four main issues:

- *Energy shortage issues*. The depleting of fossil fuels, in particular, oil, gas and coal, together with the widely believed "peak oil" which is close to or possibly already round the corner, such that our current urgent energy application question perhaps, is: "Can renewables fill the gap of the fossil fuel depletion?" Or, a better question could be: "Can alternative fuels (including renewables, energy efficiency, etc.) and nuclear energy fill the gap of the fossil fuel depletion?" Can all countries, in particular, industrialized countries, adopt a whole range of reasonable energy policies and associated policies to tackle the complicated technology issues to make their energy futures more "sustainable"? Undoubtedly, our world must continue to set realistic targets to monitor and control our energy supply in parallel with our utilization systems as closely as possible.

- *Carbon and greenhouse emissions issues.* The second important issue is linked to the goal towards global sustainability in terms of reducing greenhouse gas emissions, in particular CO_2. In 1997, the Kyoto Protocol has called for a reduction of the average emissions by at least 5% below 1990 levels in the period 2008–2012. Specifically, it aimed to reduce about 7, 6, and 8%, for the EU, Japan and United States, respectively. According to many, not only does this initiative imply a reduction in the use of fossil fuels, particularly coal, but also such emission targets cannot be achieved by burning biomass and waste, two of the most promising renewables in the short-term. A further problem is that many less developed nations, for Example China and India, foresee substantial increases in CO_2 emissions as their economies develop.

- *Social-economic and implementation issues*. Historically, the main problem for financing alternative/renewable projects has been small project size, which has discouraged financial institutions from providing loans. Problematic legal frameworks, poor tax or subsidy structures, and the dearth of local groups or retailers to develop local markets have deterred private investors as well. As an example, in the United States, only 14 new wind turbine manufacturing plants (wind energy projects are generally considered sustainability projects) were installed in 2010; the U.S. industry was hampered with late extension by the U.S. Congress of the Investment Tax Credit (ITC), low natural gas and electricity prices, and transmission access issues, so that renewable project developers managed only half the number of the renewable projects they did in 2009 [40].

- *Environmental life-cycle issues*. At the other end of the product life-cycle, burgeoning production, operation, and decommissioning processes have highlighted growing environmental and materials issues. In the solar PV sector in particular, questions about material and energy flows, environmental impacts, and the reprocessing of used components have become increasingly central. With total installed global solar PV capacity increasing sevenfold between 2005 and 2010, these practices have come under greater scrutiny, driving innovations in efficient manufacturing, new production equipment, recycling of process water and other resources, and the on-site generation of renewable process energy. Of growing importance is the recycling of solar panels that have reached the end of their service life. While current quantities of disused PV modules remain too small to fully support an extensive recycling operation, it is predicted that around 130,000 tonnes of end-of-life PV panels will be ready for disposal in Europe by 2030 [40].

The above issues in relation to energy usage and energy applications imply the following six main areas of great importance towards sustainability's target, i.e.:

1. Energy efficiency improvement

2. Energy recovery and energy saving (the opposite of energy wastage)

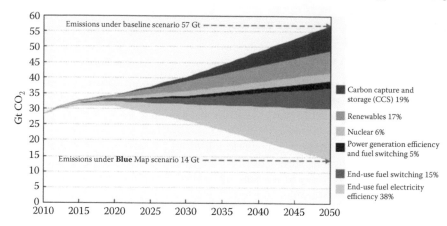

Figure 12.2

Key technologies for reducing CO_2 emissions. (From International Energy Agency, Co-generation and renewables. Solution for a low-carbon energy future, OECD/IEA, Paris, 2011. With permission.)

3. Energy priority and security

4. Techno-socio-economical and energy total management

5. Short term nuclear energy applications

6. Long-term current and future renewable energy applications.

To provide sustainability, IEA, for instance, has suggested two priorities of energy applications, namely co-generation (CHP) and renewables [35] and proposed a Blue Map strategy. The description is given in the following paragraph:

Energy efficiency and renewables are both important if a sustainable future is to be realized. This is well illustrated in Energy Technology Perspectives 2010 (ETP 2010), where the current trend represented by the Baseline Scenario (Figure 12.2) is described as unsustainable. In contrast, when appropriate actions are taken, the BLUE Map Scenario 1 is described as more sustainable. There are a number of key technologies that can help bridge the gap between these two courses of action. Both energy efficiency and renewable energy can contribute to the transition from an unsustainable energy path to a sustainable one [35].

Co-generation is considered one option that contributes substantially to energy efficiency, including through district heating and cooling networks. Such contributions are embedded in Figure 12.2 within the three lower wedges (power generation efficiency and fuel switching 5% + end-used fuel switching 15% + end-use fuel and electricity efficiency 35%), together accounting for 58% of total contributions [35].

12.3   Waste and "Waste-to-Energy (WTE)" Issues

Around the globe, waste is piling up at an alarming rate and we need an effective waste disposal schemes very urgently while at the same time making use of the unwanted waste to provide *clean* energy via smart and efficient composting or recycling methods.

At the present, the common methodologies for dealing with municipal solid wastes consists of land-filling, composting, recycling, and combustion with energy recovery commonly called "waste-to-energy" or simply WTE.

Generally, the WTE conversion methods can be divided into a few technologies, namely land-filling, incineration, pyrolysis, and gasification (Figure 12.3). Both pyrolysis and gasification can create high energy production, usually > 750 kWh/metric tons; while 500 kWh/metric tons of electricity generation could be expected through incineration methods, and about 100-125 kWh/metric tons of electrical energy can be extracted from the common land-filling methods.

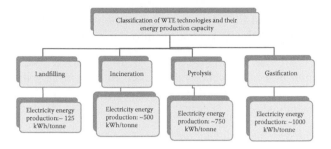

Figure 12.3
Classification of WTE technologies and electricity energy production.

According to the U.S. Environmental Protection Agency (EPA [18]), between 1996 and 2007, there were no new WTE facilities in the USA because of the issues of environmental and political pressure. The major concern has been the perceived release of hazardous toxic substances into the environment. In the past, the primary focus has been on air emissions issues, especially in relation to dioxins/furans and heavy metals. However, after the EPA implemented the maximum available control technology (MACT) regulations in the 1990s, WTE emissions have been reduced to a point such that in 2003 the EPA named WTE as one of the cleanest sources of energy production schemes. In particular, the implementation of the MACT regulations by the U.S. WTE industry has resulted in reducing mercury and other volatile metal emissions by 99% and dioxin and furan emissions by 99.9% (Psomopoulos et al., 2009 [16]). It is expected that in the future, biofuel from garbage could cut carbon emissions by 80% while replacing our need for such large amounts of petroleum. An estimated 82.9 billion liters of biofuel could be produced from the waste currently found in landfills worldwide; see, e.g., [19]).

From the energy applications perspective, the interesting key questions or issues arising from WTE that must be further addressed are (see, e.g., [17]):

1. What are the regulatory, technological, environmental, and economic market drivers for WTE methods?

2. What is the role of WTE in the mix of primary and renewable energy resources?

3. Can WTE help to mitigate GHG emissions by displacing fossil fuel use, provide renewable energy, and improve energy security? How might WTE do this?

4. How is the market structured and who are the key market players?

5. What is the size of the WTE market opportunity by region and technology category?

12.4 "Zero-energy" Building (ZEB) Issues

A "zero-energy" building (ZEB), also known as a "net zero energy" building (NZEB), is a popular term used to describe a building's use with zero net energy consumption and zero carbon emissions annually.

In America, the Energy Independence and Security Act of 2007 (EISA 2007) authorize the Net- Zero Energy Commercial Building Initiative to support the goal of net zero energy for all new commercial buildings by 2030. It further specifies a zero-energy target for 50% of U.S. commercial buildings by 2040 and net zero for all U.S. commercial buildings by 2050 [20].

The European Parliament defines a net-zero energy building as "a building where, as a result of the very high level of energy efficiency of the building, the overall annual primary energy consumption is equal to or less than the energy production from renewable energy sources on site' [21]. Within the recast of the Directive on Energy Performance of Buildings (EPBD) adopted in May 2010, the EPBD establishes the "nearly zero energy building" as the building target from 2018 for all, buildings publicly owned or occupied by public authorities buildings, and from 2020 for all new buildings [22].

Despite the clear international goals and the international attention given to the ZEBs, two major issues remain and need to be met before full integration of the ZEB concept into national building codes and international standards. They are (Marszala et al., 2011 [23]):

1. *The adaptation of a common and unambiguous definition.* In the existing literature the Zero Energy Building concept is described with a wide range of terms and expressions and a number of distinct approaches towards ZEB definitions can be distinguished.

2. *The development of a supporting methodology for computing the energy balance.* The need for a robust calculation methodology has gained attention with the growing number of ZEB projects and thus the interest in how the "zero" balance is computed.

Here, we will enumerate and summarize the main issues pointed out by Marszala et al. in relation to the available ZEBs definitions and the associated methodologies:

1. *Metric (Unit) of the energy balance.* The applied unit for the "zero" balance can be influenced by a number of measures; therefore more than one unit can be used in the definition and/or calculation methodology. These can be, i.e., the final, also called delivered, end-use, or un-weighted energy, primary energy, CO_2 equivalent emissions, exergy, the cost of energy, or other parameters defined by national energy policy.

2. *Period of the balance.* The period of time over which the building calculation is performed can vary very much. It can be an exhaustive full life cycle of a building or the operating time of the building (e.g., 50 years) or, very commonly used, annual balance or (applied in special situations) a seasonal or monthly balance.

3. *Type of energy use.* The methods adopted for computing the energy use of a building are very diverse and include various inputs. For example, type of energy demand for building operation, energy use related to users such as appliances, lighting, and/or embodied energy.

4. *Type of balance.* This issue is mostly relevant to grid connected ZEBs, because in this type of ZEB there are two possible balances between: (1) the energy use and the renewable energy generation or (2) the energy delivered to the building and the energy feed into the grid. As both balances are the same in most cases (the exception is the fossil fuel Combined Heat and Power (CHP), which is not considered in the first balance but can be taken into consideration in the second one), the main difference is the period of application. The first balance is more applicable during the design phase of the building and the second to the monitoring phase. In the off-grid ZEBs, the situation makes clear that the energy use has to be offset by renewable energy generation.

5. *Renewable energy supply options.* The renewable sources can either be available on the site, e.g., sun or wind, or need to be transported to the site, e.g., biomass. Therefore, in principle two renewable energy supply options exist: on-site supply and off-site supply, respectively.

6. *Connection with the energy infrastructure.* Many technologists and policy makers focuses either on off-grid or on-grid zero energy buildings. The main difference is the connection to the energy infrastructure. The off-grid ZEB is not connected to any utility grid and hence needs to use some electricity storage system for periods with peak loads. The issues of large storage capacity, backup generators, energy losses due to storing or converting energy, and oversized renewable energy producing systems in autonomous ZEBs result in lack of global implementation of the off-grid ZEB [24].

7. *Requirements.* The requirements that can significantly influence the design and thus the "quality" of ZEBs are (1) energy efficiency requirements; (2) indoor climate requirements (such as temperature, IAQ, lighting, etc.); and in the case of grid connected ZEBs (3) building–grid interaction requirements (type and application of renewable energy systems — i.e., are they only on-site or also external).

The American society of heat, refrigerating and Air-conditioning Engineers (ASHRAE) initially defines NZEB as buildings that "on an annual basis use no more energy than is provided by on-site renewable energy sources." But the general question can be "how do we define net zero? Is it zero cost, zero energy, or zero carbon?" What is clear, according to Holness (2011) is that net zero energy cannot be achieved by energy efficiency alone; renewable energy components must be applied. Then the next challenge or issue becomes, how do we apply renewable energy in dense urban and high-rise buildings? The answer is: "Only by a fully integrated design and construction approach addressing, among others things" [31]:

- Building orientation to suit climate zone

- Coordinated siting, landscaping, and building location

- Highly insulated building envelope

- Optimized use of day-lighting

- Low density ambient lighting (electronic dimmable)

- High efficiency task lighting (occupancy control)

- Super efficient HVAC systems

- Radiant heating and cooling systems

- High performance packaged systems, including variable refrigerant flow (VRF) systems

- Consideration of renewable energy

Zero energy buildings can use energy independently from the energy grid supply, i.e., energy can be harvested on-site in combination with available renewable energy schemes such as solar, wind, hydro, or hybrid energy while reducing the overall use of energy with extremely efficient HVAC, lighting, VFR, and other technologies. Now, we largely know the technology to achieve net zero energy.

With NZEB as the ultimate goal, the following are the main issues or challenges that we have to face, including:

- Improvement to building energy codes, and with more rigorous application and enforcement

- An integrated building design approach

- Federal, state and utility rebates and incentives

- Benchmarking, measurement, and verification

- Metering and sub-metering

- Operator and user training

- Performance measurement, tracking, and regulation of performance

---

12.5 "Biofuels from Crops" Issues

There are still several setbacks to using biofuels on a wide scale at the current stage of development worldwide. The main issues can be broadly classified into technological, socio-economic, and environmental perspectives (Figure 12.4). We shall note that over the years, however, some of the obstacles were overcome in a progressive manner.

Technology issue. Our current worldwide biofuel technology still has not yet reached a mature stage of wide applications. The biofuels generally are not competitive compared to conventional fuels, with somewhat unfavored energy balance and low (overall) energy output per energy input. This is primarily because, in general, we have not found a techno-economical way:

1. to mass produce biofuels at the same (high) quality as conventional fossil fuels;

2. to generate biofuels at a required rate of production;

3. to generate attractive revenue from biofuels currently due to both the relative low quality and the expensive biofuels that are currently available; and to find suitable and reliable supply of resources.

4. to provide efficient cultivation technology such as that of sugarcane biofuel program in Brazil, which uses modern equipment and cheap sugar cane as feedstock, with the residual cane-waste successfully recycled for CHP applications.

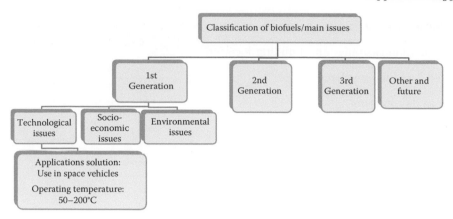

Figure 12.4
Classification of biofuels' main issues.

Production of ethanol from crops has been around in the world for a long time; but the technology and associated applications still raise many concerns. One drawback of crop-derived ethanol for, e.g., corn ethanol production is that it requires a large amount of land and fresh water, along with fertilizers and other energy usage. This results not only in potential competition with food sources for land use but also water resources as well for other industrial and commercial uses. In addition, with current technology, perhaps 50 to 70% of the overall energy value of corn ethanol production is used just to produce the fuel, and most of those energy essentially comes from fossil fuel-based electricity.

Socio-economical issue. The socio-economical problem is one of the main issues leading to the problem of biofuel applications worldwide. Currently, there is no attractive promise of good return on investment to start up the biofuel industry. Moreover, we are still lacking the biofuels research and associated development projects that are essential to carry out for practical technology, commercialization, and sound applications. Two of the general social concerns are:

1. The general social concerns of supply and demand of crop harvests used to produce biofuels

2. The associated shrinking crop sizes, the issue of which is essentially one of the worldwide concerns in the agricultural sector

The economical concerns are:

1. The primary economic concern about biofuels is that the large amount of crops needed in order to mass produce biofuels may take land away from crop farms, cattle farms, and other forms of land use that are essential to the global community for food, textiles, manufacturing, and other industries.

2. If these lands are depleted due to biofuel demands, the costs of these goods will rise as supply decreases and demand either stays static or increases. This will affect the global economy, pushing many third world countries even lower into poverty.

3. Another economic concern is the fact that until there is a true demand for biofuel, it may not be economically advisable to produce it. This means that until there is a large enough fleet of vehicles that run economically on biofuel, there will not be a large enough demand for it; and the investment involved may in the end face failures. Unfortunately, until today, only a limited number of places such as Brazil have shown a positive performance in term of ethanol production for vehicle fuel applications.

The above economical issues have further results in some of the popular social-related issues so much so that some argue that biofuels can actually contribute more problems than solutions. The popular arguments are summarized as follows:

1. *Food-based biofuel.* Food is considered by many as far more important than fuels especially because currently a large part of our world's seven billion people have scarce food for living.

2. *Increase of global crop commodity prices.* Many believe that increase in biofuel demand leads to inevitable increases in crop prices, and increases in raw material prices. However, we shall note that global crop production can also increase, to stabilize crop prices.

3. Land as well as the water needed for crops is becoming a scarce resource as a result of rising per capita consumption and globally population.

4. Some believe that growing demand for biofuels will lead to deforestation, which in turn will add to global warming.

5. Some believe that even the second generation biofuel derived from non-food sources such as microalgae will cause a reduction of algae worldwide. A reduction of algae, one of the most important oxygen producers, would possibly cause an unwanted world ecology imbalance.

Environmental issues. There are various environmental issues with regards to biofuel production and applications, summaries of which are enumerated below:

1. *Soil nutrients depletion*: One of the main environmental concerns of biofuels is that there are few crops that can actually be turned into biofuels, and at a good yields. This is a big concern because each type of crop takes different nutrients out of the soil, and unless crop rotation is implemented, the soil will quickly be depleted of these nutrients until the crop harvests are no longer suitable for consumption, either as food or fuel.

2. *Water supply:* Worldwide, many countries experience droughts yearly, in particular during hot summer days, a large percentage of which often enforce a watering ban for those critical periods. Thus for crop plantation and harvests, this may be a major setback.

3. *Environmental effect of biofuel use.* There is a lack of research into the effects of biofuel applications, which includes the process of evolution of biofuels, i.e., biofuel industry grow the and change over time patterns is not known. The study of emissions from biofuel combustion processes in particul,ar for transportation applications has also not been thoroughly studied; environmental issues such as cleanliness, safety, and the toxicity level of these biofuel emissions are not known or not established yet. According to Pimentel (1998) [38]: "Ethanol produces less carbon monoxide than gasoline, but it produces just as much nitrous oxides as gasoline. In addition, ethanol adds aldehydes and alcohol to the atmosphere, all of which are carcinogenic, as a result of which may also contribute to air pollution problems."

4. Sustainable feedstock for biofuel production.

The United Nations has said that production of biofuel from first generation biofuels such as corn has caused shortages in the world food supply.

However, for the second generation biofuels, for instance, biomass that is derived from food stalks and other by-products notably algae, switch grass, and prairie grass, there could be an advantage compared with the first generation biofuels. They are, however, under research and development. Some of the popular alternative bio-resources are briefly discussed in the paragraphs below:

12.5.1 *Algae*

Algae is a green substance that can be chemically altered to produce biofuel. Some state that it may be 7–30 times more productive per unit of land than typical biofuel plants [36]. Algae also consumes CO_2, which has led the big power plants to start attaching algae cultivation ponds to their plants. The CO_2 from the power plant is fed through to the biofuel plant. Algae can be grown in many different places, like the smokestacks of power plants, to have a dual role in the production of energy, first as a filter to clean power plant emissions, then as an energy source. Algae is extremely efficient, and yields about 1,000 gallons of oil for each acre of land. It is estimated that the amount of algae used to filter at a typical 1,000 megawatt power plant could produce 40 million gallons of biodiesel as well as 50 million gallons of ethanol, and still filter over 40% of the carbon emissions from the power plant [37]. However, it is to be noted that the harvesting of algae should not be of those already existing on our earth as this will definitely cause a worldwide reduction of algae — an oxygen contributor — as this would cause ecological imbalance. Therefore, if algae is going to be used, it must be systematically planted, and specifically only for biofuel production.

12.5.2 *Switch Grass*

Also useful as a biofuel is switch grass. Switch grass can grow up to 12 feet tall in a single season and when harvested, can produce 320 gallons of usable oil per acre. This is much more efficient than either soy or corn. Switch grass also makes an attractive biofuel alternative because it can be grown in soil that might not otherwise support any crop. Switch grass is also environmentally friendly in that it doesn't require a lot of fertilizers or insecticides to grow well and requires relatively little water [37].

12.5.3 *Prairie Grass*

According to studies done by educational institutions, the biomass that is the best for the environment is prairie grass. Prairie grass is best for energy per acre, removing carbon dioxide from the atmosphere, reviving degraded land, and possibly the best for economic and sustainability conditions [37]. Unlike other biomass, which need land clearing and other environmental sensible cultivation/harvesting techniques, prairie grass may offer some advantage.

12.5.4 Biofuel from Waste Fats (Biodiesel)

Unlike bioethanol production, which often deals with sustainable feedstock issues such as their potential competition with land and water resources, *biodiesel* can use cheaper inedible vegetable oil and animal derived triglycerides as feedstock. The issues with biodiesel production and applications, however, are:

1. Some of the low grade biodiesel feedstock contains a high level of free fatty acids (FFA) up to 25% which need further pretreatment.

2. While the low grade feedstock may be much cheaper than edible vegetable oil, the pretreatment processes may add considerable cost to the production of biodiesel (methyl or methyl ester).

Palm oil. Palm oil is a popular biodiesel feedstock due to high yield; however, palm oil has issues with relation to native rain forests and native fauna habitats.

Jatropha. Jatropha can grow on virtually barren land or infertile soil areas and is attracting much interest as a biodiesel feed stock that would not compete with food and land compared with palm oil ([15]). A typical application example is the jatropha plantation along the Mumbai and Delhi train line in India, the trains run on 15–20% biodiesel. However, it has also setbacks:

1. Currently, where jatropha harvesting is done manually; therefore it will not be economical for countries or areas that have high labor cost.

2. Unlike food crop oils, jatropha oil is fairly unstable and quickly breaks down, according to some reports.

3. Jatropha oil is high in free fatty acid (FFA), about 45%, which is far beyond the limit of the 1% FFA level that can be converted into biodiesel easily via transesterification employing an alkaline catalyst (Tiwari et al. 2007 [13]). In other words, it requires pretreatment called an *esterification* process that can add extra cost.

4. Seeds and leaves of jatropha are toxic to human beings and animals (Parawira, 2010 [14]).

Synthetic diesel fuel. There is a considerable effort taking place to use a biomass feedstock for a modified Fischer-Tropsch plant that could produce a synthetic diesel fuel; however, this process is expensive and requires huge capital investment.

Waste cooking oil. Waste cooking oil has been suggested and used by many; but currently only on a small scale.

Future trends and possibilities   The so called third (third) generation biofuels, employing regenerative savings and recycling methods with useful energy production, that is, total energy output per input, would be one of the future potentials of biofuels.

Production of ethanol from fermentation of syngas has already been demonstrated by many; however, according to Ruane et al., 2010 [10], considerable improvements could still be made: For example, isolation of micro-organisms that act well in hot temperatures is important as gasification results in syngas at high temperature, thus and approaches metabolic engineering techniques (see, e.g., [11]) could be employed. If this process could be made commercially viable, it would be particularly advantageous as, unlike biochemical conversion, the lignin in LC biomass, as well as cellulose and hemicellulose, would be converted to a liquid biofuel [12].

12.6 Adsorption Technology from Vehicle Waste Heat

Refrigeration and air-conditioning technology are required to evolve in accordance with the Montreal Protocol, adopted in 1987, and the Kyoto Protocol of 1997. This regulation concerns climate change in an attempt to phase out chlorofluorocarbons (CFCs), followed by hydro-chlorofluorocarbons (HCFCs) and then moving to 1,1,1,2-tetrafluoroethane (HFC-134a) starting in 2011. This trend leads to a strong demand for new systems of air-conditioning, especially in automobiles. Adsorption cooling systems, among other proposed cooling technologies, has a very good potential for automobile applications [9].

The employment of adsorption air-conditioning technology could be one of the best options to replace conventional compression cooling systems. The reasons follow:

1. The potential use of low-grade heat would decrease the quantity of carbon dioxide emission from combustion of the fossil fuels as the engine load is decreased, thus reducing the overall operational cost.

2. Clean refrigerants (instead of CFCs and HCFCs) that have zero ozone depletion and global warming potential can be used as a working fluid.

3. Low regeneration temperature (less than 150 °C) could be used to operate the system.

4. Fewer moving parts, low maintenance cost, and simple system structure make it attractive for many cooling applications.

However, the adsorption technology has not been applied in today's automobiles yet due to a few limitations and associated issues [9] summarized in the following paragraphs:

One of the main limitations is due to the fact that most currently available adsorbents have low adsorptive–desorptive capacity, which could lead to huge system requirements and thus be quite difficult to install into automobiles. In order to overcome this limitation, new synthetic or natural adsorbents with higher adsorptive capacity should be invented to build a compact system.

Besides that, heat transfer rate inside the adsorber normally is quite slow and thus increases the overall cycle time. Coated adsorbent technology and heat pipes could be an alternative to improve the system efficiency by reducing the cycle time and overall heat transfer rate.

Another limitation is that commonly used adsorbates have low latent heat and high boiling point, which cause a low cooling effect generated. Therefore, new types of adsorbates that have less impact to the environment shall be used.

Apart from being compact in size, a practical near-future automobile adsorption system would perhaps be able to achieve a COP range from 0.5 to 0.8, with SCP $\geq$ 1 kW/kg.

12.7 *Assignments, Questions, and Problems*

Assignments

1. Analyze and discuss the issues in relation to the following energy storage methods:

   (a) ammonia storage

   (b) solar thermal energy storage

   (c) wind energy storage

   Which one of the above applications is most applicable or suitable for use in an energy-intensive industry, and why? Your total discussions, including the executive summary, must be less than 10 pages.

2. Analyze and discuss energy efficiency and energy sustainability issues in the cement industries. Supposing that you are one of the energy policy makers in your country, spell out the policy and programs that you would like to see.

3. Many SME- related industries such as those of the food and agricultural sectors have "contributed to a vast amount of unwanted waste," as frequently reported by newspapers. Discuss how we can go about this in terms of mitigation of the waste. Also, choose a particular waste-to-energy (WTE) technique of your choice, discuss how the waste can be recycled effectively, and the issues associated with it used. You must first identify the types so waste of waste and the waste amounts (rate of waste disposal).

4. Analyze, discuss, and compare the latest issues in relation to energy utilization of biofuel and that of fuel cell applications. Which one of the applications do you favor, and why? Your total discussion, including the executive summary, must be less than 10 pages.

Questions and Problems

Multiple choice questions

1. Which of the following is *true* about energy storage from a battery?

   (A) A lead-acid battery has a low energy density, it is bulky and heavy for the amount of energy it stores and does not stand up well to repeated charge–discharge cycles.

   (B) A sodium–sulphur (NaS) battery has a lower energy density than a lead acid battery; thus it too does not exhibit good charge–discharge cycle characteristics.

   (C) A lithium-ion battery has a lower energy density and efficiency than a lead acid battery. It uses a lithium salt in an organic solution, which is environmentally friendly such that its use is widespread in the market.

   (D) The sodium–sulfur battery operates at 300–400 °C, is a stable, non-corrosive, high-temperature, and cost effective battery.

2. Which one of the following statements *correctly* describes the application advantages of pump hydroelectricity as compared to a battery as effective energy storage?

   I It is the most mature and widespread technology being used for large-scale electricity storage.

   II It can have generating capacities ranging from tens of megawatts (MW) to several gigawatts (GW).

   III Generally, pumped storage hydroelectricity requires mountains, so opportunities are limited by geography.

   IV Pumped storage hydroelectricity is not expensive, and is environmentally-friendly.

   V Pumped storage hydroelectricity is not a good match for wind power due to the energy fluctuation characteristics of wind generation.

   (A) I only;

   (B) I and II only

   (C) I, II, and III only

   (D) I, II, III, and IV

   (E) All of the above

3. Which of the following policies or programs is related to energy sustainability?

   I Bruntland Commission 1987

   II Rio 1991 Summits

   III Kyoto 1997 Protocol

IV The BLUE Map Scenario

(A) I only

(B) II and III only

(C) III and IV only

(D) IV only

(E) All of the above

4. Which of the following is *not* an element of Energy sustainability?

   (A) Providing energy security

   (B) Maintaining economic growth

   (C) Providing environmental protection

   (D) Sacrifying the needs of the present generation

   (E) Providing the energy needs of future generation

5. Which of the following is *not* an issue related to waste-to-energy (WTE) application?

   (A) Nitrogen emission

   (B) Mercury emission

   (C) Dioxin emission

   (D) Furan emission

   (E) Carbon emission

6. Which of the following statements correctly describe the present "zero energy building (ZEB)" issue?

   I There is a problem in relation to the adaptation of a common and unambiguous ZEB definition.

   II The development of a supporting methodology for computing the energy balance is not well established yet.

   III There is a lack of building policy and directives from government to implement ZEBs.

   IV There is an issue in relation to the question of whether the renewable sources can be available on site or need to be transported to the site (i.e., either on-site supply or off-site supply).

   (A) I only;

   (B) I and II only

   (C) I, II, and III only

   (D) I, II, and IV only

   (E) All of the above

7. Production of ethanol from crops has been around in the world for a long time. However, ethanol from crops raises many concerns. Which of the following issues *is* a concern of the use of jatropha for bio-ethanol production?

   I Biofuel crops require a large amount of land and fresh water, which can result in potential competition with food sources for land use and fresh water as well as for other industrial and commercial uses.

   II Biofuel crops need significant inputs of fertilizers.

   III Some biofuels derived from biofuel crops may require expensive pretreatment processes.

   IV Some biofuel feedstocks may possess some level of toxicity for humans or animals.

   (A) I only

   (B) I and II only

(C) I, II, and III only

(D) III and IV only

(E) All of the above

8. Which of the following statements is *not* a popular issue, or which does *not* support the notion that "the biofuels can contribute more problems rather than being a solution," in terms of energy sources applications?

(A) Biofuel derived from fuel is an issue since a large part of our world's population has scarce food for living.

(B) There will be an increase of global crop commodity prices.

(C) Land as well as the water needed for crops is becoming a scarce resource as a result of rising per capita consumption and population globally.

(D) The second generation biofuel derived from non-food sources such as algae will be a total solution for biofuel production, and as such this will end all the biofuel application issues.

(E) Growing demand for biofuels will possibility lead to deforestation.

9. Which of the following policy/program/statements is directly concerned with our world climate change in an attempt to phase out the use of chlorofluorocarbons (CFCs) for refrigeration and air-conditioning applications?

I Montreal Protocol adopted in 1987

II Kyoto Protocol, 1997

III The BLUE Map Scenario

IV Directive on Energy Performance of Buildings (EPBD), 2010

(A) I only

(B) I and II only

(C) I, II, and III only

(D) III and IV only

(E) All of the above

Theoretical questions

1. Discuss the problems and issues in relation to industrial ammonia tank storage. What are the measure that can be taken to ensure the integrity of an ammonia tank?

2. Why were there no new WTE facilities in the United States between 1996 and 2007?

3. Define a "zero-energy building." What are the main issues in the definitions of "zero-energy building?"

4. Pumped hydroelectricity has high storage efficiency of 70–85%. List and briefly discuss the issues in relation to pumped hydroelectricity storage.

5. Classify and discuss three main issues in relation to the use of crop-derived biofuels for worldwide energy applications.

6. What are the current problems of fuel cell and battery storage? Suggest at least three ways for improvement.

7. Discuss the current application problems of adsorption technology from vehicle waste heat. What is the possible future or outlook of such technology?

8. Currently, what are the limitations of adsorption technologies for vehicle cooling applications? Discuss ways to improve those limitations.

Answers to multiple choice questions:

1. (A) is true

2. (C) I, II, and III only

3. (E) All I through IV are policies or programs related to energy sustainability.

4. (D) should be meeting the needs of the present generation.

5. (A) Nitrogen is not an issue.

6. (D) I, II, and IV only. III is not true as there are clear directives and encouragement from various energy policy bodies for adoption of ZEB and building efficiency.

7. (D) III and IV only. Unlike other biofuel crops, jatropha can grow on virtually barren land or infertile soil areas. Only III and IV are concerns of the use of jatropha for fuel production.

8. (D) Some believe that even the second generation biofuel derived from non-food sources such as algae will cause a reduction of algae world wide. A reduction of algae, one of the most important oxygen producers, would cause the ecology imbalance.

9. (B) I and II only

Bibliography

[1] Lunde, L., Nyborg, R., (1989) SCC of Carbon Steel in Ammonia – Crack growth studies and means to prevent cracking. American Chemical Institute of Engineers 1989 Ammonia Symposium: Safety in ammonia plant and related facilities, San Francisco, CA. Paper No. 238 c.

[2] Della, R. M., Rand, D. A. J., Energy storage — a key technology for global energy sustainability. *Journal of Power Sources*, Volume 100, Issues 1-2, 30 November 2001, pp. 2–17.

[3] Lindley, D., Smart grids: The energy storage problem. *Nature* 463, 18–20 (2010) doi:10.1038/463018a.

[4] Tester, J. et al. 2006. The Future of Geothermal Energy: Impact of Enhanced Geothermal Systems (EGS) on the United States in the 21st Century. Massachussetts Institute of Technology and Idaho National Laboratory.

[5] Barbier, E., Geothermal energy technology and current status: an overview. *Renewable and Sustainable Energy Reviews*, 2002, 6, 3-65.

[6] Gutzeit, K., Batteries for Telecommunications Systems Powered by Solar Energy, Telecommunications Energy Conference. In: INTELEC'86 International, October, 19–22, 1986. pp. 73–6.

[7] Kalogirou, S., The potential of solar industrial process heat applications. *Applied Energy*, 2003; 76 (December (4)):337–61.

[8] Balaras, C. A., Grossman, G., Henning, H. M., Infante Ferreira, C. A., Podesser, E. et al. Solar air conditioning in Europe—an overview. *Renewable and Sustainable Energy Reviews*, 2007;11(February (2):299–314.

[9] Abdullah, M. O., Tan,I. A. W., Lim, L. S., *Renewable and Sustainable Energy Reviews*, 15 (2011) 2061–2072.

[10] Ruane, J., Sonnino, A., Agostini, A., Bioenergy and the potential contribution of agricultural biotechnologies in developing countries, *Biomass and bioenergy*, 34(2010) 1427–1439.

[11] Henstra, A.M., Sipma, J., Rinzema, A., Stams, A.J.M., Microbiology of synthesis gas fermentation for biofuel production, *Current Opinion in Biotechnology*, 2007;18:200-6.

[12] Larson, E.D., Biofuel production technologies: status, prospects and implications for trade and development. United Nations Conference on Trade and Development (UNCTAD); 2008. Available from: http://www.unctad.org/en/docs/ accessed February 28, 2011.

[13] Kumar Tiwari, A., Kumar, A., Raheman, H., Biodiesel production from jatropha oil (Jatropha curcas) with high free fatty acids: An optimized process, *Biomass and Bioenergy*, Volume 31, Issue 8, August 2007, pp. 569–575.

[14] Parawira, W., Biodiesel production from Jatropha curcas: A review, *Scientific Research and Essays*, Vol. 5(14), pp. 1796-1808, July 18, 2010.

[15] Berchmans, H.J., Hirata, S., (2008). Biodiesel production from crude Jatropha curcas L. seed oil with a high content of free fatty acids. *Bioresour. Technol.*, 99: 1716–1721.

[16] Psomopoulos, C.S., Bourka, A., Themelis, N.J., Waste-to-energy: A review of the status and benefits in USA, *Waste Management* 29 (2009) 1718–1724.

[17] http://www.pikeresearch.com/research/waste-to-energy-technology-markets accessed January 15, 2011.

[18] US Environmental Protection Agency, 2003, www.wte.org/docs/epaletter.pdf, accessed March 13, 2011.

[19] See e.g. http://www.eco20-20.com/One-Mans-Trash-is-Another-Mans-Biofuel.html.

[20] Crawley, D.,Pless, S., Torcellini, P., Getting to net zero, *ASHRAE Journal*, 51 (9) (2009) 18–25.

[21] European Parliament, Report on the proposal for a directive of the European Parliament and of the Council on the energy performance of buildings (recast) (COM(2008)0780-C6-0413/2008-2008/0223(COD)), Brussels, Belgium, April 16, 2009.

[22] The Directive 2010/31/EU of the European Parliament and of the Council of 19 May 2010 on the energy performance of buildings, Official Journal of the European Union, 53, 2010.

[23] Marszala, A.J., Heiselberga, P., Bourrelle, J.S., Musall, E., Vossc, K., Sartori , I., Napolitano, A., Zero Energy Building – A review of definitions and calculation methodologies, *Energy and Buildings*, 43 (2011) 971–979.

[24] Torcellini, P., Pless, S., Deru, M., Crawley, D., Zero Energy Buildings: A Critical Look at the Definition, in: ACEEE Summer Stud, Pacific Grove, California, 2006.), (K. Voss, What is Really New about Zero-Energy Homes? in: 12th International Conference on Passive Houses, Nuremberg, Germany, 2008.

[25] Schnitzer, H., Christoph, B., Gwehenberger, G., Minimizing greenhouse gas emissions through the application of solar thermal energy in industrial processes. Approaching zero emissions. *Journal of Cleaner Production*, 2007;15 (September (13–14)):1271–86.

[26] Kulkarni, G. N., Kedare, S. B., Bandyopadhyay, S., Design of solar thermal systems utilizing pressurized hot water storage for industrial applications. *Solar Energy*, 2008;82(August (8)):686–99.

[27] World Commission, *Our common future: Report on the World Commision on Environement and Development*, 1987, Oxford: Oxford University Press.

[28] Evans, A., Bieberle-Hütter, A., Rupp, J. L.M., Gauckler, L. J., Review on microfabricated micro-solid oxide fuel cell membranes, *Journal of Power Sources*, 194 (2009) 119–129.

[29] G.J. La O, H.J. In, E. Crumlin, G. Barbastathis, Y. Shao-Horn, *Int. J. Energy Res.*, 31 (2007) 548.

[30] Wanga, H.-Y., Bernarda, A., Huang, C.-Y., Lee, D.-J., Chang, J.-S., Micro-sized microbial fuel cell: A mini-review, *Bioresource Technology*, 102 (2011) 235–243.

[31] Holness, G.V.R., On the path to net zero: how do we get there from here?, *ASHRAE Journal*, June 2011, pp. 50–60.

[32] Cataldi, R., Social acceptability of geothermal energy: problems and costs. In: *Geothermal District Heating Schemes, 1997 Course Textbook*, International Summer School on Direct Application of Geothermal Energy, Ankara-Skopje, 1997, 6–1/6–15.38.

[33] Abdullah, M. O., Zen, J., Yusof, M., 2011, On the Stress Corrosion Cracking, Crack Growth Prediction and Risk-based Inspection of Industrial Refrigerated Ammonia Tanks, *Corrosion*, 67 (4) National Association of Corrosion Engineers (NACE) International, Houston, USA.

[34] Högselius, P., Spent nuclear fuel policies in historical perspective: An international comparison, *Energy Policy*, Volume 37, Issue 1, January 2009, pp. 254–263.

[35] International Energy Agency, Co-generation and renewables. Solution for a low-carbon energy future, OECD/IEA, Paris, 2011.

[36] Danielo, O., 2005, An algae-based fuel, *Biofutur*, No. 255. May 2005.

[37] http://www.eco20-20.com/Biofuels-3rd-Generation.html, accessed on October 22, 2011.

[38] Pimentel, D., 1998, Energy and Dollar Costs of Ethanol Production with Corn: *Hubbert Center Newsletter*, 98/2, M. King Hubbert Center for Petroleum Supply Studies, Golden, Colorado.

[39] World Energy Council, *2010 Survey of Energy Resources*, London, United Kingdom.

[40] *Renewables 2011 Global Status Report*, Renewable Energy Policy Network for the 21st Century, Paris.

[41] Cairns, E. J., Albertus, P., Batteries for Electric and Hybrid-Electric Vehicles Annual Review of Chemical and Biomolecular Engineering Vol. 1: 299-320 July 2010.

Epilogue

And, What's Next?

Student workdays are filled with all types of studying loads such as readings, assignments, tests and experimental works; so much so that most students, are in particular engineering students, at times less motivated to take on new horizons and generic courses. The question is, what should our students learn or do now for their elective programs?

While teaching undergraduate students, conducting research, undertaking a bit of consultancy and related community works, in relation to applied energy, all these years, I was also reading and thinking of inventions and starting to encourage student-based businesses. Making changes to the present available elective syllabus and enhanced study materials, e.g., from simply *Energy* to *Applied Energy, Innovation, and Entrepreneurship* in universities, could make students feel like spending time on increasingly interesting things. Encouraging student entrepreneurs or entrepreneurs-to-be, either individually or in groups, would, essentially, encourage the generation of incomes for the new generations apart from promoting traditional science and general engineering. Students' futures are full of opportunities, and we all look forward to an interesting educational journey.

Appendix

Some conversion factors

Length

1 ft = 2.54 cm
1 meter = 91.43 cm = 3.281 ft

Density

1 kg/m^3 = 0.06243 Ib/ft^3

Work, energy, or heat

1 J = 1 watt-sec = 0.93 cal = 2.7778 x 10^{-7}kWh
1 kJ = 10^3 Nm = 0.9478 Btu
1 hp = 745.7 W
1 Btu = 1.055 × 10^3 J
1 quad = 1.055 × 10^{18} J = 10^{15}Btu
1 TJ = 10^{12} J
1 ZJ = 10^{21} J
1 cal = 4.184 J = 1.1622 × 10^{-6} kWh
1 toe = 11.63 MWh = 41.868 GJ
1 Mtoe = 11,63GWh
1 kW = 1 kJ/s = 3412 Btu/h
1 kWh = 860 kcal = 3.6 × 10^6 J
1 MW = 1,000 kW
1 MWh = 3.6 × 10^9 J
1 TW = 10^3 GW = 10^6 MW = 10^9 kW

Pressure

1 atm =1.01325 bar
1 bar = 750.06 mm Hg = 10^{5}N/m^2
1 N/m^2 = 1 Pascal = 10^{-5} bar = 10^{-2}kg/m − sec^2

Mass

1 kg = 2.205 Ib

Thermal conductivity

1 kW/m K = 577.8 Btu/ft h R

Heat transfer coefficient

1 kW/m^{2}K = 176.1 Btu/ft^{2}h R

Specific energy

1 kJ/kg = 0.4299 Btu/Ib

Glossary

Glossary of Some Important Terms

Absorption and adsorption

Absorption is the chemical integration of one chemical into another, i.e., transfer of a volume of energy or mass into another. It occurs when one substance holds another via physical bonds. On the other hand, adsorption is a process where molecules of a gas (or liquid) contact and adhere to a solid surface, i.e., transfer of a volume of energy or mass onto a surface. (The adsorption process is always exothermic where heat is liberated during the process.)

Best available technology

The latest stage of practical technology available in terms of processes, facilities, or operation methods, in particular, for enhancing overall energy efficiency.

Biomass and bioenergy

Biomass refers to biological materials derived from living organisms such as woods and other residues from agriculture. It includes solid biomass (e.g. wood), gas and liquid biomass (e.g., industrial waste and municipal waste). In terms of applied energy, biomass is used to produce bioenergy - in the forms of electricity, heat, or both.

Capacity factor

The ratio of the actual electical production to the maximum possible generation for a given period of time. It is often expressed in percentage.

Catalytic reforming

Catalytic reforming is a chemical process where a chemical component is converted into another chemical component, using a catalyst.

Combined heat and power

Combined heat and power (CHP) or cogeneration is the use that is a heat engine or a power plant to simultaneously generate both electricity and useful energy from the waste heat that is otherwise rejected to the environment. By making use of the waste heat, CHP is a way of energy recycling to improve the eciency of the overall power plant or engine system.

Energy audit

Formal energy account in terms of efficiency energy consumption and the associatd cost of an energy system, over a period of time-usually n a yearly loasis.

Energy demand

Energy input required to provide various energy applications to sustain our socioeconomics and energy activities.

Energy intensity

Energy intensity is generally defined as the energy consumption divided by the economic output. (In other words, it is a measure of total primary energy use per unit of physical/gross domestic product). Energy efficiency is the reciprocal of energy intensity.

Energy recovery

Energy retrieval or energy revival of an energy system, which was initially thought not techno-economically viable.

Energy storage

Energy storage refers to a space or system devices to store some form of energy to perform some useful operations at a later time. For example, an oil reservoir stores petroleum while a gas reservoir stores natural gas. Our own human body also poses excellent means of storage systems, such that the food that we consumed can be utilized for useful energy production and work. It allows humans to balance the supply and demand of energy, and provide energy sustainability.

Energy supply

The production of fuels as well as other energy inputs from available resources.

Energy target

Energy target is the energy goal, plan or commitment that an entity (normally a government at the local, state, or national level) is going to achieve in the future, usually in 5 or 10 years ahead of the policy made. Some energy targets are legislated while others are set by regulatory agencies.

Hybrid energy system

An energy system consists of two or more energy sources used together, via suitable energy conversion techniques, to power that energy system.

Peak load

The point in time when energy needs are highest, and the energy system experiences the largest demand.

Renewables

Renewables refers to energy resources that are readily available and replenished constantly, and include biomass, geothermal, solar, wind, wave, hydropower, and so forth. It is to be noted that the term "alternative energy" is preferred by some as energy can not be actually be renewed but converted from one form to another.

Sankey diagram

A flow diagram that visually represents various energy inputs, outputs, lossses, and energy stored in an energy system.

Specific energy consumption

The energy consumed by an energy system per unit output, e.g., in the case of a building, it can be the energy consumption per unit area of the building.

Index

Absorption energy conversion, 349
Acid rain, 204
Adsorption energy conversion, 354
Adsorption technology from vehicle waste heat, 395
Alternative or "renewable" energy, 205
Aluminum, Iron and steel industries, 42
Applied energy modeling and simulation, 138

Basic hydro plant components, 254
Battery, 227
Battery storage issues, 380
Bioenergy or biomass energy, 309
    Biodiesel, 315
    Bioethanol, 314
Biofuels from crops issues, 391
Building industries, 60

Cement and concrete industries, 49
Chemical and petrochemical industry, 46
Coal, 198
Combined heat and power (cogeneration), 160
Combined thermionic–thermoelectric, 365

Energy, 1
    Chemical energy, 3
    Electrical energy, 2
    Electrochemical energy, 4
    Electromagnetic energy, 4
    Kinetic energy, 2
    Nuclear energy, 5
    Potential energy, 3
    Sound energy, 4
    Thermal energy, 3
Energy audit, 111
Energy conversion systems, 13
Energy efficiencies and losses, 25
Energy equation, 8
Energy life-cycle analysis, 116
Energy performance curve, 12
Energy planning and statistics, 134
Energy policy, 133
Energy policy, planning, and statistics, 133
Energy pollutants, 129
Energy power industry, 37
Energy recovery, 147
    Enhanced Oil Recovery (EOR), 147
    Oil and gas (O&G) industry, 147
Energy resources, 6, 80

comparison, 80
Energy resources, supply and demand, 79
Energy storage, 23, 167
    Ammonia storage, 167
    Capacitor, 175
    Flywheel storage, 170
    Fuel cell storage, 173
    Hydrogen storage, 174
    Inductor, 175
    Nuclear storage, 168
    Pump hydroelectric storage, 172
    Solar energy storage, 171
    Supercapacitor or Ultracapacitor, 175
    Thermal energy, 174
    Wind energy storage, 172
Energy storage comparison, 178
Energy storage issues, 379
Energy supply *and* energy demand, 83
    Denmark, 92
    India, 90
    Japan, 90
    Jordan, 95
    Kuwait, 96
    Malaysia, 91
    United Kingdom, 93
Energy supply *and* energy demand
    United States, 97
Energy sustainability issues, 387
Energy use and fuel consumption study, 112
Energy, environment, and health, 125
Energy-intensive industry, 41

First Law of Themodynamics, 8
Fuel cell, 232
    Alkaline fuel cell (AFC), 233
    Direct methanol fuel cell (DMFC), 234
    Microbial fuel cell (MFC), 237
    Molten carbonate fuel cell (MCFC), 236
    Phosphoric acid fuel cell (PAFC), 233
    Proton exchange membrane fuel cell (PEMFC), 235
    Solid oxide fuel cell (SOFC), 236
Fuel cells and hydrogen storage issues, 385
Fuel consumption study, 115

Gasification, 311
General mechanical energy equation, 9

Geothermal energy, 273
Gibbs free-energy relation, 12

Hybrid electric vehicles (HEVs), 335
Hybrid energy, 327
Hybrid energy systems for rural application, 337
Hybrid fuel cell/turbine and fuel cell/microturbine
        system, 332
Hybrid geothermal/fossil system, 327
Hydro energy, 253

Impact of energy use, 131
Impacts of transport on energy consumption, 64
Integrated gasification combined-cycle (IGCC), 331

Magnetohydrodynamic (MHD) power conversion, 366
Metal processing industry, 52
Mining industry, 53

Natural gas (gaseous petroleum), 194
Nuclear power plant, 295
Nuclear storage issues, 380

Ocean energy, 262
Ocean thermal energy, 266
Oil based mud drilling (OBM), 316
Other fossil fuels, 201
        Natural bitumen, 202
        Oil shales, 201

Petroleum, 187
Power, 5
Process integration and pinch technology, 155
Pyrolysis, 312

Renewable energy storage issues, 380
Residential energy applications, 68

Sankey diagram, 99
Second Law of Thermodynamics, 9
SME energy industry , 56
Solar photovoltaic (PV), 215
Solar thermal energy, 218

Textiles, Pulp and paper industries, 50
Thermionic energy conversion, 363
Thermoelectric energy conversion, 360
Tidal energy, 262
        Application advantages, 262
Transportation industries, 63
        Climate change and emissions, 67
        Energy consumption, 64

Volumetric equation of oil in place, 188

Waste-to-energy (WTE), 388
Wave energy, 265

Application advantages, 266
Wind energy, 268
        Offshore windmills, 272

Zero-energy building (ZEB) issues, 389

T - #0136 - 250919 - C12 - 279/216/21 - PB - 9780367380731